T0315523

Pharmaceutical Quality by Design: A Practical Approach

ADVANCES IN PHARMACEUTICAL TECHNOLOGY

A Wiley Book Series

Series Editors:
Dennis Douroumis, University of Greenwich, UK
Alfred Fahr, Friedrich–Schiller University of Jena, Germany
Jürgen Siepmann, University of Lille, France
Martin Snowden, University of Greenwich, UK
Vladimir Torchilin, Northeastern University, USA

Titles in the Series

Hot-Melt Extrusion: Pharmaceutical Applications
Edited by Dionysios Douroumis

Drug Delivery Strategies for Poorly Water-Soluble Drugs
Edited by Dionysios Douroumis and Alfred Fahr

Computational Pharmaceutics: Application of Molecular Modeling in Drug Delivery
Edited by Defang Ouyang and Sean C. Smith

Pulmonary Drug Delivery: Advances and Challenges
Edited by Ali Nokhodchi and Gary P. Martin

Novel Delivery Systems for Transdermal and Intradermal Drug Delivery
Edited by Ryan Donnelly and Raj Singh

Drug Delivery Systems for Tuberculosis Prevention and Treatment
Edited by Anthony J. Hickey

Continuous Manufacturing of Pharmaceuticals
Edited by Peter Kleinebudde, Johannes Khinast, and Jukka Rantanen

Pharmaceutical Quality by Design: A Practical Approach
Edited by Walkiria S. Schlindwein and Mark Gibson

Forthcoming Titles:

In Vitro Drug Release Testing of Special Dosage Forms
Edited by Nikoletta Fotaki and Sandra Klein

Characterization of Micro- and Nanosystems
Edited by Leena Peltonen

Therapeutic Dressings and Wound Healing Applications
Edited by Joshua Boateng

Process Analytics for Pharmaceuticals
Edited by Thomas de Beer, Jukka Rantanen and Clare Strachan

Pharmaceutical Quality by Design

A Practical Approach

Edited by

WALKIRIA S. SCHLINDWEIN
De Montfort University
Leicester
United Kingdom

MARK GIBSON
A M PharmaServices Ltd
United Kingdom

WILEY

This edition first published 2018

Registered Offices
John Wiley & Sons, Inc., 111 River Street, Hoboken, NJ 07030, USA
John Wiley & Sons Ltd, The Atrium, Southern Gate, Chichester, West Sussex, PO19 8SQ, UK

Editorial Office
111 River Street, Hoboken, NJ 07030, USA
350 Main Street, Malden, MA 02148-5020, USA
9600 Garsington Road, Oxford, OX4 2DQ, UK
The Atrium, Southern Gate, Chichester, West Sussex, PO19 8SQ, UK
Boschstr. 12, 69469 Weinheim, Germany
1 Fusionopolis Walk, #07-01 Solaris South Tower, Singapore 138628

For details of our global editorial offices, customer services, and more information about Wiley products, visit us at www.wiley.com.

Wiley also publishes its books in a variety of electronic formats and by print-on-demand. Some content that appears in standard print versions of this book may not be available in other formats.

Library of Congress Cataloging-in-Publication Data

Names: Schlindwein, Walkiria S., 1961– editor. | Gibson, Mark, 1957– editor.
Title: Pharmaceutical quality by design : a practical approach / edited by Dr. Walkiria S. Schlindwein, Mark Gibson.
Description: First edition. | Hoboken, NJ : John Wiley & Sons, 2018. | Series: Advances in pharmaceutical technology | Includes bibliographical references and index. |
Identifiers: LCCN 2017030338 (print) | LCCN 2017043153 (ebook) | ISBN 9781118895221 (pdf) | ISBN 9781118895214 (epub) | ISBN 9781118895207 (cloth)
Subjects: LCSH: Drugs–Design. | Drugs–Quality control.
Classification: LCC RS420 (ebook) | LCC RS420 .P47 2018 (print) | DDC 615.1/9–dc23
LC record available at https://lccn.loc.gov/2017030338

Cover design by Wiley
Cover image: (Background) © ShutterWorx/Gettyimages; (Graph) Courtesy of Walkiria S. Schlindwein and Mark Gibson

Set in 10/12pt Times by SPi Global, Pondicherry, India

10 9 8 7 6 5 4 3 2 1

Contents

List of Figures

List of Tables

List of Contributors

Noel Baker, AstraZeneca, Macclesfield, United Kingdom

Claire Beckett, Associated with OSIsoft, London, United Kingdom

Andreas Berghaus, ColVisTec AG, Berlin, Germany

Paul A. Butterworth, AstraZeneca, Macclesfield, United Kingdom

Brian Carlin, De Montfort University (visiting Professor) Lawrenceville
New Jersey, United Kingdom

Alan Carmody, Pfizer, Canterbury, United Kingdom

Ian Cox, JMP Division, SAS Institute, Manchester, United Kingdom

Bruce Davis, Bruce Davis Ltd, Haslemere, United Kingdom

Joe de Sousa, AstraZeneca, Macclesfield, United Kingdom

Lennart Eriksson, Formerly associated with MKS and now associated
with Sartorius Stedim Data Analytics AB, Sweden

Mark Gibson, Formerly associated with AstraZeneca and now associated with
AM PharmaServices Ltd, Congleton, United Kingdom

David Holt, AstraZeneca, Macclesfield, United Kingdom

Erik Johansson, Formerly associated with MKS and now associated with Sartorius Stedim Data Analytics AB, Sweden

Line Lundsberg-Nielsen, Lundsberg Consulting Ltd and NNE, London, United Kingdom

Martin Owen, Formerly associated with GSK and now associated with Insight by Design Consultancy Ltd, Stevenage, United Kingdom

Walkiria S. Schlindwein, De Montfort University, Leicester, United Kingdom

Siegfried Schmitt, PAREXEL Consulting, PAREXEL International, London, United Kingdom

Gerry Steele, PharmaCryst Consulting Ltd, Loughborough, United Kingdom

Conny Wikström, Formerly associated with MKS and now associated with Sartorius Stedim Data Analytics AB, Sweden

Roger Weaver, Formerly associated with Pfizer and now associated with Weaver Pharma Consulting, Canterbury, United Kingdom

Mustafa A. Zaman, PAREXEL Consulting, PAREXEL International, London, United Kingdom

Advances in Pharmaceutical Technology: Series Preface

The series *Advances in Pharmaceutical Technology* covers the principles, methods and technologies that the pharmaceutical industry uses to turn a candidate molecule or new chemical entity into a final drug form and hence a new medicine. The series will explore means of optimizing the therapeutic performance of a drug molecule by designing and manufacturing the best and most innovative of new formulations. The processes associated with the testing of new drugs, the key steps involved in the clinical trials process and the most recent approaches utilized in the manufacture of new medicinal products will all be reported. The focus of the series will very much be on new and emerging technologies and the latest methods used in the drug development process.

The topics covered by the series include the following:

Formulation: The manufacture of tablets in all forms (caplets, dispersible, fast-melting) will be described, as will capsules, suppositories, solutions, suspensions and emulsions, aerosols and sprays, injections, powders, ointments and creams, sustained release and the latest transdermal products. The developments in engineering associated with fluid, powder and solids handling, solubility enhancement, colloidal systems including the stability of emulsions and suspensions will also be reported within the series. The influence of formulation design on the bioavailability of a drug will be discussed, and the importance of formulation with respect to the development of an optimal final new medicinal product will be clearly illustrated.

Drug Delivery: The use of various excipients and their role in drug delivery will be reviewed. Among the topics to be reported and discussed will be a critical appraisal of the current range of modified-release dosage forms currently in use and also those under development. The design and mechanism(s) of controlled release systems including macromolecular drug delivery, microparticulate controlled drug delivery, the delivery of biopharmaceuticals, delivery vehicles created for gastrointestinal tract targeted delivery, transdermal delivery and systems designed specifically for drug delivery to the lung will all be reviewed and critically appraised. Further site-specific systems used for the delivery of drugs across the blood–brain barrier including dendrimers, hydrogels and new innovative biomaterials will be reported.

Manufacturing: The key elements of the manufacturing steps involved in the production of new medicines will be explored in this series. The importance of crystallization; batch and continuous processing, seeding; and mixing including a description of the key engineering principles relevant to the manufacture of new medicines will all be reviewed and reported. The fundamental processes of quality control including good laboratory practice, good manufacturing practice, Quality by Design, the Deming Cycle, regulatory requirements and the design of appropriate robust statistical sampling procedures for the control of raw materials will all be an integral part of this book series.

An evaluation of the current analytical methods used to determine drug stability, the quantitative identification of impurities, contaminants and adulterants in pharmaceutical materials will be described as will the production of therapeutic bio-macromolecules, bacteria, viruses, yeasts, moulds, prions and toxins through chemical synthesis and emerging synthetic/molecular biology techniques. The importance of packaging including the compatibility of materials in contact with drug products and their barrier properties will also be explored.

Advances in Pharmaceutical Technology is intended as a comprehensive one-stop shop for those interested in the development and manufacture of new medicines. The series will appeal to those working in the pharmaceutical and related industries, both large and small, and will also be valuable to those who are studying and learning about the drug development process and the translation of those drugs into new life-saving and life-enriching medicines.

Dennis Douroumis
Alfred Fahr
Jürgen Siepmann
Martin Snowden
Vladimir Torchilin

Preface

The Quality by Design (QbD) concept is not a new one, but it is only in recent years that it has been adopted by the pharmaceutical industry as a systematic approach to the development and control of drug products and their associated manufacturing processes. It became, and still is, a 'hot topic' after the US Food and Drug Administration (FDA) observed diminished drug approval rates, and recognized that something needed to be done to address the numerous quality manufacturing issues occurring post approval due to poorly developed products and manufacturing processes. As a result, the FDA outlined its initiative to address these concerns in its report 'Pharmaceutical Quality for the 21st Century: A Risk-Based Approach', and subsequently, in collaboration with major pharmaceutical companies, established a series of pharmaceutical QbD guidance documents adopted by ICH (International Conference on Harmonization) to help streamline the drug development and regulatory filing process.

The term 'Quality by Design' with respect to pharmaceutical development is defined in ICH guideline Q8(R2) 'Pharmaceutical Development' as 'a systematic approach to development that begins with predefined objectives, emphasizes product, process understanding and process control, based on sound science and quality risk management' (ICH, 2009). A key principle is that quality should be built into a product with a thorough understanding of the product and the manufacturing process. This includes establishing a knowledge of the risks involved in manufacturing the product and how best to mitigate those risks. This new paradigm for pharmaceutical product development employing QbD varies a great deal from the traditional approach, which was extremely empirical; should result in better control of parameters and variables; and reduce the emphasis on end-product testing. There are also potential opportunities to operate within a broader "design space," post approval, without the need for additional regulatory scrutiny.

Due to the combination of regulatory pressure, the carrot of regulatory flexibility, and an acceptance by many pharmaceutical companies that QbD is an improved way of working for the development of new products, it has been adopted as the preferred way of working by many companies. However, implementation has often been found challenging because the FDA and ICH guidance documents are written at a fairly high level, and it is up to each company to interpret them. Some leading pharmaceutical companies have found a way forward and have obtained approval of QbD applications, but there are many others who still struggle to implement QbD in practice and are still feeling their way.

A new group specializing in Pharmaceutical Quality by Design was initiated in 2010 at De Montfort University, Leicester School of Pharmacy, led by Walkiria S. Schlindwein, in recognition of an opportunity to fill a perceived gap in education and learning associated with QbD. A distance learning postgraduate certificate programme was launched in 2010 and expanded into a full validated MSc programme in 2012. This programme has been specifically designed with the needs of the pharmaceutical and allied industries in mind and has been created through a unique collaboration between industry and academia. A wide range of experienced industrial and regulatory experts have been engaged in preparing and delivering pre-recorded lectures that form the core of the programme. They represent manufacturing and development companies, excipient suppliers, process equipment suppliers, data analysis software suppliers, consultancies and regulators. In addition, DMU has developed a dedicated online platform to deliver online short courses tailored to the needs of industry.

The most significant departure in the creation of a QbD training for industry programme, however, has been the completion of a dedicated laboratory to simulate the conditions for QbD in action. Here, programmes can be designed to address the specific training needs of companies from the United Kingdom and the rest of the world, to give support to institutions of all sizes within the industry.

This book, *Pharmaceutical Quality by Design: A Practical Approach*, is intended to complement the DMU QbD teaching materials already available to support the MSc course and distance learning modules, and also, to be consistent in terms of interpretation of the principles, approach and their application in practice.

Pharmaceutical Quality by Design: A Practical Approach includes 12 different chapters covering a broad scope of QbD aspects that are considered important by the editors and contributors. Each of the subsequent chapters are written by "experts" in their field and provide relevant, up-to- date and tailored information. Each part will stand alone, but it is the sum of these individual parts that makes Quality by Design whole and provides the compelling story that will ultimately benefit patients and give clarity of understanding in what is important when designing, manufacturing and supplying products to our customers.

The content is applicable to development scientists, manufacturing specialists and those in supporting roles, such as quality, analytical, engineering, validation and more. It is intended to be helpful, practical and wide-ranging and for use by novices, experienced practitioners or those who want to expand their current knowledge.

Walkiria S. Schlindwein
Mark Gibson

1

Introduction to Quality by Design (QbD)

Bruce Davis[1] *and Walkiria S. Schlindwein*[2]

[1] *Bruce Davis Ltd, United Kingdom*
[2] *De Montfort University, Leicester, United Kingdom*

1.1 Introduction

The aim of this chapter is to introduce the principles of Quality by Design (QbD) to those who want to understand pharmaceutical QbD, and that may include readers from industry, academia, regulators or indeed anyone interested in finding out more about this important subject.

The content is applicable to development scientists, manufacturing specialists and those in supporting roles, such as quality, analytical, engineering, validation and more. It is intended to be helpful, practical and wide-ranging and for use by novices, experienced practitioners or those who want to expand their current knowledge.

Each of the subsequent chapters is written by experts in their field and provide relevant, up-to-date and tailored information. Each part will stand alone, but it is the sum of these individual parts that makes QbD whole and provides the compelling story that will ultimately benefit patients and give clarity of understanding in what is important when designing, manufacturing and supplying products to our customers.

So, who are these customers? Some may be our family or friends or colleagues, but most will be individuals we do not know and will never meet. A customer may choose a generic medicine from a shelf in the pharmacy, their choice perhaps being influenced by the descriptions on the packaging or by marketing and advertising, or, alternatively, they may

Pharmaceutical Quality by Design: A Practical Approach, First Edition.
Edited by Walkiria S. Schlindwein and Mark Gibson.

have their medicine prescribed and administered by healthcare professionals. Some may be supporting others, for example, a parent helping their child, or an adult helping an elderly relative or colleague.

But no matter what the circumstances are in which someone takes a medicine, there is one overriding principle: that every patient, healthcare professional, parent or career has to trust the pharmaceutical industry to provide what is intended and that the medicine will be safe, efficacious and of the required quality.

So it is important that we value this trust we have been given. History says that most of the time the pharmaceutical industries have delivered on this trust, but there have been occasions when the industries have not, and such mistakes, albeit small in number compared with all the medicines that are taken, have sadly damaged the trust that customers put in the industry.

So how does this impact the development, manufacturing and distribution of pharmaceuticals? First, we should recognise that we in industry have the detailed technical knowledge, and the customers usually do not. Second, we should ensure strong linkages across the product lifecycle, from development to manufacturing to supply. Third, we not only have to understand the underlying science, what risks there might be and mitigate these risks proactively before products reach the patient, but we have to communicate these risks effectively. So, for example, if there are a million tablets in a batch, we have to be sure, to the best of our ability, that these tablets have been produced of the appropriate quality, as each one may go to a different customer.

And this is where the term QbD comes in – sometimes referred to as 'a science and risk-based approach', and this set of words gives a little more insight into what QbD is about.

The definition of QbD [1] is:

> A systematic approach to development that begins with predefined objectives and emphasizes product and process understanding and process control, based on sound science and quality risk management.

The term 'Quality by Design' was first used by Juran in 1985 when the first draft of his book [2], published in 1992, was available for consultation by 50 senior representatives of industry. The Juran Trilogy stated: 'Managing for quality is done by use of the same three managerial processes: planning, control and improvement' [3].

This book will provide detail to expand this same idea into a practical reality, but for this introduction, we can consider QbD as helping the pharmaceutical industry to continue to take a science- and risk-based approach to enable safe and efficacious medicines to be developed and produced over the product lifecycle, that lifecycle being from the time the product is being conceived to the time it is finally withdrawn from the market, including managing the impact of any changes that may occur during this period.

1.2 Background

Historically, an ultra-compliant approach had dominated the way the pharmaceutical industry operated, perhaps even threatening, wrongly, to potentially swamp the underlying science, rather than compliance being seen as a partnership with science. Fear of seeking change for already approved regulatory documents, even when new enhanced science or technology developments came to light, has meant that industry continued to operate within compliance-driven, historically established boundaries. One example might be where manufacturing limits had been approved in a regulatory submission. Yet, over time, as more product and process knowledge accumulated to support widening – or maybe tightening – the original limits, industry had a fear of discussing this new knowledge with regulators. Sadly, this fear still partly exists today, though it is not as prevalent.

So why did compliance come to have such an overbearing role? Maybe it was a perceived fear of regulators? Or business pressure to have approvals in place to meet rapid launch of products ahead of competitors? Or was it to assess matters as either 'right' or 'wrong', rather than recognising there is a continuum of risk in regard to pharmaceuticals – or indeed in any product that is designed, manufactured and supplied to customers? It could be none, any or all of these reasons; one will never know.

This compliance-focussed mindset did not mean that quality problems were not occurring. Indeed, one could sense, looking back, some frustration within industry and also for regulators, particularly when things did go wrong, as they occasionally did. Regulators started to impose increasingly large fines, but this did not seem to resolve recurrence of quality issues.

This point was captured in a *Wall Street Journal* article of 3 September 2003 [4]:

> Dr. Woodcock had been among the architects of an FDA crackdown under which the agency fined drug makers as much as $500 million for manufacturing failures in recent years. Yet 'we still weren't seeing acceptable levels of quality', she says, because 'production techniques were outmoded. Just refining procedures and documentation wasn't going to change that'.

This article was significant as it gave a reflection of what was happening at that time both within the industry and for regulators, but it also recognised that the public were beginning to take a keener interest in the pharmaceutical industry and expecting regulators to continue to play their part in ensuring quality.

One other important factor was that the pharmaceutical market was becoming increasingly international, yet regulators, who were normally based in one country, were, understandably, focussed on ensuring their own particular local interest was protected, both for imports and for products made within their national boundaries.

Gradually it became apparent that a less local perspective would be beneficial, and in 1990 the International Conference on Harmonisation (ICH) was created [5].

ICH is an interesting concept as it brings together regulators, industry associations and observers from different parts of the world to meet and jointly write guidance documents. The members include US, EU and Japanese regulators and industry bodies, as well as observers from the World Health Organisation (WHO) and others.

As is stated on the ICH web site [5]:

> The ICH Parties comprise representatives from the following regulatory parties:
>
> - European Union, the Regulatory Party is represented by the European Commission (EC) and the European Medicines Agency (EMA).
> - Japan, the Regulatory Party is the Ministry of Health, Labour and Welfare (MHLW).
> - USA, the Regulatory Party is the Food and Drug Administration (FDA).
> - Canada, the Regulatory Party is the Health Products and Food Branch (HPFB).
> - Switzerland, the Regulatory Party is the Swissmedic.

As well as from the following industry parties:

- Europe, the European Federation of Pharmaceutical Industries and Associations (EFPIA).
- Japan, the Japan Pharmaceutical Manufacturers Association (JPMA).
- The United States, the Pharmaceutical Research and Manufacturers of America (PhRMA).

ICH produces guidelines under headings of Safety, Quality and Efficacy. It has eminent and broad-ranging groups of experts involved in producing these guidelines, so these guidances are as near to global as one can obtain, even though they are neither global nor mandatory, unless – as happens in some cases – regulatory agencies include them in their national Good Manufacturing Practices (GMPs). Most can be considered, to all intents and purposes, as internationally applicable.

It was ICH that first brought the term QbD to the pharmaceutical industry when in 2009 it published ICH Q8 (R2) [1], 'Pharmaceutical Development'.

This was a watershed moment for the industry, as, after this, the importance of taking a science- and risk-based approach moved to front stage, and even terms like 'manufacturing science' began to be heard.

1.3 Science- and Risk-Based Approaches

Science, of course, has always been a fundamental element of the development of pharmaceuticals and, historically, innovative application of science has been core to producing the many life-saving and life-enhancing drugs that the industry has produced over time.

So why all this supposedly new approach? What has changed?

Well, the fundamentals driving the need to understand pharmaceutical science remain the same, but perhaps the following are factors that influenced a change in perspective:

- The application of science is becoming more complex; for example, biotechnology-based drugs are more complicated to understand, make and analyse compared to small molecules; specialised therapies such as advanced therapeutic medicinal products, gene therapies, etc. are beginning to emerge.
- The use and application of more sophisticated tools, for example, process analytical technology (PAT), [6] has become more commonplace – although this tool is relatively new for the pharmaceutical industry, it has been in use by other industries for many years.
- More powerful data processing is now available to enable such tools to be used. An example is design of experiments (DoE) (see Chapter 7) and multivariate analysis (MVA) (see Chapter 8) can now be used for more sophisticated analysis than was possible previously.
- The industry has become more global, often with many differing countries involved in the supply chain, which has made it necessary to maintain quality across various international boundaries and cultures.
- The supply chain has become more fragmented and diverse, with many more parties involved, including contract research organisations (CROs), contract manufacturing organisations (CMOs) and external suppliers. 'Virtual' companies are now emerging, a role that did not exist a few years ago.
- Internal organisations are being re-structured. An 'over the wall' attitude for technology transfer, development and manufacturing is being heavily discouraged. Business benefits are being seen in having closer working internal partnerships.

- Knowledge management is becoming increasingly important, from development to manufacturing to supply. Knowledge is not just about knowing the pure science but is also about embracing the application of that science through technology, manufacture, engineering, materials science and many other disciplines.
- Regulatory pressures are continuing. The recent US Food and Drug Administration Safety and Innovation Act (FDASIA) law [7] and the current discussions about quality metrics is an example, as is the stronger emphasis on data integrity, lifecycle, process validation and many other topics.
- Public pressure continues to grow. The pharmaceutical industry is expected to not only do things right but to also reduce costs.
- Location of manufacture is moving east. India, China and other countries have more dominant and extensive pharmaceutical industries.

All these factors have reinforced the need to not only understand pharmaceutical science, but to also understand where any potential risks lie, to mitigate and control these risks, and to ensure clear communication over the lifecycle of the drug to deliver safe and efficacious products to patients.

So industry has always recognised the importance of science and that risks should be quantified.

But there is probably one major factor that has been a significant gap, and that is the rationale has not been as clearly articulated as it could have been.

So QbD is not just about doing the science- and risk-based work; it is also about explaining the story clearly, both verbally and in written form. Everyone who is a part of a product's lifecycle has to be made aware of their role in ensuring product quality is in place at all stages – be they a development scientist, an operator on the manufacturing plant, a business leader, a regulatory department, a third party supplier, an equipment vendor or anyone else involved in this, at times, complicated supply chain.

1.4 ICH Q8–Q12

The following published ICH Guidelines set forth the principles about how a science- and risk-based approach should be delivered.

The key ICH documents that make up the QbD 'family' are:

- ICH Q8 (R2) – 'Pharmaceutical Development' [1].
- ICH Q9 – 'Quality Risk Management' [8].
- ICH Q10 – 'Pharmaceutical Quality System' [9].
- ICH Q11 – 'Development and Manufacture of Drug Substances (Chemical Entities and Biotechnological/Biological Entities' [10].
- ICH Q12 – Concept paper (at the time of writing this) – 'Technical and Regulatory Considerations for Pharmaceutical Product Lifecycle Management' [11].

Taking each of these in turn, in brief:

ICH Q8 (R2) lays out the principles of using science for development of a drug product. It was the first ICH document to use the term 'enhanced, Quality by Design' approach. It includes two Parts and two Appendices. Part 1 is Pharmaceutical Development; Part 2 gives the Elements of Pharmaceutical Development, also introducing the terms laid out in the next section of this chapter, and details for Submissions; Appendix 1 is about differing approaches and gives examples of 'minimal' and 'enhanced, Quality by Design' approaches; Appendix 2 is Illustrative Examples. The development and manufacture of drug product with the application of ICH Q8 (R2) is discussed in detail in Chapter 6 of this book by Mark Gibson.

ICH Q9 lays out a framework on approaches for quality risk management, including risk initiation, assessment, control, review, communication and the tools to use. This framework is explained in more detail later in this book in Chapter 2 (Noel Baker). ICH Q9 has two Annexes: the first is on methods and tools to use, and the second is about applications. Importantly, ICH Q9 uses clear terms and definitions, and this author recommends being consistent and rigorous in using these terms, as they enable clearer communication both within a company and externally, be this to third parties or regulators.

ICH Q10, 'Pharmaceutical Quality Systems', lays out the fundamentals of what a quality system should cover, including management responsibility, and continual improvement of process performance and product quality and also of the quality system itself. Quality systems and knowledge management are discussed further in Chapter 3 of this book by Siegfried Schmitt.

ICH Q11, 'Development and Manufacture of Drug Substances (Chemical Entities and Biotechnological/Biological Entities)', is a partner to Q8, and is based on similar principles, with extensive use of the term 'enhanced'. Significantly, it covers API (active pharmaceutical ingredient) for both large and small molecules. It covers selection of starting materials, control strategy, process validation, submission and lifecycle and gives some examples. See Chapter 4 of this book by Gerry Steele for more details on the development and manufacture of small molecule drug substances.

ICH Q12, at the time of writing this book, is being drafted. The concept paper indicates it will cover the regulatory dossier, the quality system and lifecycle change management. Further reference to this and other regulatory guidance is given in Chapter 12 of this book.

All these guidelines form the basis of taking a science- and risk-based approach to cover the product and process lifecycle.

1.5 QbD Terminology

The ICH Q8–Q11 documents have helped bring great clarity to terms and definitions. The pharmaceutical industry is complex and does not help itself when companies or individuals use different language to describe in essence the same thing. Indeed, there are examples where the regulators have been concerned when there has been a lack of clarity.

The following are some of the key terms on which QbD is founded:

- Quality target product profile (QTPP).
- Critical quality attribute (CQA).
- Critical process parameter (CPP).
- Critical materials attribute (CMA).

- Design space (DS).
- Control strategy (CS).
- Lifecycle.

The following chapters in this book will expand, bring alive and give context to these terms.

Use of ICH terminology enables industry and regulators to use a common language both in-house as well as, say, during regulatory applications. No longer it is acceptable to sloppily mix terms such as *parameter* and *attribute*. They mean completely different things and using terms precisely enables clarity of communication within development and on to manufacturing and, indeed, over the full product lifecycle. So, it is strongly recommended that ICH terminology is used wherever possible.

1.6 QbD Framework

The following 'QbD framework' given in Figure 1.1 will be used for this book.

The following chapters will expand on and explain the elements of this diagram.

1.7 QbD Application and Benefits

QbD fundamentally links patient requirements to drug product and then drug substance. It is used to understand product specific requirements, which can then be supported by GMP [12].

QbD normally starts in development and progresses through to manufacturing, with the intent of producing a control strategy for commercial-scale production. Sometimes, say,

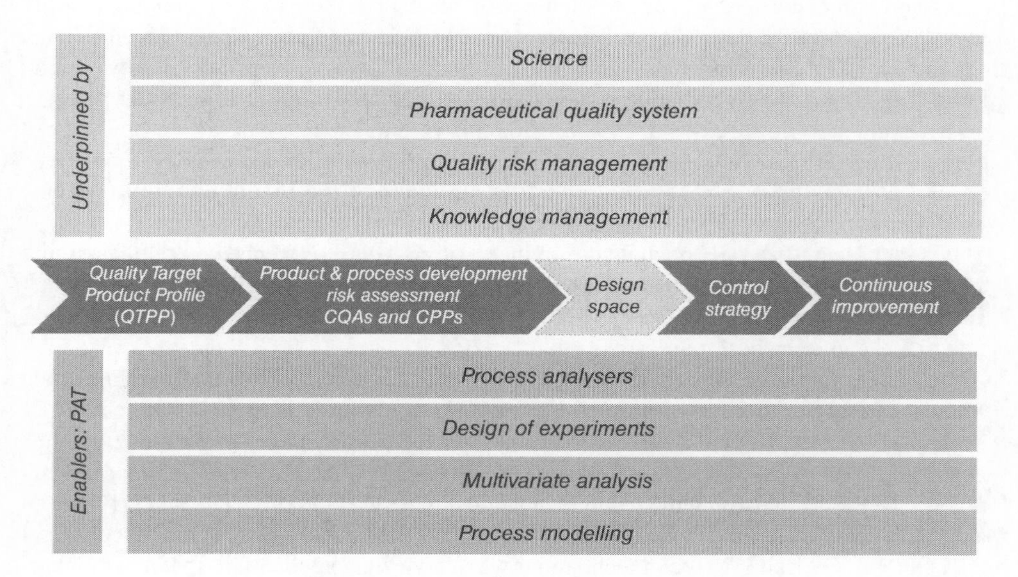

Figure 1.1 *A framework for QbD.*

with a legacy product, QbD may start with an existing manufacturing process, for example, where a rich history of product and process knowledge is available.

QbD can be applied to small and large molecules, to drug substance and drug product, to vaccines, to combination products, to all or parts of a process, to novel drugs or to generics. It can be used by leading companies, by contract research or contract manufacturing companies or 'virtual' companies. It is up to the particular organisation to decide an appropriate level and application of QbD. QbD can be applied nationally or internationally.

Understanding the science underpinning a product and its associated risks helps prioritise what is important for manufacturing and so normally leads to efficiency gains and cost benefits. On the basis of a survey of several companies, Reference [13] concludes that companies found strong business benefits in using QbD. Part of the Concluding Remarks stated the following:

> QbD seems strongly embedded in the companies interviewed. The benefits realized have met the expectations set by companies when they embraced QbD.... improved product and process understanding; a more systematic approach across the development portfolio; to continue to improve patient safety and efficiency; to improve manufacturing efficiency; and to improve development efficiency.

1.8 Regulatory Aspects

QbD is not mandatory, but product and process understanding is an expectation of regulators. So how does one obtain such understanding without considering QbD principles? With difficulty, is the answer!

As indicated earlier, legacy products with an established history can provide a wealth of qualitative data to confirm, for example, that ranges and acceptance criteria set during manufacturing have produced products of the appropriate quality. Such knowledge is extremely valuable, as regulators are not insisting that companies should go back and start doing experiments afresh to provide quantitative evidence (unless, of course, there is demonstrably a lack of product and process understanding), but they do expect an assessment has been made that products of the required quality can be produced.

It is significant that ICH documents are increasingly being referenced by regulators, in areas where compliance is expected. For example, validation is a regulatory requirement, and it is interesting that ICH Q8, Q9 and Q10 are referenced in guidelines for the United States, EU and elsewhere.

One other matter of importance relative to this area is the US FDA internal guideline Manual of Policies and Procedures, MAPP 5016.1, Applying ICH Q8(R2), Q9, and Q10 Principles to CMC Review [14]. As it says: 'This MAPP outlines and clarifies how the chemistry, manufacturing, and controls (CMC) reviewers in the Office of Pharmaceutical Science (OPS) should apply the recommendations in the ICH Q8 (R2), Q9, and Q10 guidances to industry'. It also interestingly says: 'OPQ product quality reviewers will consider ICH Q8(R2), Q9, and Q10 recommendations when reviewing applications that may or may not include QbD approaches'.

1.9 Summary

This book includes chapters on quality risk management, quality systems, knowledge management, development and manufacture of drug product and drug substance, the role of excipients, DoE, multivariate analysis, process analytical technology, manufacturing and process controls, analytical QbD and regulatory guidance. Within each of these chapters is a wealth of information written by practitioners who have been and are actively involved in this important subject.

So, in summary, QbD is about gaining product and process understanding, communicating it and delivering it, such that patients can continue to benefit from the medicines they may take.

Hopefully this book will help provide some tools to enable this to be put into practice.

1.10 References

[1] ICH Q8 (R2) (2009) *Pharmaceutical Development*, http://www.ich.org/fileadmin/Public_Web_Site/ICH_Products/Guidelines/Quality/Q8_R1/Step4/Q8_R2_Guideline.pdf (accessed 30 August 2017).

[2] Juran, J.M. (1986) The Quality Trilogy: A Universal Approach to Managing for Quality, *Quality Progress*, Volume **19** (8), pp. 19–24.

[3] Juran, J.M. (1992) *Juran on Quality by Design: The New Steps for Planning Quality into Goods and Services*, 1st ed., The Free Press, New York, USA.

[4] Abboud, L. and Hensley, S. (2003) *New Prescription for Drug Makers: Update the Plants*, http://www.wsj.com/articles/SB10625358403931000 (accessed 30 August 2017).

[5] ICH (1990) *The International Council for Harmonisation of Technical Requirements for Pharmaceuticals for Human Use*, http://www.ich.org/home.html (accessed 30 August 2017).

[6] FDA (2004) *Guidance for Industry PAT – A Framework for Innovative Pharmaceutical Development, Manufacturing, and Quality Assurance*, http://www.fda.gov/downloads/Drugs/…/Guidances/ucm070305.pdf (accessed 30 August 2017).

[7] FDA (2012) *Safety and Innovation Act (FDASIA)*, http://www.fda.gov/RegulatoryInformation/Legislation/SignificantAmendmentstotheFDCAct/FDASIA/ucm20027187.htm (accessed 30 August 2017).

[8] ICH Q9 (2005) *Quality Risk Management*, http://www.ich.org/fileadmin/Public_Web_Site/ICH_Products/Guidelines/Quality/Q9/Step4/Q9_Guideline.pdf (accessed 30 August 2017).

[9] ICH Q10 (2008) *Pharmaceutical Quality System*, http://www.ich.org/fileadmin/Public_Web_Site/ICH_Products/Guidelines/Quality/Q10/Step4/Q10_Guideline.pdf (accessed 30 August 2017).

[10] ICH Q11 (2012) *Development and Manufacture of Drug Substances (Chemical Entities and Biotechnological/Biological Entities)*, http://www.ich.org/fileadmin/Public_Web_Site/ICH_Products/Guidelines/Quality/Q11/Q11_Step_4.pdf (accessed 30 August 2017).

[11] ICH Q12 (2014) *Concept paper: Technical and Regulatory Considerations for Pharmaceutical Product Lifecycle Management*, http://www.ich.org/fileadmin/Public_Web_Site/ICH_Products/Guidelines/Quality/Q12/Q12_Final_Concept_Paper_July_2014.pdf (accessed 30 August 2017).

[12] MHRA (2014) *Good Manufacturing Practice and Good Distribution Practice*, https://www.gov.uk/guidance/good-manufacturing-practice-and-good-distribution-practice (accessed 30 August 2017).

[13] Kourti, T. and Davis, B. (2012) The Business Benefits of Quality by Design (QbD), *Pharmaceutical Engineering – The Official Magazine of ISPE* **32** (4), 1–10.

[14] FDA (2011) *Manual of Policies & Procedures (CDER) – Office of Pharmaceutical Science MAPP5016*.**1**, http://www.fda.gov/downloads/AboutFDA/CentersOffices/OfficeofMedicalProductsandTobacco/CDER/ManualofPoliciesProcedures/UCM242665.pdf (accessed 30 August 2017).

2

Quality Risk Management (QRM)

Noel Baker
AstraZeneca, United Kingdom

2.1 Introduction

Risk management principles have been established for several decades and are utilised by many business and government sectors to control and mitigate harm to the consumer. Examples of these sectors – though this is not an exhaustive list – include finance, insurance, occupational safety and public health as well as the government and independent agencies regulating these sectors [1].

Risk in the pharmaceutical sector is defined as the combination of the probability of occurrence of harm and the severity of the harm. For example, the well-established law for Control of Substances Hazardous to Health (COSHH) implemented by the Health and Safety Executive (HSE) [2] in the United Kingdom requires businesses that undertake activities with hazardous substances to prevent or reduce exposure of workers to risks to health. The health risks associated with these hazardous substances are typically detailed in a Chemical Safety data sheet, which businesses can use to:

- Find out what the health hazards are.
- Decide how to prevent harm to health (perform a risk assessment).
- Provide control measures to reduce harm to health.
- Make sure they are used.
- Keep all control measures in good working order.
- Provide information, instruction and training for employees and others.
- Provide monitoring and health surveillance in appropriate cases.
- Plan for emergencies.

Pharmaceutical Quality by Design: A Practical Approach, First Edition.
Edited by Walkiria S. Schlindwein and Mark Gibson.

It is this shared understanding of COSHH, to determine and control risk to harm, which has enabled both industry and the regulators to effectively communicate and maintain appropriate awareness of, and control and prevent, risks to health.

A reason for using COSHH as an example here is to demonstrate that risk management is more than performing a one-off risk assessment, this is important to understand the principles of quality risk management (QRM) described later in this chapter.

Specific definitions to help consolidate the differences between risk management and risk assessment are as follows:

Risk management is a systematic and methodical approach to developing an understanding of the variability of a process or procedure, including all associated hazards and failure modes, and implementing means of controlling or eradicating the risk in a given process or procedure.

Risk assessment is a one-off activity that utilises an appropriate tool to capture perceived risks to a process or procedure and then with appropriate expertise to determine options to control or mitigate the risk. The output from the assessment will inform the business on appropriate mitigation plans and associated controls.

It was agreed by the International Conference on Harmonisation (ICH) members in 2003 that a shared understanding of risk management was generally absent across the pharmaceutical industry and associated regulating agencies. In addition, the ICH members felt that the pharmaceutical industry underutilised both risk management and risk assessment through a medicinal products lifecycle. Maintenance of the quality of a medicinal product supplied to the consumer, in this case a patient requiring a medicine to address an urgent need, is of critical importance to understanding sources of harm and ultimately prevent harm from occurring.

It was expected that without this shared understanding of QRM, resources and time could be consumed inefficiently by both Pharma and regulators while discussing potential and/or realised causes of harm, even as supply of product to patients declines. Achieving a shared understanding of the application of risk among stakeholders is extremely important, otherwise each stakeholder might perceive different potential harms, place a different probability on each harm occurring and attribute different severities to each harm.

The perceived impact of not having a standardised approach to risk management in the pharmaceutical industry is not solely restricted to reduced supply to patients. As disclosed in ICH Q9 Final Concept paper [3], the impact was seen to be much wider:

- Product may not be available to patients, when needed.
- May increase the potential for the release of unacceptable product to the market;
- New product introductions to the marketplace may be delayed.
- Delays may occur during implementation of changes and improvements to processes.
- Safe and effective drugs may be discarded or recalled from the market.
- Manufacturers may be reluctant to implement new technologies or continuous improvements to the products or processes.
- Scarce resources may not be optimally allocated.
- Lack of appropriate data to evaluate risk most effectively.

It is the problems disclosed in this section that prompted the ICH members to form an expert working group in 2003 whose remit was to develop and establish a harmonised approach to risk management across the pharmaceutical sector. The output of this effort was the ICH Q9 'Guidance on Quality Risk Management' [4].

2.2 Overview of ICH Q9

ICH Q9 Quality Risk Management was introduced on 9 November 2005 [4], its goal being a harmonised approach to risk management. ICH Q9 was one of the new guidelines released by the ICH to support the launch of Quality by Design (QbD). As described in the introduction to ICH Q9:

> The manufacturing and use of a medicinal product, including its components, necessarily entails some degree of risk. The risk to its quality is just one component of its overall risk. It is therefore, important to understand that product quality should be maintained throughout the product lifecycle such that the attributes that are important to the quality of the medicinal product remain consistent with those used during clinical studies. [4]

It is therefore no surprise that QRM should be regarded as a cornerstone of QbD and the development of a medicinal product.

The definition of QbD from ICH Q8R2 [5] 'Pharmaceutical Development' further reinforces this conviction:

> A systematic approach to development that begins with predefined objectives and emphasises product and process understanding and process control, based on sound science and *quality risk management*.

It is recognised across the pharmaceutical industry that ICH Q9 provides an excellent high-level framework for the use of risk management in pharmaceutical product development [4]. The ICH Q9 guideline, like each of its companion ICH guidelines, is well laid out and clear in content and purpose. The guideline discloses regulators' expectations on scope, principles, responsibilities, QRM process overview (a graphical representation of the process is presented later in this section; see Figure 2.2) and gives examples of risk management methodology and tools.

One key message to reflect on from the guideline is that the medicinal product development effort can be linked to the level of determined risk to the patient, as the amount of effort used for risk control should be proportional to the significance of the risk. In other words, Pharma have an opportunity to use development resources to focus on what risks are critical rather than all identified risks, and there is a general agreement that risks carrying a low probability of harm can be accepted if they can be justified, thus driving development efficiency and potentially shortening development timelines and increasing speed of product launch to the patients in need.

With the inception of each ICH guidance, the expectation is that each ICH guideline, inclusive of QRM, will not be routinely used in isolation. It is clearly stated in the definition of QbD that ICH Q9 should also be used in conjunction with ICH Q8(R2) 'Pharmaceutical Development' [5], ICH Q10 'Pharmaceutical Quality System' [6] and ICH Q11 'Development and Manufacture of Drug Substance' [7]. A pictorial representation of the application of the four ICH guidelines versus a typical medicinal product development is presented in Figure 2.1. The purpose of the implementation of the new guidance was not to change the traditional route to medicinal product development. Instead, the intention was to offer a supportive structure to cater to the needs of patients via a focus on what is critical to patient safety and product performance.

ICH Q9 specifically addresses the management and assessment of risk; however, neither risk management nor an assessment can be performed in the absence of clear goals. ICH Q8(R2) 'Pharmaceutical Development – Annex' [5] elaborates on possible approaches to forming the basis of a design brief, for example, a quality target product profile (QTPP) for the development of a medicinal product based on the un-met need of the patient. In addition, ICH Q8(R2) 'Pharmaceutical Development – Annex' also provides illustrative examples of how to determine criticality of a process or product.

Terms commonly used by both Pharma and regulators for assigning criticality are critical material attributes (CMAs), critical process parameters (CPPs) and critical quality attributes (CQAs). It is through linking of material attributes and process parameters to the product CQAs which is of paramount importance to assuring patient safety and product performance at the point of use. A more comprehensive review of the identification of CMAs and CPPs for both drug substance and drug product and application of design of experiments (DoE) is presented by Gerry Steele (Chapter 4, 'Quality by Design (QbD) and the Development and Manufacture of Drug Substance') and Mark Gibson *et al.* (Chapter 6, 'Development and Manufacture of Drug Product').

As presented in Figure 2.1, the purpose of the implementation of the new guidance was not to change the traditional route to medicinal product development. Instead, the intention was to offer a supportive foundation to cater to the needs of patients via a focus on what is critical to patient safety and product performance.

The process detailed in ICH Q9 (Figure 2.2) is structured to support identification of what is critical, what should be controlled and what level of risk is acceptable to the patient.

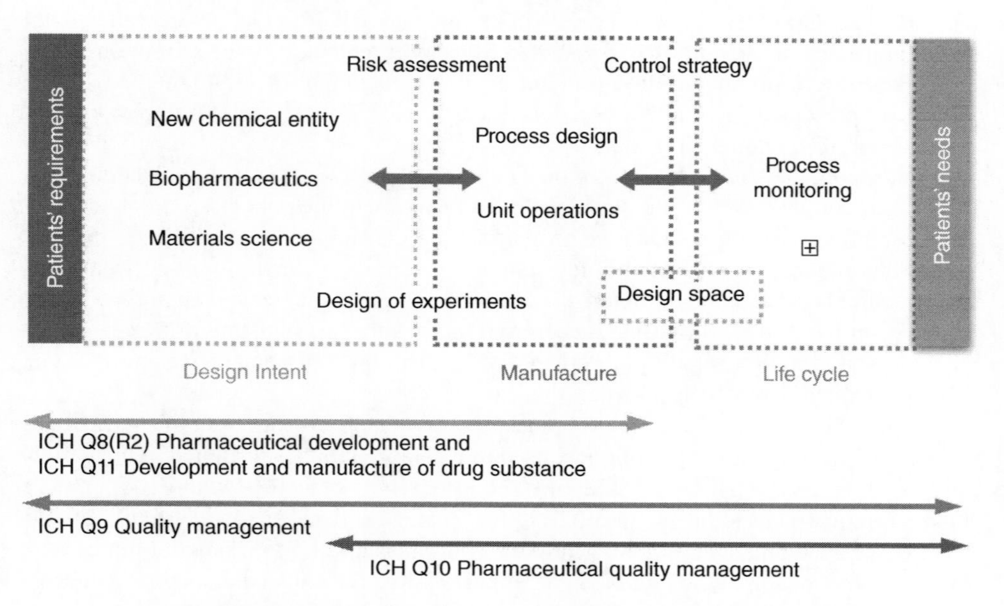

Figure 2.1 *Medicinal product development flowchart.*

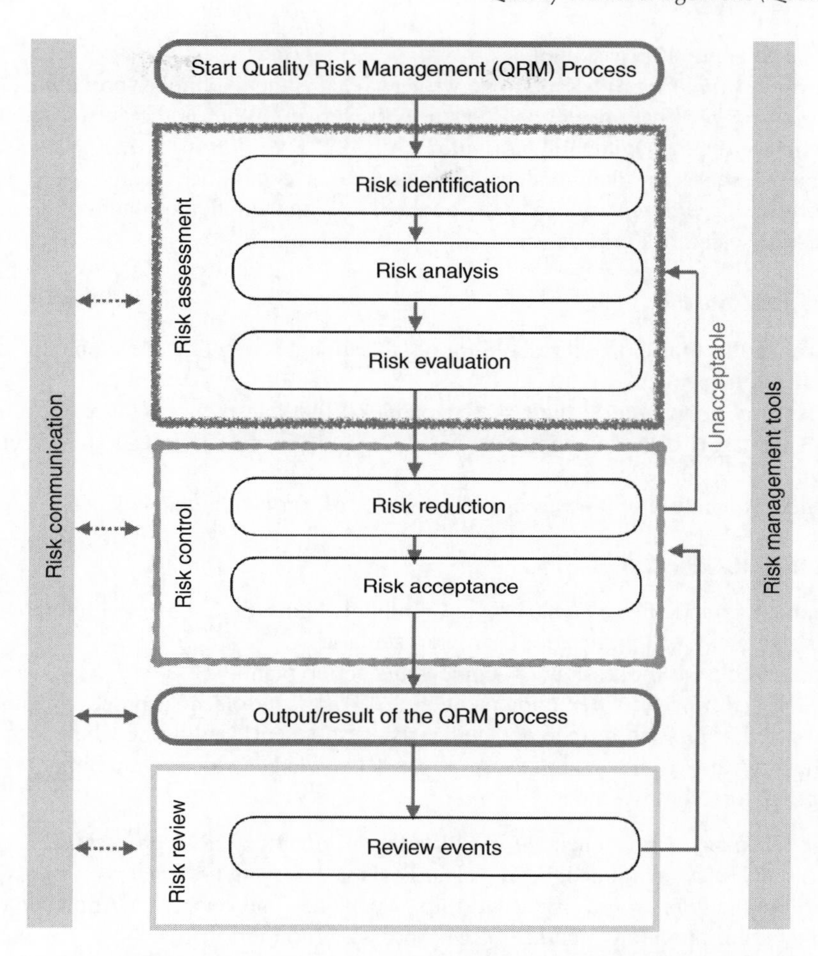

Figure 2.2 *Overview of a typical quality risk management process (ICH Q9).*

ICH Q9 does a very good job of defining each stage in the process presented in Figure 2.2; however, to help the reader, additional clarifications and considerations for each stage are summarised in the following sub-sections.

2.2.1 Start QRM Process

- Determine the scope/focus area of the QRM process and the risk question (e.g. risks to achieving QTPP, specification, either generally or for specific supply threats).

2.2.2 Risk Assessment

- Collect and organise prior knowledge of the subject.
- Perform *risk identification*, which is the identification of threats to quality and ultimately threats to patient safety.

- Choose an appropriate risk tool.
- Perform *risk analysis*: What might go wrong? What is the likelihood (probability) that it will go wrong? What are the consequences (severity)? What is the current method of detection of the defined quality attribute?
- Define risk scales for likelihood, consequence and detectability.
- Perform *risk evaluation*, which is the qualitative or quantitative evaluation of each risk versus the scaled risk criteria.

2.2.3 Risk Control

- Determine the appropriate threshold for organisational action (e.g. development studies; specification change).
- Perform *risk reduction*: Is the risk acceptable? What can reduce, control or eliminate risk? Can earlier detection be implemented (e.g. on-line measurements) to detect threat sooner?
- *Risk acceptance*: Determine acceptable level of risk for patient and business.

2.2.4 Risk Review

- Document, communicate and approve: Communications might be through internal business or with external parties, for example, regulators.
- Maintain living risk assessment document and action plan.
- Ongoing *risk review*: After implementation of risk controls, if a process changes, or unexpected process shift, then update the risk assessment. Determine whether there are any new threats to quality, and review any new knowledge which might impact the original regulatory control strategy.

There are many dependencies to performing an effective QRM process; these stretch from, as previously mentioned, having clear goals of assessing potential risks (e.g. QTPP), to instilling a culture where risk ownership, acceptance and review are embedded in the team and organisation (e.g. internal QRM process). A non-exhaustive list of points to consider for a QRM process are the following:

- Internal guidance, procedures on application of QRM, for example, pharmaceutical business standard operating procedure (SOP).
- Internal training QRM and associated tools.
- Clearly defined quality attributes/targets, for example, QTPP, clinical or proposed commercial specification for the medicinal product or defined problem including pertinent assumptions.
- Available and easily accessible knowledge on the subject, for example, development reports and analytical data.
- Leader and multi-disciplinary team identified with technical expertise in the subject of assessment, for example, materials, formulation, process, analytical and statistical scientists.
- Appropriate tools to deliver the required tasks, for example, risk assessment failure mode and effects criticality analysis (FMECA) and hazard and operability study (HAZOP).
- Cultural acceptance and internal agreement on what level of risk is tolerable to help drive development and prioritisation of higher-risk components.

- Recognition that QRM is a cyclical process; for example, risk assessments should be re-visited at an appropriate time when more knowledge has been obtained, such as prior to and during technical transfer, or prior to commencement of a significant change programme.
- Clear accountability of ownership of risk and clearly defined lines of communication.

A significant output from QRM, if utilised appropriately during the development of a medicinal product, is a regulatory submission containing a sound, robust regulatory control strategy backed up by technical justification and evidence that the controls applied are appropriate for patient safety and efficacy and can meet the defined CQAs. This does not, however, mean that QRM for a medicinal product stops at approval; change should be expected post approval (e.g. a material supplier changes, manufacturing site changes, unexpected events) which will warrant reflection and re-assessment of existing controls to maintain the quality to the patient. All prior development knowledge and more recently acquired process performance metrics should be utilised as part of the re-assessment. In fact, it is the ICH's expectation that QRM will be used throughout the lifetime of the product and utilised as part of change management [8].

One final note for this section is about the increased utilisation and recognised importance of the use of statistical tools to monitor process performance and the understanding of sources of variation. The expectation that continued process verification (CPV) will be applied to commercialised medicinal products has enhanced the quality and amount of data generated both during development and commercial manufacturing. A more comprehensive review of CPV is presented by Mark Gibson in Chapter 11, 'Manufacturing and Process Controls'.

This on its own is very valuable for re-assessing risks, but now with the more routine application of control charts and capability metrics, the pharmaceutical sector is in a better position than ever to escape testing in quality (a reactive paradigm). Through application of statistical tools, a greater understanding is achieved of sources of variation associated with materials, processes and analytical methodology during development and post market approval, and this additional information will support and inform risk management.

2.3 Risk Management Tools

As defined in ICH Q9 [4]:

> Quality risk management supports a scientific and practical approach to decision-making. It provides documented, transparent and reproducible methods to accomplish steps of the quality risk management process based on current knowledge about assessing the probability, severity and sometimes detectability of the risk.

Therefore, it should not be forgotten that a risk assessment is not just 'a one-off event which once performed is filed away and forgotten'; a risk assessment is part of a process designed to identify the materials and manufacturing process parameters critical to the product. A risk assessment should be performed at key stages of medicinal product development and as part of post-approval change control and is an iterative process where any previous associated risk assessment should be used as the basis of performing the next risk assessment. For example, it is good practice to perform a risk assessment prior to establishment of the product and prior to and post validation/commercialisation. The risk assessments can be used to inform and direct development and technical studies and assure development and maintenance of a robust control strategy at both the commercial manufacturing site and disclosure in regulatory submissions. The usefulness of risk assessments for any significant change programme during the commercial phase – for example, site-to-site manufacturing process transfers – and for assessing threat to patient safety for any unexpected quality event should also not be underestimated.

A key early step in the execution of an effective risk assessment is to clearly define the risk question to provide context, scope and focus to the objective of the QRM exercise. The next step is to determine the appropriate risk assessment tool or methodology. There is generally no single or restricted set of tools applicable to any given assessment process, and the selection of the appropriate risk methodology should be based on the depth of analysis required, complexity of the subject and familiarity with the assessment tool. This section will provide a non-exhaustive list of risk assessment tools currently in use within the pharmaceutical industry.

ICH Q9 provides an example of risk management methodologies which could be used and adapted based on the needs of the risk assessment. The list is provided as an example, and the ICH does not intend the list to be prescriptive, as long as an appropriate approach to risk management is taken and an appropriate tool or methodology is used to assess risk. To be honest, there is no wrong methodology, and simple tools can be very effective for quickly identifying risks associated with simple risk questions. The more complex tools are usually retained, for example, for more a generalised risk review of manufacturing processes; however, again, both basic and advanced tools are interchangeable.

A useful review was undertaken in 2008 by a working group of industry and FDA representatives commissioned by the Pharmaceutical Quality Research Institute Manufacturing Technology Committee (PQRI-MTC). The group was set up to seek out good case studies of actual risk management practices used by Pharma and bio-pharmaceutical companies for the purpose of sharing with the industry at large [9]. The working group generated an extremely useful table detailing the most commonly used risk assessment tools. It is no surprise that the tools presented in Table 2.1 are commensurate with the tools described in ICH Q9, and demonstrates some success of the intended harmonisation of QRM approaches by the ICH members.

Of the case studies evaluated by the PQRI-MTC, the most commonly used basic risk assessment tools were found to be flowcharts, check sheets (see Table 2.2) and risk ranking. For more sophisticated analysis needs, failure mode effect analysis (FMEA) and hazards analysis and critical control points (HACCP) were found to be the most common tools used. In addition, since the case study described, there has been a demonstrable preference in the pharmaceutical industry for FMECA when judging risks associated between the manufacturing process and medicinal product quality (CQAs), typically performed under the umbrella term quality risk assessment (QRA).

It should come as no surprise that FMECA is an extension of FMEA. The key difference is the inclusion of the degree of severity of the consequence of the effect.

A fundamental component of risk ranking, FMEA and FMECA tools is the need to have general agreement on the risk scales for probability (likelihood) and detectability (and consequence (severity) for FMECA alone). These must be defined and agreed prior to undertaking a risk assessment to prevent any disagreement during risk discussion which

Table 2.1 Commonly used risk assessment (RA) tools (adapted from [9]).

	RA Tool	Description / Attributes	Potential Applications
Basic Tools	Diagram analysis • Flowcharts • Check sheets • Process mapping • Cause/effect diagrams	• Commonly used to gather/organise data, structure risk management and facilitate decision making	• Compilation of observations, trends or other empirical information to support a variety of less complex deviations, complaints, defects or other circumstances
	Risk ranking and filtering	• Method to compare and rank risk • Typically involves evaluation of multiple diverse quantitative and qualitative factors for each risk, and weighting factors and risks scores	• Prioritise operating areas/sites for audit/ assessment • Useful for situations when the risks and underlying consequences are diverse and difficult to compare using a single tool
Advanced tools	Fault tree analysis (FTA)	• Method used to identify all root causes of an assumed failure or problem • Used to evaluate system/sub-system failures one at the time, but can combine multiple causes of failure by identify casual chains • Relies heavily on full process understanding to identify casual factors	• Investigate product complaints • Evaluate deviations
	Hazard operability analysis (HAZOP)	• Tool assumes that risk events are caused by deviations from the design and operating intentions • Uses a systematic technique to help identify potential deviations from normal use or design intentions	• Access manufacturing processes, facilities and equipment • Commonly used to evaluate process safety hazards

(Continued)

Table 2.1 *(Continued)*

RA Tool	Description / Attributes	Potential Applications
Hazards analysis and critical control points (HACCP)	• Identify and implement process controls that consistently and effectively prevent hazard conditions from occurring • Bottom-up approach that considers how to prevent hazards from occurring and/or propagating • Emphasises strength of preventive controls rather than ability to detect • Assumes comprehensive understanding of the process and that critical process parameters (CPPs) have been defined prior to initiating the assessment. Tool ensures that critical process parameters will be met	• Better for preventive applications rather than reactive applications • Great precursor or complement to process validation • Assessment of the efficacy of CPPs and the ability to consistently execute them for any process
Failure mode effects analysis (FMEA)	• Assesses potential failure modes for processes, and the probable effect on outcomes and/or product performance • Once failure modes are known, risk reduction actions can be applied to eliminate, reduce or control potential failures • Highly dependent upon strong understanding of product, process and/or facility under evaluation • Output is a relative 'risk score' for each failure mode	• Evaluate equipment and facilities; analyse a manufacturing process to identify high-risk steps/critical parameters

could undermine the whole process. Risk scales will typically contain textual or numerical levels to help separate risks with a textual definition/description. In the case of FMEA, the risk tool may be used in conjunction with a risk ranking table (where the x scale = probability, and the y scale = detectability (see Table 2.3)), to help the organisation undertaking the QRM process to communicate risk visually. In the case of FMECA, a numerical risk scale is likely to be used, and the numerical score for probability, consequence and detectability will be multiplied to provide a risk priority number (RPN) to facilitate risk ranking and focusing of effort.

The final and no less important component of the QRM process is having a general internal agreement on the 'Threshold for Action'. Without this in place, there is a chance the organisation will be compelled to investigate every risk, no matter how unlikely. For example, if using the risk ranking table, the organisation must agree on the risk placement acceptable for no further investigation or, alternatively, that warrants investigation. For FMEA or FMECA tools, the organisation must agree on the priority or RPN score considered acceptable for no further investigation or, alternatively, that warrants investigation.

A pictorial overview of each of the commonly used risk assessment tools, with appropriate risk scales, is provided in Figure 2.3 and Figure 2.4 and Table 2.1, Table 2.3, Table 2.4, Table 2.5 and Table 2.6. A more in-depth case study on the use of FMECA is provided in Section 2.4.

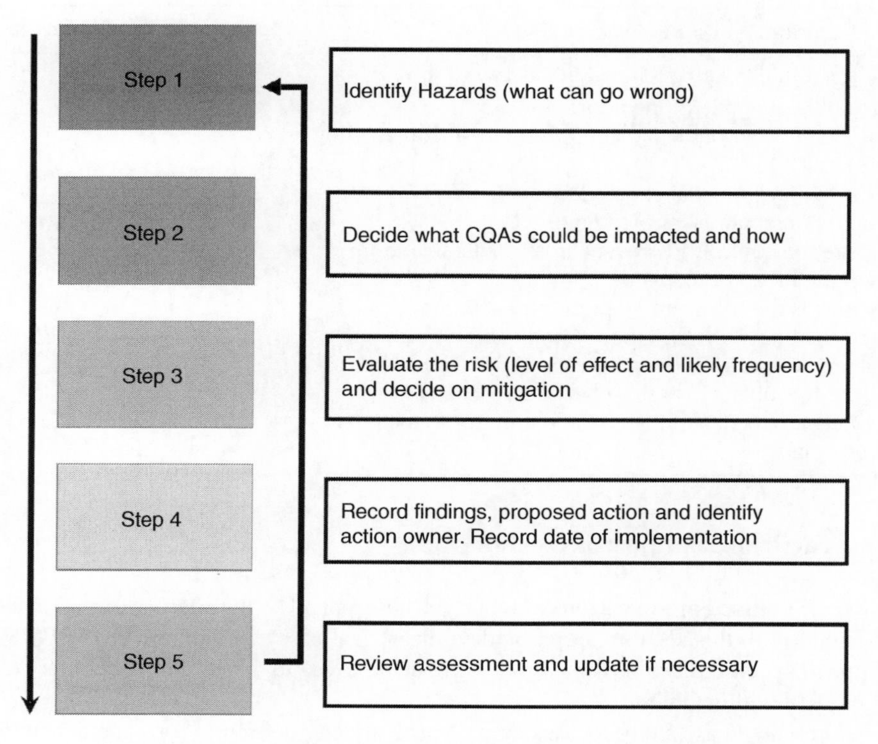

Figure 2.3 *Example of risk assessment tool: flowchart (basic tool).*

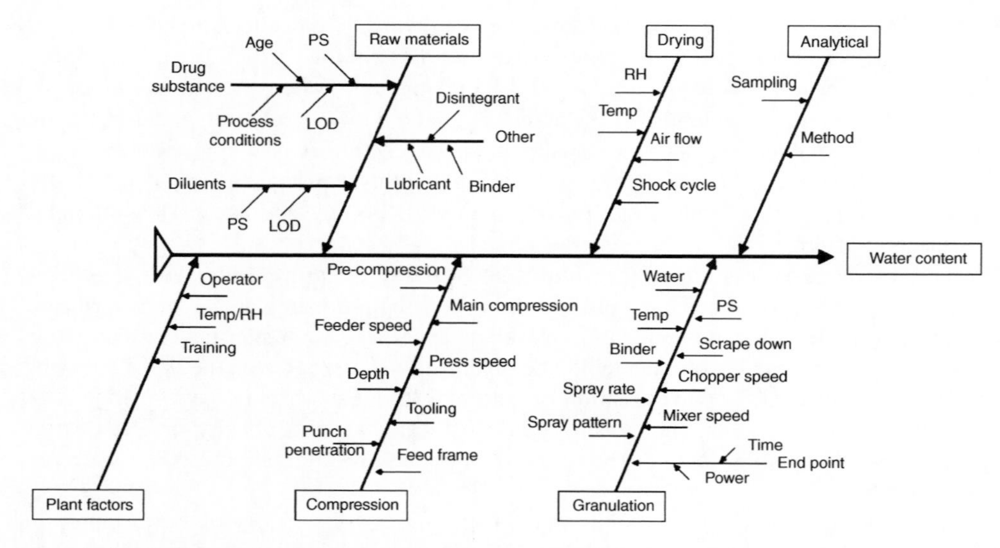

Figure 2.4 *Example of a cause and effect diagram (Ishikawa fishbone diagram).*

Table 2.2 *Example of risk assessment tool: check sheet (basic tool).*

Manufacturing (oral solid dosage OSD)	Comments and observations
How is the raw material handled between the dispensing area and the manufacturing floor?	
What are the environmental controls on the manufacturing floor?	
How is the raw material charged to the blender?	
What are the controls for blending?	
What are the controls to prevent cross-contamination?	
How is the blend discharged?	
How long is the blend stored prior to next process stage?	
How is the blend charged to the compression machine?	
What are the controls for compression?	
How are the tablet cores stored in after discharge?	
How long are the tablet cores stored for prior to the next process stage?	

2.4 Practical Examples of Use for QbD

Section 2.3 outlined risk management tools typically applied by the pharmaceutical industry. The intention of this section is to provide a theoretical case study on the use of QRM. In addition, a specific risk assessment tool will be employed to add additional context to the use of QRM within QbD.

The presented case study focuses on a control strategy definition for a medicinal product. A QRA is at its core, and the expected output from the development team will include

Table 2.3 *Example of risk assessment tool: risk ranking (basic tool).*

Example Risk Ranking Table:

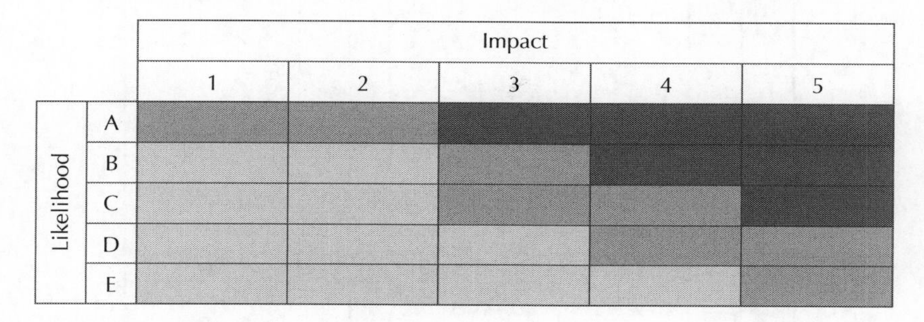

Example Risk Ranking Scoring:

A = Almost certain	(1:2)	1= Insignificant
B = Likely	(1:10)	2= Stop of process flow
C = Possible	(1:100)	3= Batch deviation
D = Unlikely	(1:1000)	4= Does not meet specification
E = Rare	(1:10000)	5= Only detected by patient

Table 2.4 *Example of risk assessment tool: failure mode effect analysis (FMEA) (advanced tool). Note: red = dark grey; amber = light grey and green = medium grey*

Example FMEA Table:

Failure mode	Failure effect	L	D	P	Risk mitigation	Owner
Compression force too high	Tablet capping				Ascertain max. force which can be applied	J. Bloggs
Poor blend homogeneity	Poor tablet content uniformity				Determine appropriate blending parameters to obtain suitable blend uniformity	J. Bloggs

L = likelihood; D = detectability; P = priority

Example FMEA Scoring:

Likelihood	*Detectability*	*Priority*
Red (1:2)	Red – By patient/on stability	Organisations may weight L and D differently to achieve a prioritisation score
Amber (1:100)	Amber – Release testing	
Green (>1:10000)	Green – Before/during manufacture	

additional studies to complement the commercial process design and a series of control items, which is likely to include a combination of CMAs and CPPs and an understanding of method capability via analytical method measurement system analysis (MSA). A more comprehensive review of the identification of CMAs and CPPs for both drug substance and drug product and application of DoE is presented by Gerry Steele (Chapter 4, 'Quality by

Table 2.5 *Example of risk assessment tool: failure mode effect and criticality analysis (FMECA) (advanced tool).*

Example FMECA Table:

Quality attribute	Control target	Failure mode	Failure effect	S	L	D	RPN	Risk mitigation	Owner
Description	Meets AQL	Compression force too high	Tablet capping	5	5	1	75	Ascertain max. force which can be applied	J. Bloggs
Tablet content uniformity	Meets BP acceptance limit	Poor blend homogeneity	Poor tablet content uniformity	10	1	5	150	Determine appropriate blending parameters to obtain suitable blend uniformity	J. Bloggs

Example FMECA Scoring:

Severity	*Likelihood*	*Detectability*	*Priority*
10 Patient affected	10 (1:2)	10 – By patient/on stability	**RPN = S x L x D**
5 Batch failure	5 (1:100)	5 – Release testing	
1 Stop in production flow	1 (>1:10000)	1 – Before/during manufacture	

Table 2.6 *Example of risk assessment tool: hazards analysis and critical control points (HACCP) (advanced tool).*

Process	Critical parameter	Hazard	Control	Limit	Monitoring (what, how, when)	Person	Record	Corrective action
Compression	Compression force	Over-compression can cause tablet capping	Maintain force below 14 kN	Experience has linked capping to compression forces in excess of 15 kN	**What:** max compression force **How:** compactor integrated force transducer **When:** check every 10 min	Facility engineer	Batch record	Perform experimental studies and in-process monitoring to build body of data to qualify 14 kN is appropriate in-process limit
Blending	Blending time	Inadequate blending leads to poor tablet	Blending time greater than 10 min	Experience has linked blending times greater than 10 min	**What:** min and max blend time controlled by manufacturing recipe **How:** calibrated timer with auto shut-off of blender **When:** annual calibration	Facility engineer	Annual calibration certificate	Perform experimental studies and in-process monitoring to build body of data to qualify 5 to 10 min as an appropriate in-process range.

Design (QbD) and the Development and Manufacture of Drug Substance') and Mark Gibson, Alan Carmody, and Roger Weaver (Chapter 6, 'Development and Manufacture of Drug Product'). The case study also refers to the analytical method MSA, which is described in greater detail by Joe de Sousa, David Holt and Paul Butterworth (Chapter 10, 'Analytical Method Design, Development, and Lifecycle Management').

During the presentation of the theoretical case study in the following sections, Notes are provided to enhance the readers' understanding.

A single comprehensive example of QRM in use is presented in the following sections; however, there are many good examples of the application of QRM in the literature. The most comprehensive demonstration of the application of QRM during QbD development is the regulatory submission for the fictitious drug molecule Acetriptan ('ACE') [10]. Another good alternative source of QRM examples is contained in a presentation provided on the Pharmaceuticals and Medical Devices Agency (PMDA) web site provided by Pfizer (Author unknown) [11].

2.4.1 Case Study

A medicinal product has been identified by a pharmaceutical business as a likely candidate for commercialisation. The product development team has been given the remit to define a regulatory control strategy (formulation, process and analytical controls) for the medicinal product. The medicinal product is an immediate release tablet using conventional excipients and is manufactured using a conventional direct compression process train. The team performs a QRA to determine the capability of the current process to achieve the CQAs both presented in the QTPP and clinical specification and identify any priority areas for further assessment to achieve a control strategy definition.

Note: The QRA functions as an iterative process, similar to that outlined in Figure 2.1, and is an assessment of the risk of the input materials, process parameters and analytical methodology to the CQAs of the medicinal product.

A QRA had previously been performed by the project team, and product development and analytical method robustness studies had been undertaken using designs of experiments (DoEs) to enable the project team to undertake an informed risk assessment of the controls for the medicinal product and associated process and analytical methodologies.

The team undertook the QRA process as described in Figure 2.5. Additional details on the pre-work and scoring step are provided later in this section.

Note: As you will see, the process offers a lot of similarity to Figure 2.2, the overview of a typical QRM process (adapted from ICH Q9).

2.4.2 Pre-work

In order to prepare for an effective FMECA risk review, a risk matrix containing a list of failure modes and failure effects was prepared by the development team. The risk matrix was generated prior to performing the FMECA to utilise the knowledge of the development team and assessed versus the medicinal product QTPP. The finalised risk matrix was then used to pre-populate the FMECA tool. The risk matrix considered all known/potential relationships between every raw material and/or process parameter (failure mode) and every quality

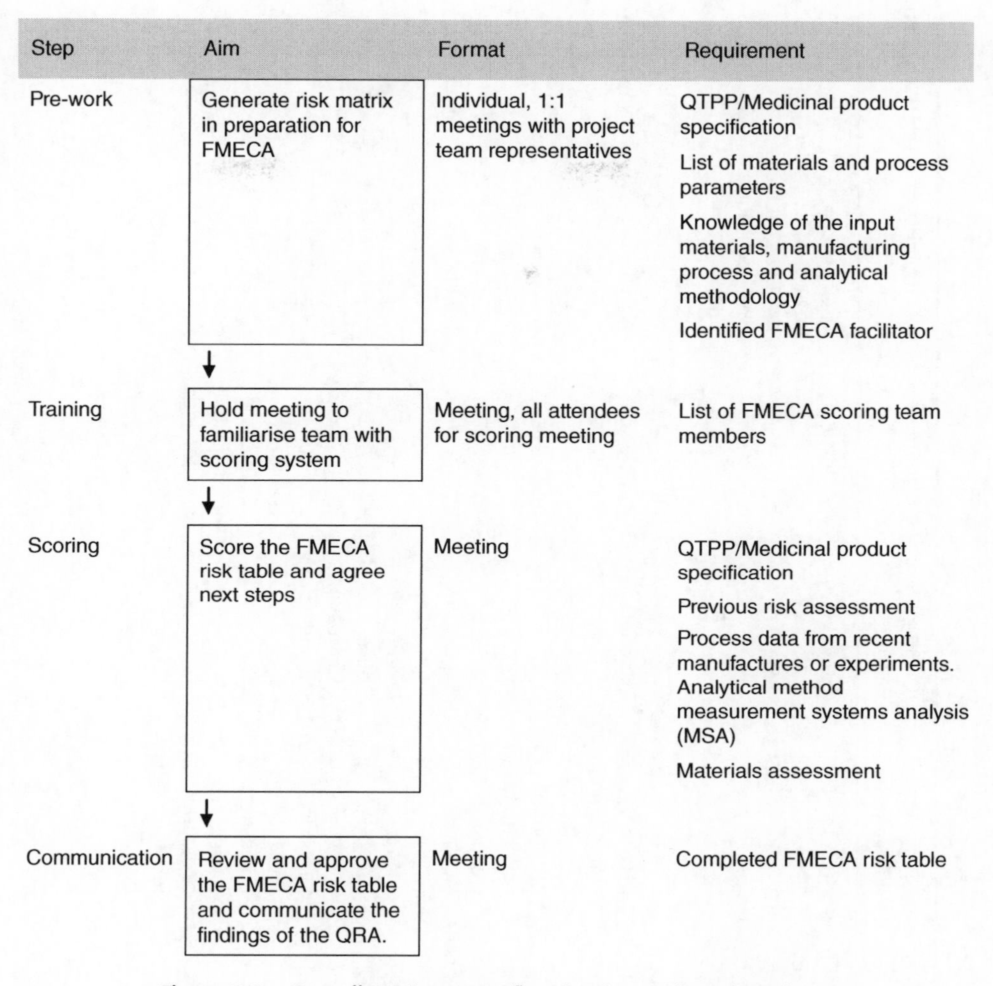

Figure 2.5 *Overall QRA process flow for the medicinal product.*

attribute (failure effect) taken from the QTPP for the medicinal product. A risk matrix already existed for the medicinal product, produced during the previous QRA, therefore the project team updated the matrix based on knowledge obtained since the previous version. An extract from the risk matrix is presented in Table 2.7 for exemplification of the approach.

Note:

- Assessing every raw material attribute and process parameter against every quality attribute ensures that no combinations are missed.
- Documentation of 'no effect' where there is considered to be no viable cause-effect relationship between a process parameter and a quality attribute may be useful in answering regulatory questions. Where an assessment of 'no effect' is perhaps unusual, it would be good practice to add a justification.

Table 2.7 *Risk matrix for medicinal product.*

Material/ unit operation	Attribute/process parameter	Assay	Uniformity of dosage units	Degradation products	Dissolution	Microbiological quality	Description
Drug substance	Moisture content (Proven range: 3.5 to 5.5% w/w)	Moisture content will contribute mass to charge amount	No effect	No effect	No effect	No evidence to suggest water levels will achieve water activation limit	No effect
Drug substance	Particle size distribution (Proven range: D90 35–75 um)	No effect	Risk of segregation if particle size distribution poor match to major excipients	No effect	Particle size distribution	No effect	No effect
Sodium starch glycolate	Supplier/grade (used Type A to date)	No effect	No effect	Type B or increase in acidity can cause degradation	No effect	No effect	No effect
Blending 1	Number of revolutions	No effect	No effect due to presence of downstream mixing operations; >20 revolutions has been used to date	No effect	Low number of revolutions could cause inadequate mixing, and slow dissolution of some tablets due to poor distribution of disintegrant	No effect	No effect

Blending 1	Volume fill %	No effect	No effect due to presence of downstream mixing operations. Trials have been performed between 40 and 60% fill	No effect	Under/over fill could cause inadequate mixing, and slow dissolution of some tablets due to poor distribution of disintegrant	No effect	No effect
Screening	Screen size	No effect	No effect	No effect	Screen of 1.6 mm square hole has demonstrated adequate de-agglomeration of magnesium stearate	No effect	No effect
Blending 2	Number of revolutions	No effect	Revolutions >60 have demonstrated no impact on dosage uniformity. Shorter blend times have not been investigated	No effect	Short time could cause inadequate mixing, and slow dissolution of some tablets due to poor distribution of disintegrant	No effect	No effect

(Continued)

Table 2.7 *(Continued)*

Material/ unit operation	Attribute/process parameter	Assay	Uniformity of dosage units	Degradation products	Dissolution	Microbiological quality	Description
Blending 2	Volume fill %	No effect	No effect due to presence of downstream mixing operations. Trials have been performed between 40 and 60% fill	No effect	Under/over fill could cause inadequate mixing, and slow dissolution of some tablets due to poor distribution of disintegrant	No effect	No effect
Hold time Blend 2	Up to 30 days	No effect	No effect	No effect observed up to 30 days	No effect	No effect	No effect
Compression	Turret speed RPM	No effect	No effect	No effect	No effect	No effect	Speeds >80 RPM has contributed to tablets with low hardness and tablet chipping during coating
Compression	Main compression kN	No effect	No effect	No effect	Forces >25kN produces tablet with hardness >200 N and demonstrate a slower release	No effect	No effect

Compression	Tablet hardness N	No effect	No effect	No effect	The desirable target is 160–180 N to achieve desired tablet robustness and dissolution performance	No effect	The desirable target is 160–180 N to achieve tablet robustness appropriate dissolution
Hold time tablet cores	Up to 30 days	No effect	No effect	No effect observed up to 30 days	No effect	No effect	No effect
Coating	Coat amount %	No effect	No effect	No effect	No effect	No effect	>2% by weight required to provide adequate core coverage
Primary packaging	Up to 1 year	No effect	No effect	No effect observed up to 1 year	No effect	No effect	No effect

- An assumption made is that all scenarios for each failure mode are reasonable and within those likely to be attainable for routine manufacture.
- In performing this assessment, it may be helpful to consider what would happen if the process parameters were at the extremes of the normal operating range.
- The risk matrix is a pre-work tool to drive FMECA meeting efficiency. There is opportunity to add additional risks during the FMECA.

2.4.3 Scoring Meeting

To achieve a balanced interpretation of the risks for the medicinal product, a multi-disciplinary team was convened. The team included the following:

- Project team members – from both development and commercial manufacturing consisting of representation from analytical, formulation, quality assurance and process engineering disciplines.
- Independent experts – senior process engineer, senior materials scientist and senior formulation scientist.
- Facilitator – Independent of the project but an expert in the application of the FMECA tool.

Note: A leader and multi-disciplinary team, with experience of the product, and specialists naive to the specific product but with expertise in the associated formulation and process type must be identified.

2.4.4 FMECA Tool

A scoring system based on the different aspects of a relationship between a failure mode and a failure effect was used. The system is based on two scores:

- Risk score (incorporates an assessment of likelihood and severity).
- Detectability score.

The two scores were then multiplied together to provide a risk prioritisation number between 0 and 100. An extract from the FMECA risk table is presented in Table 2.8 as an example of the approach. The two scores are described in more detail below.

Note: The FMECA tool used by the team is considered a very good example of a semi-quantitative assessment of the current state and is ideally suited to definition of a control strategy for regulatory submission. The scoring reflects the assessment of the appropriateness of the existing formulation, process and analytical controls on the basis of their influence on reliably attaining the target for each CQA.

2.4.5 Risk Score

Risk scenarios were constructed on the basis of variation in input material properties or process parameters. The risk score was broken down into an assessment of up to three different aspects, as presented in Table 2.8. For each risk scenario, the team picked the highest score for a given risk scenario (failure mode and failure effect). For example, for a change in a parameter, consideration was made against any intended future changes in scale, prior

Table 2.8 *Risk score definitions.*

Score	Risk associated with future change to input material supplier or technical grade	Risk associated with a future equipment change[a]	Risk associated with operating space, including physical properties of input materials (accounts for likelihood and impact)[b]
1	No change to proposed commercial suppliers/material grades, or functionality-related characteristics well understood.	No change to equipment type or operating principles required. Operating at proposed commercial scale or future scale changes required, but highly predictable scaling behaviour and assessed as having no impact	$DF >> EF$ (or $AC >> BC$) Distance between average result and acceptable limit[c] is greater than variability of results at these conditions (i.e. buffer zone AB sufficiently large)
5	Future supplier or material grade change required/desired to reduce commercial supply costs. Some knowledge of functionality-related characteristics, therefore no significant impact expected.	Equipment type remains the same, but change in operating principles or geometry expected to be impactful (e.g. change of blending vessel shape from bin to double cone) or future scale changes required ($<10\times$) to achieve commercial scale and empirical knowledge or poorly defined scaling behaviour	$DF \approx EF$ (or $AC \approx BC$) Distance between average result and acceptable limit[c] is of a similar magnitude to the variability of results at these conditions (i.e. buffer zone AB only just large enough)
10	Future supplier or material grade change required/desired to reduce commercial supply costs. Currently no or poor knowledge of functionality-related characteristics. Expect negative impact of change.	Change of equipment type and/or operating principle required (e.g., tray dryer to fluid bed dryer, impact mill to screening mill) and effect unknown or large future scale changes required (e.g., $\geq 10x$) to achieve commercial scale and scaling behaviour unpredictable or not known	$DF << EF$ (or $AC << BC$) Distance between average result and acceptable limit[c] is less than variability of results at these conditions (i.e., buffer zone AB is too small)

[a] Equipment type = SUPAC class, operating principles = SUPAC sub-class.
[b] Refer to Figure 2.6 for visualisation.
[c] For example, specification limit from QTPP, or other accepted limit based on stage of development.

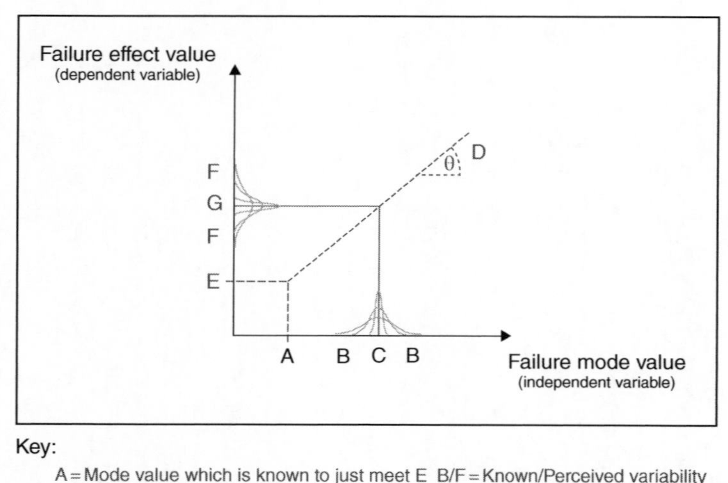

Key:

A = Mode value which is known to just meet E
C = Target input/Set point
E = Target output/Specification
B/F = Known/Perceived variability
D = Relationship (mode and effect)
G = Measured output

Figure 2.6 *Risk score approach (high level) based on current operating space.*

to commercialisation, for that parameter and associated unit operation. If a risk scenario was scale dependent, the information from laboratory scale studies were of little relevance, so the score given to the risk scenario reflected this. If no change to the process or materials was envisaged prior to commercialisation, then the risk scoring reflected the current operating space and understanding of variability (95% confidence limits) associated with the process and analytical methodology. A comment was recorded in the FMECA tool to clarify which risk scenario was scored highest and why. A score of zero was assigned if experimental evidence or the literature existed to confirm no effect of this failure mode on this failure effect, and a link to experimental evidence was supplied to justify this position. A high-level graphical representation of this scoring approach is presented in Figure 2.6 and a more detailed view in Figure 2.7.

Note: It is an expectation if performing a QRA at the point of defining the control strategy for registration that studies will have been performed at commercial scale (e.g. commercial site establishment trials) and on commercially relevant analytical apparatus (methods system analysis trials), thus enabling the development team to apply the semi-quantitative scoring described in the third column of Table 2.8. The initial two columns are self-explanatory and can be used as part of change control.

2.4.6 Detectability Score

The detectability score reflects the mode or effect as measured during manufacture such that a corrective change may be applied if necessary. The detectability aspects of the risk were scored using values from 0 to 10, with lower numbers reflecting higher levels of reassurance of product quality. A summary of detectability scores is presented in Table 2.9.

An extract from the applied FMECA tool is presented in Table 2.10 as an example of this approach.

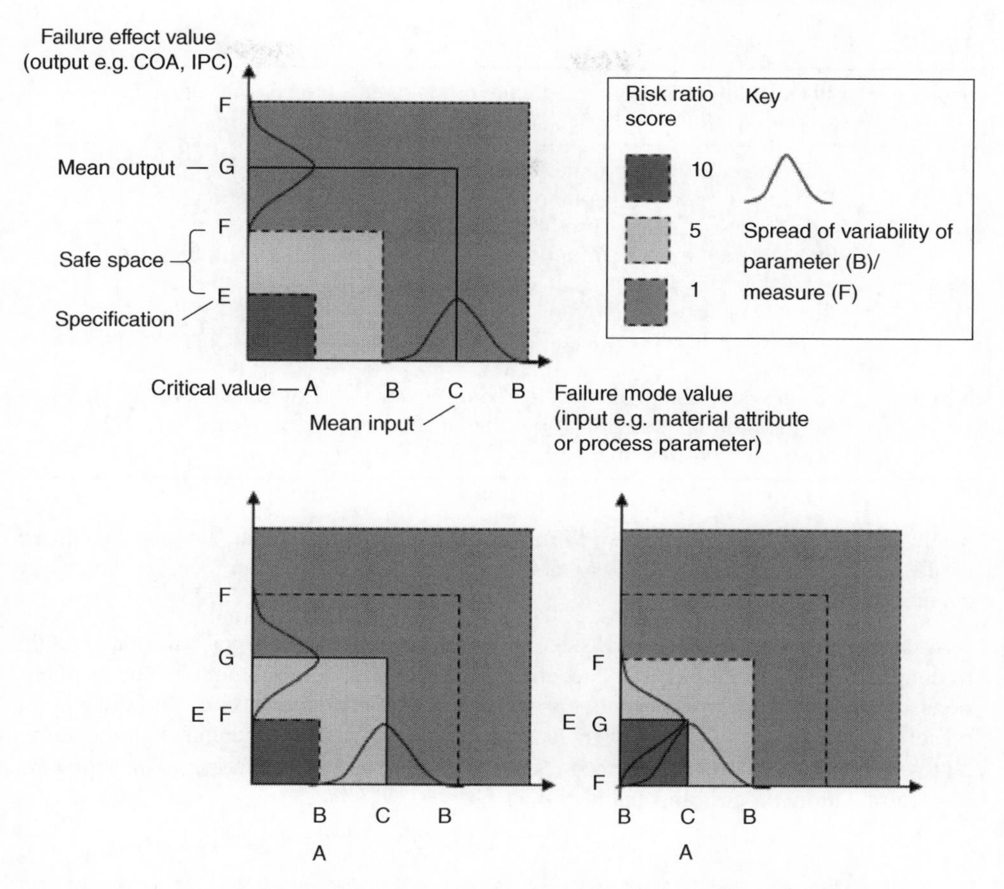

Figure 2.7 *Risk score approach (detailed level) based on current operating space.*

2.4.7 Communication

The 'Threshold for Action' set by the organisation undertaking the risk assessment was set at ≥35 to accommodate failure modes which may only be discovered during end-product testing with process and analytical contributions to variability (95% confidence limits) at or close to the specification limit. Therefore, as summarised in the action column in Table 2.10, further studies were recommended by the team undertaking the QRA.

To facilitate both communication of existing project risks and agreement on resourcing, after the scoring meeting the FMECA tool was accommodated in a report and reviewed and approved by business stakeholders. This led to an agreement on the described actions, which led to successful accommodation of the manufacturing process at the commercial manufacturing site and regulatory agreement and approval of the control strategy within the regulatory submission.

The team also agreed to perform an additional QRA either one year after the launch of the medicinal product, since it was expected that sufficient data from CPV would be

Table 2.9 Detectability score.

Score	Detectability (monitoring of failure mode or effect to ensure acceptable product quality)
0	Mode detected before unit operation
1	Continuous monitoring of mode or effect during unit operation (e.g. temperature probe in stirred tank)
3	Intermittent measuring/monitoring of mode or effect during unit operation (e.g. tablet weight)
5	Testing of effect at end of unit operation or intermediate step in the manufacturing process (e.g. impurity testing at end of stage, granule moisture content)
7	End-product testing of effect (e.g. drug substance assay, tablet dissolution)
10	No monitoring, measuring or testing of effect at time of release (e.g. tablet dissolution drop on stability)

available to perform a comprehensive re-assessment of the process and analytical controls, or alternatively, if a change control was raised (e.g. change to materials or process), whichever came soonest.

Note: It is critical that organisational stakeholder communication and approval is undertaken to demonstrate the organisation's acceptance of risk and/or agreement on the implications of risks with scores beyond the acceptable threshold to assure commitment to resources. The stakeholders may include development and commercial manufacturing representation from project management, quality assurance and heads of technical departments (e.g. formulation, process and analytical).

2.5 Concluding Remarks

To conclude, it should be clear that the use of risk assessment tools is only one component of a QRM process. The QRM process is cyclical, and risk review, assessment and control should continue during the life of a medical product: development > commercialisation > termination of supply. It is also of paramount importance that once a risk assessment is complete and risks ranked that the risks perceived to be acceptable and of unacceptable risk to quality need to be communicated. Through this dialogue, agreed and appropriate action can be taken by the organisation to ensure development of a robust regulatory control strategy, appropriate steps taken as part of change control and agreement on timings arrived at for the next risk review and to ultimately assure product quality to the patient. It is worth referring back to Figure 2.2 in this chapter for a visualisation of this cyclical process. A good summary of the tacit benefits of QRM are also exemplified by a diagram taken from the ICH Q9 training slide pack [11] as shown in Figure 2.8.

Finally, within the presentation on the PMDA web site provided by Pfizer (author unknown) (http://www.pmda.go.jp/files/000152079.pdf) [11], a very useful and thought-provoking summary of the requirements to implement QRM within a pharmaceutical

Table 2.10 *FMECA for medicinal product.*

Unit operation/ process step	Material attribute, process parameter, failure mode or independent variable	Quality attribute, failure effect, response or dependent variable	Risk description	Risk score	Detectability	RPN	Reasoning for scores assigned	Action	Owner	Timings
Drug substance	Moisture content (proven range 3.5 to 5.5% w/w)	Microbiological	Combined moisture content of tablet reaches and/or exceeds water activation level	1	0	0	Commercial-scale API manufacture centring around 4.1% w/w with a range of 3.5 to 4.6% w/w. Moisture content determined routinely as a non-specification test	Risk acceptable No action	N/A	N/A

(Continued)

Table 2.10 (Continued)

Unit operation/ process step	Material attribute, process parameter, failure mode or independent variable	Quality attribute, failure effect, response or dependent variable	Risk description	Risk score	Detectability	RPN	Reasoning for scores assigned	Action	Owner	Timings
Drug substance	Particle size distribution (Proven range D90 35–75 μm)	Uniformity of dosage units (AV ≤15)	D90 drops below 35 μm taking outside proven range leading to risk of insufficient mixing	1	7	7	Commercial-scale API manufacture centring on 40 μm, with a range of 35 to 45 μm. Upper specification of 75 μm in place to aid dissolution. No experience below 35 μm, and API process has demonstrated reduction in D90 over time. Acceptance value centred on 8 with a range of 4 to 10	Risk acceptable No action	N/A	N/A

Sodium starch glycolate	Supplier/ grade (used Type B to date)	Degradation products (Named degradant DEG 1 < 0.4% w/w)	Increase in acidity of SSG could induce degradation	5	10	50	a. Excipient compatibility studies and forced degradation studies demonstrate API sensitivity to acidic excipients. b. Low method capability for quantification of associated degradation product ±10% at 0.3% w/w	Combined manufacturing and analytical studies to assess supplier variability and improve method sensitivity	Formulator, analyst	<3 months

(Continued)

Table 2.10 *(Continued)*

Unit operation/ process step	Material attribute, process parameter, failure mode or independent variable	Quality attribute, failure effect, response or dependent variable	Risk description	Risk score	Detectability	RPN	Reasoning for scores assigned	Action	Owner	Timings
Blending 1	Number of revolutions	Dissolution (Q =70% in 30 min)	Insufficient mixing can lead to poor distribution of disintegrant, leading to variable dissolution	5	7	35	a. No DoE studies performed to assess influence of blending 1 and blending 2 on dissolution performance. b. Studies under way to identify process influence on dissolution performance	Combined manufacturing and analytical studies to assess influence of blending parameters on dissolution performance	Formulator, analyst	<6 months

Blending 1	Volume fill %	Uniformity of dosage units (AV ≤15)	Insufficient mixing can lead to poor distribution of API leading to variable content uniformity	1	7	7	No DoE studies performed to assess influence of blending 1 and blending 2 on blend homogeneity. Acceptance value centred on 8 with a range of 4 to 10	Risk acceptable No Action	N/A	N/A
Blending 1	Volume fill %	Dissolution (Q =70% in 30 min)	Insufficient mixing can lead to poor distribution of disintegrant, leading to variable dissolution	5	7	35	a. No DoE studies performed to assess influence of blending 1 and blending 2 on dissolution performance. b. Studies under way to identify process influence on dissolution performance	Combined manufacturing and analytical studies to assess influence of blending parameters on dissolution performance	Formulator, analyst	<6 months

(Continued)

Table 2.10 *(Continued)*

Unit operation/ process step	Material attribute, process parameter, failure mode or independent variable	Quality attribute, failure effect, response or dependent variable	Risk description	Risk score	Detectability	RPN	Reasoning for scores assigned	Action	Owner	Timings
Compression	Turret speed RPM	Description (appearance)	Low dwell time can contribute negatively to tensile strength of tablet	5	5	25	Edge of failure reached at 80 RPM on proposed commercial press. Desire to run as fast as practically possible to maximise plant efficiency	Risk acceptable Agreement to run at 75 RPM	N/A	N/A
Compression	Main compression kN	Dissolution (Q =70% in 30 min)	Increased tablet hardness can increase disintegration time and reduce dissolution rate	5	7	35	A maximum force of 22 kN and a target hardness range of 160–180 N has been established, ensuring desired tablet hardness and dissolution performance are achieved. Typical force used is 20 kN to achieve a hardness of 165 N	Implement plan to monitor compression force and tablet hardness during routine manufacture due to close proximity of likely edge of failure >22kN	Commercial manufacturing	<1 Month

Coating	Coat amount %	Description (appearance AQL)	Inadequate core coverage due to too little coat	10	5	50	Coating optimisation DoE not performed in commercial-scale coater	Manufacturing optimisation study	Formulator, commercial manufacturing	<3 Months
Coating	Coat amount %	Description (appearance AQL)	In-fill of debossing due to too much coat	10	5	50	Coating optimisation DoE not performed in commercial-scale coater	Manufacturing optimisation study	Formulator, commercial manufacturing	<3 Months

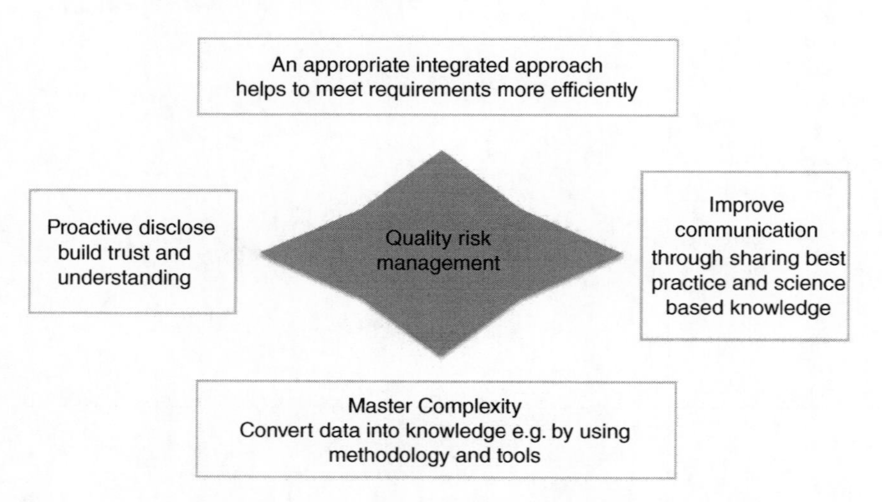

Figure 2.8 *Tacit benefits of QRM.*

organisation is presented, and it is recommended that before undertaking a QRM exercise, each of the following items are used as guidance to fulfilling the ICH principles of QRM and to underpin QbD:

- Have leadership support/accountability for QRM.
- Need all functions involved – not just the quality department.
- Provide clear guidance to your colleagues on where and how QRM should be used.
- Incorporate philosophy into existing SOPs.
- Provide training in tools and methodology.
- Identify site QRM champions.
- Across all functions/departments, not just the quality department.
- Have a proactive approach to finding opportunities to use QRM.
- Communicate QRM analysis and outcomes.
- 'Listen' to your QRM analysis.
- Take active decisions and actions.
- Bad decisions cannot be 'made right' by applying QRM.
- QRM works best with multi-disciplinary perspectives.
- Even if the tool is simple (informal), you must document how you reached your decision.
- There are no strict rules for applying tools, and adapting tools is acceptable.
- Each situation is different, but all situations should relate back to consumer and compliance risk.

2.6 References

[1] Lis, Y., Roberts, M.H. and Kamble, S. *et al.* (2012) Comparisons of Food and Drug Administration and European Medicines Agency Risk Management Implementation for Recent Pharmaceutical Approvals: Report of the International Society for Pharmacoeconomics and Outcomes Research Risk Benefit Management Working Group, *Value in Health*, **15**, 1108–1118.

[2] HSE (2016) *Control of Substances Hazardous to Health (COSHH)*, http://www.hse.gov.uk/coshh/basics.htm (accessed 30 August 2017).
[3] ICH Q9 (2003) *Final Concept Paper Quality Risk Management*, http://www.ich.org/fileadmin/Public_Web_Site/ICH_Products/Guidelines/Quality/Q9/Concept_papers/Q9_Concept_Paper.pdf (accessed 30 August 2017).
[4] ICH Q9 (2005) *Quality Risk Management*, http://www.ich.org/fileadmin/Public_Web_Site/ICH_Products/Guidelines/Quality/Q9/Step4/Q9_Guideline.pdf (accessed 30 August 2017).
[5] ICH Q8(R2) (2009) *Pharmaceutical Development*, http://www.ich.org/fileadmin/Public_Web_Site/ICH_Products/Guidelines/Quality/Q8_R1/Step4/Q8_R2_Guideline.pdf (accessed 30 August 2017).
[6] ICH Q10 (2008) *Pharmaceutical Quality System*, http://www.ich.org/fileadmin/Public_Web_Site/ICH_Products/Guidelines/Quality/Q10/Step4/Q10_Guideline.pdf (accessed 30 August 2017).
[7] ICH Q11 (2012) *Development and Manufacture of Drug Substances (Chemical Entities and Biotechnological/Biological Entities)*, http://www.ich.org/fileadmin/Public_Web_Site/ICH_Products/Guidelines/Quality/Q11/Q11_Step_4.pdf (accessed 30 August 2017).
[8] ICH Q8, Q9 and Q10 Questions & Answers (R4) (2010) *Quality Implementation Working Group*, http://www.ich.org/products/guidelines/quality/article/quality-guidelines.html (accessed 30 August 2017).
[9] Frank, T., Brooks, S., Creekmore, R. *et al.* (2008) Quality Risk-Management Principles and PQRI Case Studies, *Pharmaceutical Technology*, Volume **35**, Issue 7 http://www.pharmtech.com/quality-risk-management-principles-and-pqri-case-studies (accessed 30 August 2017).
[10] Conformia CMC-IM Working Group (2008) *Pharmaceutical Development Case Study ACE Tablets* https://www.ispe.org/pqli/case-study-ace-tablets.pdf (accessed 03 June 2016).
[11] PMDA (presentation provided by Pfizer) *Quality Risk Management for Quality system* http://www.pmda.go.jp/files/000152079.pdf (accessed 30 August 2017).

3

Quality Systems and Knowledge Management

Siegfried Schmitt

PAREXEL Consulting, PAREXEL International, United Kingdom

3.1 Introduction to Pharmaceutical Quality System

3.1.1 Knowledge Management – What Is It and Why Do We Need It?

Knowledge management is needed to exploit intangible assets such as process understanding, patents and customer relationships in order to repeat successes, share best practices, support continuous improvement and prevent knowledge loss after organisational changes.

The classical definition of knowledge as 'justified true belief' in the context of the pharmaceutical industry can be described, for example, as 'having appropriate experimental/scientific data, validated models and continuous verification'. The vast amount of data and knowledge makes it impossible to have anything other than partial or incomplete knowledge. Thus, it is necessary to make decisions as to which data and knowledge are needed (e.g. which experiments will be performed prior to process validation). This decision-making process is risk based.

Pharmaceutical Quality by Design: A Practical Approach, First Edition.
Edited by Walkiria S. Schlindwein and Mark Gibson.

3.2 The Regulatory Framework

3.2.1 Knowledge Management in the Context of Quality by Design (QbD)

In the traditional context of Good Manufacturing Practice, knowledge management is related to process and product understanding from a product quality perspective. In the context of QbD, knowledge management looks at process and product understanding in the context of quality, safety and efficacy, that is, the quality target product profile (QTPP). This does not necessarily mean that different parameters or attributes have to be addressed in the QbD context; it may, however, require different levels of understanding and collation of data to obtain the necessary information and, ultimately, knowledge.

For example, it is necessary to determine the critical quality attributes (CQAs) for a specific product so that they can be included in totality in the submission documents. Whereas a non-QbD approach would probably have the required information in the process validation set of documents (i.e. process validation including analytical validation), this may contain only some of the information for the QbD dossier. Certainly, various CQAs will be validated during process validation, whereas some may have been validated previously as part of development and/or during engineering runs. In a QbD context, it is equally important to have an understanding of acceptable, as well as undesirable or unacceptable ranges or parameters. In a normal process validation approach, the latter would not be deliberately included.

Therefore, knowledge management for a QbD submission will require linking information (and thus the supporting data), such as process parameters (and their associated controls) and input attributes (e.g. for raw materials) to output attributes (i.e. the quality attributes), and ultimately to the QTPP. There are various approaches as to how this can be done, for example, by process steps, or across the entire process (typically divided into upstream and downstream for biological processes).

In a correlation matrix, the CQAs are linked to the parameters and input attributes, depicting what influences what. The process view will allow one to demonstrate where these input attributes are controlled (e.g. the quality of a raw material may be controlled several steps before it is being used in the process step, where it impacts the quality of the product). In a traditional submission dossier, there may simply be a table of CQAs and the established process parameters, and a table of the CQAs with their established specifications. The knowledge, that is, the understanding of how these were established will only be partially included in the dossier, and probably not well summarised. QbD requires bringing together development information with upscaling and validation information in a logical and interlinked manner.

Knowledge management as a prerequisite to compliance is explicitly mentioned in the EU regulations:

- Part 1, Chapter 1, Pharmaceutical Quality System of EudraLex, Volume 4, 1.4 (ii): Product and process knowledge is managed throughout all life-cycle stages.
- Part 1, Chapter 7, Outsourced Activities of EudraLex, Volume 4, 7.15: The contract should describe clearly who undertakes each step of the outsourced activity, for example, knowledge management.
- Part 3, Pharmaceutical Quality System (ICH Q10) of EudraLex, Volume 4, 1.6.1: Knowledge management.

The US Code of Federal Regulations has not been changed to include knowledge management; however, ICH Q8 and ICH Q10 have been adopted by the United States. Furthermore, the US FDA have specified the need for QbD elements in Abbreviated New Drug Applications (ANDAs), and thus implicitly acknowledge the need for knowledge management.

The following is taken from the LinkedIn Group Quality-by-Design posting from August 2, 2012:

> So what does the Agency say? I wrote [to] the OGD, and got this response today from Lisa Kubaska, PharmD, a CDER spokesperson:
>
> 'Over the recent years, the Office of Generic Drugs (OGD) has presented our Quality by Design (QbD) expectations at various meetings, workshops and roundtables including the interview as addressed in your inquiry. Throughout this time, the generic industry has been engaged and allowed to provide feedback to assure that these concepts are clear. Our expectations are highlighted within the Immediate Release and Modified Release QbD examples posted on OGD's Information for Industry webpage noted below.
>
> Beginning this year, OGD has communicated to ANDA sponsors QbD expectations with the following text:
>
> We encourage you to apply Quality by Design (QbD) principles to the pharmaceutical development of your future original ANDA product submissions. A risk-based, scientifically sound submission would be expected to include the following:
>
> - QTPP.
> - CQAs of the drug product.
> - Product design and understanding including identification of critical attributes of excipients, drug substance(s), and/or container closure systems.
> - Process design and understanding including identification of critical process parameters and in-process material attributes.
> - Control strategy and justification.
>
> Examples illustrating QbD concepts can be found online at FDA's Generic Drugs: Information for Industry webpage [9].
>
> As January 1, 2013 draws closer, OGD will continue to communicate such expectations. The ANDA checklist will be updated to include QbD elements. As such, ANDAs will not be accepted for filing without these QbD elements starting January 1, 2013. For further inquiries or clarification, individuals are asked to contact OGD via email using the following: GenericDrugs@FDA.HHS.GOV'.

The key phrase may be: 'ANDAs will not be accepted for filing without these QbD elements starting January 1, 2013'. This suggests a sort of mandate (to be spelled out in the checklist) and that, indeed, starting in 2013, filing requirements for generics manufacturers will be very different from in the past.

3.2.2 Roles and Responsibilities for Quality System

The quality system (QS) is owned, maintained, communicated and its implementation verified by the quality unit (i.e. Quality Assurance). It is the quality unit's task to assure that knowledge management is correctly (as intended by the regulators) and appropriately (as befits the organisation) embedded in the QS. Quality is the responsibility of everyone within the organisation, with the guiding principles and requirements laid out in the applicable QS.

All too often the misconception persists in companies that the quality unit is responsible for anything related to quality. The quality unit's responsibilities are clearly laid out above. Individuals outside the quality unit cannot pass on their responsibilities to the quality unit, or be absolved of their duties under the QS.

3.2.3 Roles and Responsibilities for Knowledge Management

A chief knowledge officer (CKO) is an organisational leader responsible for ensuring that the organisation maximises the value it achieves through 'knowledge'. The CKO is responsible for managing intellectual capital and is the custodian of knowledge management practices in an organisation. The designation CKO is not just a re-labelling of the title 'chief information officer' – the CKO role is much broader. CKOs can help an organisation maximise the returns on investment in knowledge (people, processes and intellectual capital), exploit their intangible assets (know-how, patents and customer relationships), repeat successes, share best practices, improve innovation and avoid knowledge loss after organisational restructuring.

CKOs must have skills across a wide variety of areas. They must be good at developing/understanding the big picture, advocacy (articulation, promotion and justification of the knowledge agenda, sometimes against cynicism or even open hostility), project and people management (oversight of a variety of activities, attention to detail and ability to motivate), communications (communicating clearly the knowledge agenda, having good listening skills and being sensitive to organisational opportunities and obstacles), leadership, team working, influencing and interpersonal skills. The CKO who successfully combines these skills is well equipped as an excellent agent of change for their organisation.

Other terms for CKO include knowledge manager, director of intellectual capital, director of knowledge transfer and knowledge asset manager.

3.2.4 Implicit and Explicit Knowledge

As with QSs, knowledge within an enterprise is managed for the benefit of the business, allowing the production of quality products in a compliant manner and achieving commercial success. There is understanding and know-how that is being shared; that is, it is being described. This is referred to as explicit knowledge. However, humans do have implicit knowledge; for example, they have abilities that they and the organisation are unaware of, but which manifests itself in organisational routines and mindsets. This is referred to as implicit knowledge. Companies will wish to leverage this implicit (hidden) knowledge in order to make it accessible to others within the company.

From a QS standpoint, it is essential that the data, information and knowledge relevant to product quality and compliance is captured, documented and acted on. Clearly, there are many aspects or elements of a company's systems, processes and operations that do not fall into the categories covered by the QS. For example, an operator may be able to judge the load of a centrifuge from the noise made by it. This understanding will not be part of the batch record. Thus, in order to capture such knowledge, systems and tools need be provided that are intentionally maintained outside the QS. Such systems may be designed similar to Wikipedia, or they can be based on an existing document and record management system.

What can be seen from the above is that not all of knowledge management falls under the QS, and from a historical perspective, QSs have (formally) been in place longer than their knowledge management system counterparts. Therefore, it is not unusual to see the two systems being developed independently of each other. Although there is no generally accepted model to gauge the maturity of either system, applying the Capability Maturity Model Integration model [1] is a possibility. It is desirable to have similar maturity levels for either system. Such models, which were developed for non-pharmaceutical applications and environments, can be extremely useful in improving systems and processes. Especially in the area of QbD statistical methodologies, design of experiments, lean and other methodologies have been adapted and adopted.

3.3 The Documentation Challenge

QSs are documented in a controlled manner, that is, with versioned, approved and dated documents. A typical QS has a pyramidal structure, where the top document may be a quality manual, a quality policy, a mission and vision statement or something similar. This tells the organisation and anyone else interested the purpose of the QS within the context of the company's business. Depending on the type, size and structure of an enterprise, the next level of QS documents may be corporate, divisional, site or department-wide guidances or instructions. The bottom layer of documents typically consists of work instructions, log books, forms, templates and so on. There can be, of course, as many layers as anyone wishes to create; the essential message here is that the QS is embedded and managed in a formal hierarchical documentation system.

Whereas the top documents describe the reasons (the why) and the scope (the what) of quality management, the lower levels detail the actual activities this entails (the how). The content is largely driven by the regulations, guidelines, inspection and audit observations and industry best practices. The purpose of the QS is to assure a product with the right quality attributes.

It is worthwhile taking a look back in time, maybe 30 years, as this will help understand better the need for knowledge management in the context of a modern QS. Most pharmaceutical companies were vertically integrated; that is, research and development (R&D), scale-up and commercialisation were all managed in-house. A large number of staff would work within the same company almost their entire working life. The absence of electronic communication tools, such as mobile phones, e-mail or even personal computers necessitated strong communication face-to-face, via telephone or in writing. Short communication channels were preferred due to the nature of the circumstances. If, for example, a production manager had a question about a particular process step during commercialisation, it would be possible to go and see the R&D chemist and ask questions directly. Knowledge was often passed on verbally and through daily practice. Instructions were often minimal in content (the author remembers a one-page instruction for a step in a synthesis that took one week to complete in the plant – the same instruction now runs to over 80 pages). Thus, process and product understanding was largely concentrated in one place. In addition, plant automation was in its infancy; though equipment controls became increasingly available (e.g. process logic controllers (PLCs), manufacturing resource planning (MRP), and supervisory controls and data acquisition (SCADA) systems), actual data and records management still posed significant challenges. Even documentation, such as standard operating procedures (SOPs), was either written manually, on a typewriter or using central application systems (like the IMB AS/400) [2].

What followed was the digital revolution, which brought unprecedented computing power, access for everyone from anywhere and a globally connected business environment. With this came opportunities and also challenges. Opportunities in the form of being able to control processes to a much greater extent, to acquire, manipulate and manage vast amounts of data, to share and to collaborate and ultimately to gain greater insight into processes and product quality. On the other hand, pharmaceutical companies have become much more virtual, with many operations and activities outsourced, and with a workforce that has become increasingly mobile. With that, knowledge within a company has been lost as staff move on, as companies reorganise and as know-how had not been written down. Also, the sheer explosion of data resulted more in an information overload than a more learned workforce. Added to this was a misguided belief that computers could replace human thinking and ingenuity.

In Figure 3.1, the situation described so far is labelled 'The Validation Age' and 'The Age of the Three Batches'. This is because from a regulatory perspective, process validation for both drug substance and drug product became commonplace in industry. And probably because the regulators put the number 'three' into the regulations and the associated guidance, three validation runs became the globally accepted standard. Worse, at the same time, industry understood, and the regulatory agencies further enforced the adage that processes had to be operated within tight operating margins that were difficult, if not impossible to alter, once defined in a drug application/submission. This meant that while R&D was evaluating optimum processing parameters, commercial production would merely attempt their best to operate within these, irrespective of any further process or product knowledge gained.

Technology transfer packages (TTPs) typically consisted of summary reports from the R&D (this is meant to include the scale-up operations) to the commercial teams. In these TTPs, the process parameters and controls determined by R&D were described, often without providing rationales or supporting information. Thus, it is not unusual to find processes

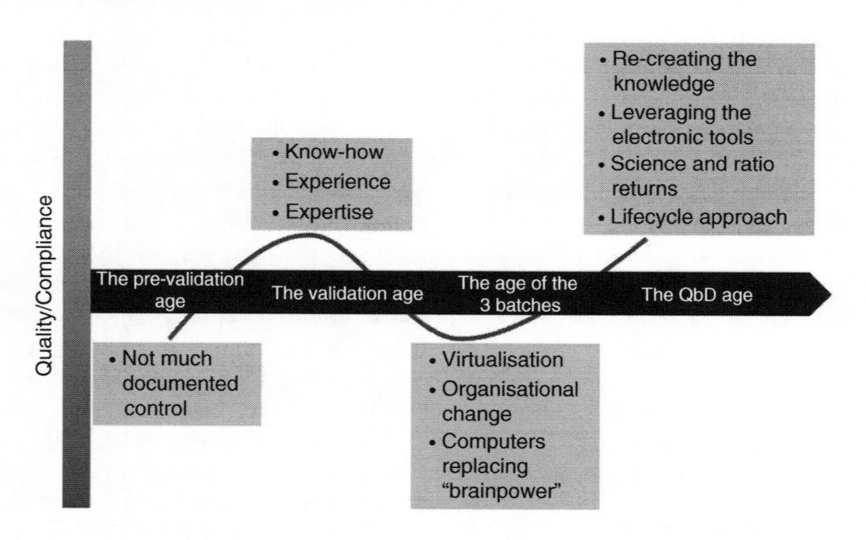

Figure 3.1 Through the ages.

that need stirring at room temperature for 12 to 13 h. This is probably because the laboratory technician let the reaction run overnight, not realising that it had completed within 15 min. Although this is of little consequence to the laboratory, on a commercial scale, this is an enormous waste of resources.

Probably the most obvious effect the digital revolution had on QSs is their management and the creation of documents in electronic format. This has many advantages, such as the ability to share/distribute easily or the option to update and revise with ease whenever needed. Unfortunately, all too often the benefits are wasted when more emphasis is put on the font for the footer than on the quality of the content. Of course, more information is stored and detailed in instructions, such as SOPs, which helps describe processes in more clarity and detail, and ultimately helps create higher-quality products. If we, however, take a holistic look at QSs, little except for the format has changed over the past few decades. That means that QSs are still largely onerous, cumbersome and extremely difficult to maintain current and compliant. They are thus not a blueprint for best practices in knowledge management.

One also has to recognise the fact that the generation of managers and senior managers in the pharmaceutical industry have grown up with documents, not records. Thus, much information is still created and stored as unstructured information, that is, in documents, rather than, for example, in records in relational databases. Let us take the example of TTPs, which are highly unstructured. The only search functionality is by the table of contents and a full-text search (if the document exists in a searchable electronic format). Neither is particularly useful or efficient. In many cases, those in need of information do not even know that such a TTP exists, where it is stored and who has access to it. Not only is it unlikely that we can go to the R&D building to talk to the chemist who developed the process, it is now also difficult to find documentation about it. The fact is, knowledge is getting lost.

Not only industry but also the regulatory agencies have noticed that this situation is untenable. Modern tools and technologies, plus advancements in pharmaceutical science, cried out for a modern science-based quality paradigm for the twenty-first century. Enter the 'QbD Age'. Developing such a new paradigm required time, collaboration and vision. Figure 3.2 shows the key milestones in this journey, which was driven by the major regulatory agencies and the International Council for Harmonisation of Technical Requirements for Pharmaceuticals for Human Use (previously known as the International Conference on Harmonisation of Technical Requirements for Registration of Pharmaceuticals for Human Use) (ICH) [3].

One of the many facets addressed by this modern approach to process understanding and product quality is knowledge management. ICH Q10 [4] describes knowledge management in paragraph 1.6.1 as follows:

> Product and process knowledge should be managed from development through the commercial life of the product up to and including product discontinuation. For example, development activities using scientific approaches provide knowledge for product and process understanding. Knowledge management is a systematic approach to acquiring, analysing, storing and disseminating information related to products, manufacturing processes and components. Sources of knowledge include, but are not limited to prior knowledge (public domain or internally documented); pharmaceutical development studies; technology transfer activities; process validation studies over the product life cycle; manufacturing experience; innovation; continual improvement; and change management activities.

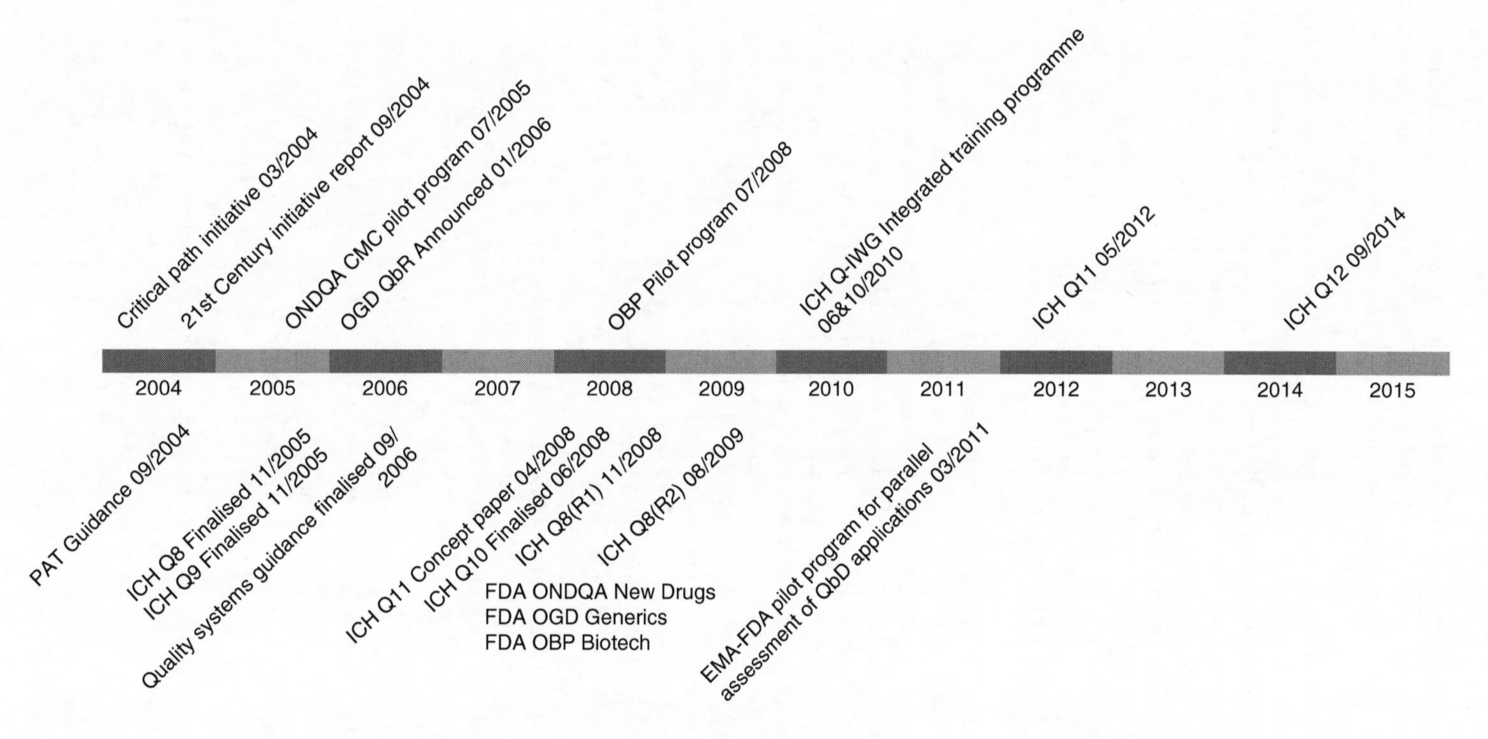

Figure 3.2 *ICH QbD timeline.*

Clearly, knowledge needs to be created, disseminated/made available, updated, used/applied and maintained throughout the life cycle of a product (please note that, for example, aspirin has been on the market for over 100 years – how is that for a life cycle?). To clarify this, let us look again at the TTPs. From a life-cycle perspective, there will be initial information on the process and the product from work performed by R&D and from the literature. There will be information on successful and unsuccessful experiments, both of which must be documented. Process parameters and controls, and also the product quality attributes will be established on small-scale studies. With scale-up and change of equipment, there will be changes at the very minimum to the process parameters and associated controls. For example, the fluid dynamics are different in equipment of different dimensions, the addition of solids to a reaction vessel is different than when adding a solid with a spatula to a glass flask or the temperature controls on a commercial size reactor are very different from those on a small-scale reactor. In fact, some parameters may be much easier to control on a large scale than on a laboratory scale. Thus, the understanding about the correlation between process inputs, process outputs, process controls and product quality attributes is evolving along the product life cycle. If all this information is maintained in reports rather than in a "live" record/repository, it will be difficult if not impossible to present the evidence and rationales for the current process and product specifications.

In inspections, one is required to go backwards in order to present the evidence that has led to the present set of operating and testing instructions in support of the product specifications. Many an exercise trying to establish why, for example, a pH parameter in a batch record set at a particular range has failed miserably. Typically, the best one can provide is a small-scale experiment where the pH was within the range. However, no information may be found whether the chosen range is in fact a proven range (i.e. supported by evidence). That is not to say that such evidence does not exist, but it may not have been considered for inclusion in the documentation. In a traditional setting, each department (R&D, quality control, manufacturing, etc.) maintains their own documentation; that is, the records do not stay with the product when it is handed over from one department to the next. Only summaries are usually provided. This means that data, information and knowledge are getting lost along the way because of the traditional documentation structures.

Thus, the way forward is to maintain a product knowledge file or record (of course, any other title may be given to this). That is not to say that all and everything should or has to be stored in the same place; it means that there is one place where both the data and the records are located or can be located from. If there is literature, for example, its location reference has to be maintained within this product knowledge file. Such a product-specific repository has enormous value, be it, as already mentioned, in inspections, or for troubleshooting when issues or deviations are observed, or simply to help optimise and continually improve the processes. Where changes are needed for processing or for product quality reasons, all the relevant supporting information should be easily accessible from this file. Only when the product is discontinued and the retention period for controlled documents has expired should the product knowledge file either be deleted (no further use), archived (for business rather than regulatory reasons) or integrated in other product knowledge files (e.g. where similar processes or controls are required).

Although, of course, resources and effort are required to collate, maintain and improve the product knowledge file or record, it vastly reduces repetitive and redundant activities, such as retrieving process and product information again and again in audits and inspections, in investigations (e.g. deviations) or for staff training.

3.4 From Data to Knowledge: An Example

From data, we want to create new knowledge and identify old and existing knowledge, though we only want to do this for knowledge that is relevant to our organisation. This is an ongoing and iterative process. Once knowledge is available, it needs to be disseminated and used for its benefits to be reaped.

Figure 3.3 depicts the process in general. This is best explained using a practical example.

The basis is data, or to be more precise, the right data. For example, a chromatographic system, such as a high performance liquid chromatography (HPLC) system, collates data. It is information that adds context to these data. For example, a time stamp needs information (meta data), whether this was a start time or an end time. Merely having this data and information is pointless, unless we have knowledge how to use these, that is, how we can interpret the chromatogram, the audit trail, and so on. The next steps beyond knowledge are understanding (i.e. why something works as it does) and wisdom (i.e. what could and

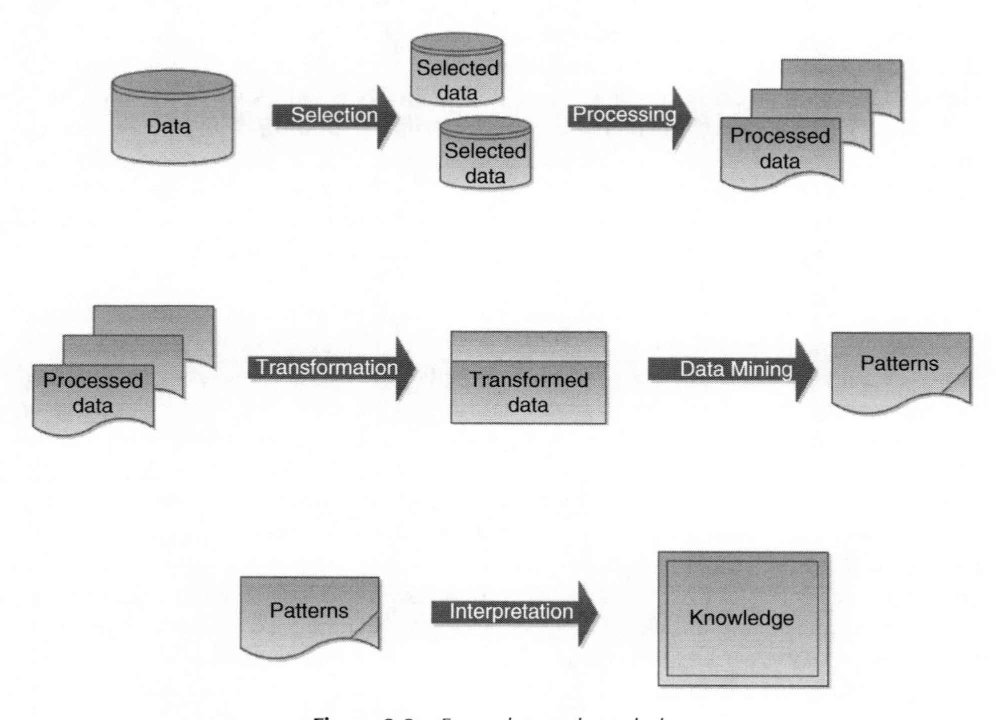

Figure 3.3 From data to knowledge.

should be done in the future). However, this chapter will only go as far as knowledge, as this in itself still is a formidable challenge for the healthcare industry.

Running a chromatographic analysis results in raw data (zeroes and ones), which in itself are meaningless. Data is the information derived from the raw data, such as an analytical result. The chromatography software provides metadata that give context and meaning to the data. In our HPLC example, the data may be 97.3. Adding metadata, this becomes: compound A, batch 57, 97.3%, analyst John Doe, Feb 2, 2014, 08:14 am GMT. In order to obtain the percentage value for compound A, only a select number of data are integrated, the integration (the processing) delivering the percentage value (the processed data). This processed data is evaluated against the specifications and is then given a pass or fail result, which is recorded in the certificate of analysis (CoA). At this point, we have transformed data.

Thus, if we or auditors/inspectors examine a CoA and wish to understand how the result was obtained and if the data and information are trustworthy, we need to have all the information on how the raw data were obtained, by whom, when and how, and how these were then processed and transformed. If, for example, the information on the integration is not available, a re-integration would not be possible. The same is true if the raw data were no longer available. This chain of custody (traceability) is referred to as data integrity by the regulatory authorities. A company's QS has to be set up to cover data integrity; that is, the QS is an enabler of knowledge management.

The transformed data, for example, the 97.3% recorded for compound A in the CoA, is of interest in comparison with the results obtained for the other batches. That is, we are interested in trend data from a business and a regulatory perspective. An example where the business perspective is applied is with regards to yields. If we find that yields are below expectations, though still within acceptable ranges, then there is a commercial concern as yield directly relates to profits. The regulatory perspective is one of preventing out of specification (OOS) results. If, for example, a trend is observed whereby an impurity is found in increasing amounts, then an investigation can be started and a possible root cause identified. Addressing the root cause can then prevent a further increase in the impurity, ensuring continued product of the specified quality. This activity is described as data mining in Figure 3.3, and leads to the identification of patterns. In short, we are continually learning more about the process and the product.

The next step is to interpret these patterns, that is, to apply the learning. Let us look at an example to explain how this can be done. Assume there are three providers of the same raw material for a chemical synthetic product. All three raw materials meet the specifications and give a product of the appropriate quality and within the expected yield range. Trending yields (patterns) showed a sudden change in yield (consistently higher yield) for product made from raw materials by one of the suppliers. There had been no change to the specification for the raw materials, and no change in patterns from raw material CoA data was observed. So how come that we see a change in yield despite a seemingly unchanged raw material? There must have been a change to some aspect of the raw material which was not captured by the specifications. In this example, the particle size of the raw material had not been specified, but it turned out to be the reason for the increased yield in the reaction. Further investigation showed that the improved particle size distribution had been achieved through a change in the drying of the raw material. The other suppliers were then asked to apply the same change, resulting in an improved product yield all around.

This is how knowledge is created: through the interpretation of patterns and the application in practice. In a traditional setting, the mere conformance to specifications would not have led to a (real-time) trending of data/records and an investigation, which would have meant that opportunities for increasing process and product understanding would have been missed.

3.5 Data Integrity

As mentioned earlier, data integrity is an essential element for creating trustworthy records that form the basis for knowledge management. Data integrity is a prerequisite for the regulated healthcare industry as decisions and assumptions regarding product quality and compliance with the applicable regulatory requirements are made on the basis of data. Drug and medical device manufacturers or (service) providers, healthcare organisations, regulators and other government organisations, and users, that is, patients and healthcare professionals, rely on data. Breaches in data integrity can have negative consequences and may lead to patient injury, or even death. Whereas in the past data integrity was relatively easy to prove using forensic methods analysing ink and paper, the advent of computerised systems has brought with it a different level of complexity. Identifying whether there could have been undocumented or even malicious changes to electronic data or records requires additional tools and expertise. As it is much easier to change electronic data and records than it is to change a paper or other physical record, there is a much higher chance of such changes being effected. The regulatory authorities have put much emphasis on data integrity in recent years, not least because they have uncovered serious cases of data integrity breaches.

The main criteria for data integrity are the following:

- Accurate – No errors or editing without documented amendments.
- Attributable – Who acquired the data or performed an action and when?
- Available – For review and audit or inspection over the lifetime of the record.
- Complete – All data is present and available.
- Consistent – All elements of the record, such as the sequence of events, follow on and are dated or time-stamped in expected sequence.
- Contemporaneous – Documented at the time of the activity.
- Enduring – On proven storage media (paper or electronic).
- Legible – Can you read the data?
- Original/reliable – Written printout or observation or a certified copy thereof.
- Trustworthy – The data and the record have not been tampered with.

In summary, the regulators' expectation is that pharmaceutical manufacturers, importers and contract laboratories, as part of their self-inspection programme, must review the effectiveness of their governance systems to ensure data integrity and traceability [5–8].

3.6 Quality Systems and Knowledge Management: Common Factors for Success

As was discussed earlier, QSs and knowledge management are complementary and mutually supportive. There are some common factors that enable successful implementation and operation of both. These include an organisational culture that supports and encourages innovation and change (for the better), collaborative relationships, curiosity to learn new skills and to acquire know-how and that enjoys senior management commitment. Although this seems to be logical and should be industry standard, this is not always the case. The fast-moving business environment that sees many reorganisations (e.g. mergers and acquisitions) has a detrimental effect on continuity, as well as staff identification with a company (and its culture). This is just one challenging aspect, and there are many others, such as differing cultures. In some countries or regions. elders and superiors are seen as infallible, and one would never challenge their decisions or opinions. It is then difficult to encourage a work environment that is against these deeply felt beliefs.

Although computers seem to have encroached on every aspect of life, automation is not as ubiquitous in the healthcare industry as one might imagine. The paperless plant is still utopia. This does hamper progress in compliance and in knowledge management in particular. Best-in-class companies therefore have computerised/automated systems and tools in place that permit electronic data transactions and data manipulation, or data mining across the organisation. Interoperability is a key aspect here as every time a manual transcription is required, the system comes to a halt, a major source of errors is introduced and the benefits of an automated system are nearly wiped out. The other aspect here is to have systems that are shared across the organisation, across divisions and borders.

Quality, compliance and knowledge need to be embedded in and aligned with the business strategy to satisfy both the regulators and the business. All too often, compliance and quality are seen as add-ons forced upon the company by the authorities, without tangible benefits. Instead of only talking about the cost of compliance, companies need to address the benefits of compliance, which can indeed be measured in hard cash. In any case, the status and the trends in all these aspects need to be verified through metrics. For example, as much as the business defines what increase in market share is targeted for a financial period, they must define the expected improvements in knowledge management, compliance and quality.

And yes, things can and will go wrong. Thus, companies need to have a framework that supports failure and deviation investigations, which is followed by corrective and preventive actions that result in continuous improvement, based on knowledge and risk. And here it is important to say that these systems must be efficient and effective. Merely having this framework is not enough. Too many companies are struggling with thousands (really!) of unresolved deviations, investigations overdue by years and wholly ineffective root cause analyses. Without such systems being of the desired quality and the operators having the necessary skills, no usable knowledge can be gleaned from the outcomes here.

3.7 Summary

There are real and practical limitations to what can be achieved in a pharmaceutical enterprise in regard to knowledge and process understanding. For example, not everyone is capable or even willing to document all their tacit know-how (experiences, expertise, insights, etc.), staff turnover always results in the loss of some knowledge and all resources are limited, including time to manage explicit (easily documented and transferred) knowledge.

Nonetheless, in most companies there still is a vast potential for improving knowledge management, be it from a contents, capabilities, program or adoption standpoint. There are many reasons for the current state of affairs in knowledge management, not least the many changes to the pharmaceutical industry with increased trends towards outsourcing, mergers and acquisitions leading to significant numbers in redundancies and the far from fully integrated automated/electronic systems landscape.

The regulations which govern quality and compliance for the pharmaceutical industry now specifically require companies to adopt and apply knowledge management as part of the modern quality paradigms.

Let us be clear: unless the tools and systems in place within an enterprise are designed, set up and continually adapted to handle data and information in a manner conducive to knowledge management, a company simply will not be able to fully exploit its business potential.

3.8 References

[1] Chrissis, M.B., Konrad, M. and Shrum, S. (2003) *CMMI: Guidelines for Process Integration and Product Improvement*, Addison-Wesley Longman Publishing Co., Inc. Boston, MA, United States.

[2] A Brief History of the IBM AS/400 and iSeries (1988) https://www-03.ibm.com/ibm/history/documents/pdf/as400.pdf (accessed 30 August 2017).

[3] ICH (1990) *The International Council for Harmonisation of Technical Requirements for Pharmaceuticals for Human Use*, http://www.ich.org/home.html (accessed March 28, 2016).

[4] ICH Q10 (2008) *Pharmaceutical Quality System*, http://www.ich.org/fileadmin/Public_Web_Site/ICH_Products/Guidelines/Quality/Q10/Step4/Q10_Guideline.pdf (accessed 30 August 2017).

[5] FDA (2016) *Code of Federal Regulations Title 2 Part 11*, **211**, 803 – Electronic Records; Electronic Signatures, http://www.accessdata.fda.gov/scripts/cdrh/cfdocs/cfcfr/cfrsearch.cfm (accessed 30 August 2017).

[6] EMA (2003) *EudraLex Vol 4 Chapter 4 and Annex 11*, http://ec.europa.eu/health/documents/eudralex/vol-4/index_en.htm (accessed 30 August 2017).

[7] Directive (1999)/93/Ec of *The European Parliament and of the Council on A Community Framework for Electronic Signatures*, *Journal of the European Communities*, December 13 http://eur-lex.europa.eu/legal-content/EN/TXT/PDF/?uri=CELEX:31999L0093&from=EN (accessed 30 August 2017).

[8] WHO Notices of Concern http://apps.who.int/prequal/assessment_inspect/info_inspection.htm (accessed 30 August 2017).

[9] FDA (2011) *Quality by Design for ANDAs: An Example for Modified Release Dosage Forms*, http://www.fda.gov/downloads/Drugs/DevelopmentApprovalProcess/HowDrugsareDevelopedandApproved/ApprovalApplications/AbbreviatedNewDrugApplicationANDAGenerics/UCM286595.pdf (accessed 30 August 2017).

4

Quality by Design (QbD) and the Development and Manufacture of Drug Substance

Gerry Steele

PharmaCryst Consulting Ltd, Loughborough, United Kingdom

4.1 Introduction

The production of small molecule candidate drugs (CDs) and/or active pharmaceutical ingredients (APIs) is usually through a synthetic organic chemistry procedure that should be safe, robust and cost-effective and be as environmentally sparing as possible. In this respect, it is different from the medicinal chemistry synthetic route applied during drug discovery, whose driver is to produce larger numbers of compounds for early screening after which most are discarded. In this situation, therefore, almost any route will suffice, since the amount of material to be produced is generally very small [1]. Once nominated into development, the process research and development (PR&D) or chemical development departments are then employed to deliver increasingly large amounts of drug substance and then transfer the optimized process into commercial production. However, even during production, the technical issues are often still addressed by R&D (known as technical stewardship) throughout the lifetime of the drug.

A PR&D department typically employs a range of scientific and engineering skills, including process chemists and engineers as well as analytical chemists. Specialist roles in, for example, process safety and crystallization may also be found in this department. Importantly, since the materials being generated by PR&D are destined to be clinically

Pharmaceutical Quality by Design: A Practical Approach, First Edition.
Edited by Walkiria S. Schlindwein and Mark Gibson.

tested, they have to be manufactured according to the requirements of current Good Manufacturing Practice (cGMP) and International Committee for Harmonization (ICH) Q7. Quality assurance (QA) professionals are therefore integral members of the process. A useful paper describing a typical PR&D organization and the typical challenges they faced in the early development of compounds at Wyeth was reported by O'Brien *et al.* (2011) [2].

This chapter will discuss how QbD aspects are applied to the production of pharmaceutical solid drug substances in the R&D arena. Although there are sound scientific reasons favouring the implementation of QbD, there are many potential regulatory benefits to be gained such as regulatory relief for the following scenarios:

- Site changes within a company.
- Outsourcing of company method to an external company.
- Facility change within site.
- Change of equipment.
- Change of batch size.
- Modifications of process parameters.
- Changes of solvent or reagent.
- Optional operations, for example, seeding.
- Alternative test methods for in-process controls (IPC) removals or changes.

In summary, there are significant advantages and incentives for a company to adopt QbD.

4.2 ICH Q11 and Drug Substance Quality

ICH Q11 (Development and Manufacture of Drug Substance) links drug substance quality to drug product. The guideline states that the drug substance manufacturing process development should include, *as a minimum*, identification of critical quality attributes (CQAs) associated with a drug substance, so that those characteristics having an impact on drug quality can be studied and controlled. As pointed out by Mehta (2013) [3], ICH Q11 describes the process of drug substance development on the basis of six principal quality principles:

- Drug substance quality linked to drug product.
- Process development tools.
- Approaches to development.
- Drug substance CQAs (defined below).
- Linking material attributes and process parameters to CQAs.
- Design space.

A CQA is a physical, chemical, biological or microbiological property or characteristic that should be within an appropriate limit, range or distribution to ensure the desired product quality. They can, of course, guide process development of drug substance manufacture.

From these principles, conditions for appropriate manufacturing processes and control strategies should be defined such that drug substance quality, the formulation performance and patient safety will be ensured.

4.2.1 Enhanced Approach

According to ICH Q11, there is also an enhanced, systematic approach, whereby the manufacturing process is evaluated, understood and refined such that, in addition to any prior knowledge, experimentation and risk assessments, the material attributes of, for example, raw materials, starting materials, reagents, solvents, process aids, intermediates and any process parameters that may affect the CQAs of drug substance are identified. Furthermore, the functional relationships that link material attributes and process parameters to drug substance CQAs can also be determined. In combination with quality risk management (QRM), to establish an appropriate control strategy, it can lead to the proposal for a design space.

It is important that the identification of CQAs should be carried out jointly by the chemical development scientists and those responsible for the API's formulation and analysis. Unfortunately, in the early stages of development, there is probably insufficient information available to determine whether some of the drug substances attributes are critical or not (e.g. particle size). If this is the case, it is usual for all potential CQAs to be included in quality risk assessments (QRAs). However, these should be reviewed as the project progresses and, as more data becomes available, the true CQAs can be agreed upon by the development scientists at the final QRA. Evidence suggests that the use of QbD is now well embedded (at least in larger pharmaceutical companies) [4].

4.2.2 Impurities

According to ICH Q11, regulatory authorities will assess the controls employed by API manufacturers, 'including those needed on how impurities are formed in the process; how changes in the process could affect the formation, fate, and purge of impurities'. The purity of a drug substance is an important potential CQA because of the potential impact of impurities on drug product safety, that is, intermediates, degradation products and products from side reactions may cause deleterious pharmacological and/or toxicological activity. This could be of particular concern if the drug is for chronic, long-term dosing, for example, diabetes [5]. Impurities are also important from a physicochemical point of view, since it is well known that they can affect the size, shape and polymorphic form a compound, which in turn can lead to batch-to-batch variation and hence the production of variable quality APIs [6, 7]. Impurities that are present or produced early in the synthesis can often be removed by, for example, the crystallization of intermediates, and subsequently be less problematic from a quality perspective. On the other hand, impurities created later in the synthesis may have a significant influence on the quality of the drug substance. Due to their potential effect on the quality of the API, it is essential that key impurities are identified in a timely manner and controlled [8].

One type of impurity that is of particular interest to the regulatory authorities are the so-called potential genotoxic impurities (PGIs) [9–11]. Since these compounds have the potential to damage DNA and lead to its mutation, their presence needs to be reduced to sub-ppm levels. This is perhaps best achieved by changing the synthetic chemistry route to substantially reduce or eliminate the presence of the PGIs altogether. Alternatively, the PGIs may be removed by adjusting the process conditions or, if this is not possible, showing from a toxicological perspective that the PGI(s) are not harmful at the levels generated by the synthesis. Of course, the last strategy is perhaps the least attractive option from a company perspective, since this would involve additional time and cost in toxicology testing of the impurities during the R&D phase.

The use of QbD to identify those factors that influence the risks and bring the process under control can be achieved through the use of multivariate analysis to develop the design space and spiking experiments to support specifications [12]. As an example of the application of QbD in an impurity control context, the development of a control strategy for a number of genotoxic impurities in the production of a fluoroaryl-amine compound nominated by GSK as a central nervous system (CNS) agent has been reported [13]. The process understanding generated by the QbD approach adopted for this compound ensured that the final testing for the genotoxic impurities could be eliminated. As a further example, Burt *et al.* (2011) [14] used the generation of an impurity and its control to exemplify the use of QbD in practice.

4.2.3 Physical Properties of Drug Substance

The physical properties of a drug substance are important depending on the dose range, drug product formulation and process selected. An early understanding of drug substance properties and their variability allows the rational, risk-based selection of drug product formulation and manufacturing process to be made. In turn, lower-cost, more efficient processes can be selected where possible, for example, direct compression or roller compaction manufacture rather than wet granulation batch processes. Knowledge of these factors should help reduce development times, complexity and risk in late-stage development and full-scale manufacturing. Overall, the relationship between risk and the number of steps from the end of the manufacturing process is the result of two main factors, physical properties and impurities. The physical properties of a drug substance are often determined during the final crystallization step and any subsequent operations, for example, filtration, drying, milling or micronization.

Assessment of a drug substance, both milled and unmilled, allows a link to be made between its properties and its processing attributes [15, 16]. The CQAs of the drug substance should be recorded and the rationale for designating these properties or characteristics as CQAs should also be provided. The aim here is to understand inter- and intra-campaign variability for these properties and to assess and mitigate risk. It should be noted that, in an R&D context, a drug substance is often manufactured in campaigns (batches) associated with the clinical phase of the project. These campaigns usually increase in size as the compound moves through the development cycle, and within a campaign there may be more than one batch produced to support, for example, toxicology, the ongoing clinical trials and formulations being tested. Therefore, there is the risk of batch-to-batch variation of the drug substance if its manufacture is not fully understood and controlled.

Typically for a solid dose formulation, for example, tablets, the physical properties that are considered important are the following:

- Surface properties – for example, wettability, cohesion and adhesion properties.
- Powder flow.
- Particle shape.
- Solid-state form (e.g. polymorph and hydrate).

- Particle size distribution – this can affect drug homogeneity in the final dosage form/dissolution/bioavailability.
- Compression properties (plastic/elastic/brittle).
- Bulk density – this affects the amount of compound that can be fitted into, for example, a granulator.

Thus, the manufacturing process development programme should identify which material attributes (e.g. of raw material and starting materials) and processes should be controlled. Furthermore, the material attributes and process parameters found to be important to drug substance quality should be addressed by the control strategy. For example, when assessing the link between an impurity in a raw material or intermediate and a drug substance CQAs, the ability of the drug substance manufacturing process to remove the impurity or its derivatives should be considered in the assessment (can it be controlled by specification or purging downstream, e.g. by crystallization?). During drug substance development, it is important to understand the formation, fate and purge (e.g. by crystallization or extraction) of impurities, as well as their relationship to the impurities that are in the resulting drug substance as CQAs. As a *basic* approach this can be done by, for example, setting material specifications and process parameter ranges based on batch history and univariate experiments. Alternatively, applying the *enhanced* approach leads to a better understanding of the connection between material attributes and process parameters to CQAs and the effect of the interactions between them. The enhanced approach is as follows (ICH Q11):

- Identification of the potential sources of variability.
- Identification of the material attributes and process parameters likely to have the largest effect on drug substance quality.
- Designing and conducting the studies to identify and confirm the relationships of material attributes and process parameters to drug substance CQAs.
- Analyzing and assessing the data to establish appropriate ranges and, if desired, establishment of the design space.

ICH Q8 (R2) defines the design space as 'the multidimensional combination and interactions of input variables (e.g. material attributes) and process parameters that have been demonstrated to provide assurance of quality'. It is therefore accepted that changes made within the design space are not considered a change. However, working outside the design space *is* considered a change and would therefore need regulatory approval. A design space might be determined per unit operation (e.g. reaction, distillation and crystallization) or a combination of unit operations which, from a regulatory perspective, can provide additional flexibility.

4.3 Linear and Convergent Synthetic Chemistry Routes

Although most of the discussion is around the production of the drug substance and its physical properties/purity, it is worth remembering that the drug substance is typically the outcome of a synthetic organic chemistry procedure. In this respect, a brief overview of the preceding chemistry procedures can help build an overview of the entire production process for the drug substance.

An analysis of the chemistry used in process development of compounds in a number of large pharmaceutical companies was carried out by Carey *et al.* [17]. In their survey of 128 drug candidates, they found that, on average, eight synthetic steps were required to assemble an API. In a synthetic procedure, it is normal to define a *stage* as being from one isolated intermediate to another isolated intermediate. For each stage, the synthesis of an intermediate involves identification of solvent(s), reagents, conditions, and so on. The length of a stage can be quite variable, that is, from one step to many steps and may involve 'telescoping' of multiple transformations into a longer sequence (without isolating the intermediates). Telescoping a reaction may help avoid issues with toxic, irritant, noxious or oily intermediates. As an example, Anson *et al.* (2011) [18] fully telescoped the production of GSK1842799, a S1P1 agonist. They were forced to produce multi-kilogram quantities of the hemi-fumarate salt of this compound by telescoping the reactions because all of the synthetic intermediates could not be crystallized, which they thought was due to the lipophilic side chain of the molecule. Note, however, that the European Medicines Agency (2014) have stated that because of the higher risk of variability of telescoped or 'one pot' reactions, regulators would expect to see a high level of process understanding and control when these are used in process development.

Another concept in the design of API synthetic routes is that of a linear versus a convergent synthesis. A linear synthesis is quite simply the addition of one intermediate after another in a linear sequence until the API is reached. Although this is conceptually straightforward, the big disadvantage of a linear synthesis is that the overall yield will drop with an increasing number of steps. For example, suppose that for each step the yield is only 50%, then the *overall* yield of the final molecule in the sequence from initial molecule is only 0.39%! In general, the linear synthesis of an API is not the preferred synthetic method in the pharmaceutical industry, and chemists tend to devise a so-called convergent synthesis to produce the desired molecule. Figure 4.1 shows an example of the concept of a convergent synthesis [19]. Employing a convergent synthesis is a more efficient method of working and has the advantages of producing a higher overall yield, less waste and greater flexibility with regard to scheduling, plant throughput and the use of contractors.

The decision to manufacture a compound by a particular synthetic route is usually based on a number of factors. For example, from a choice of four routes devised by the

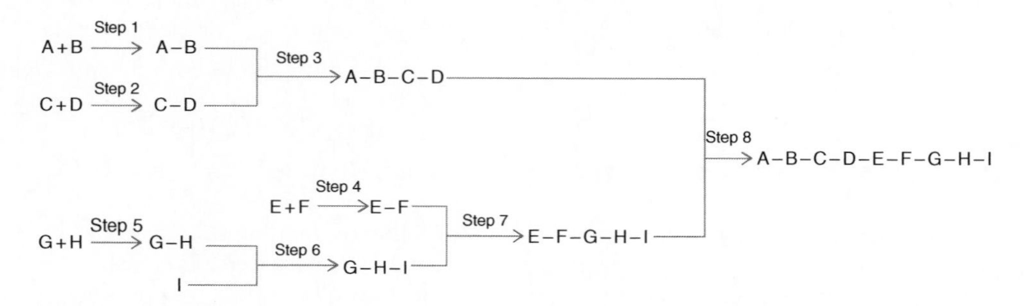

Figure 4.1 *Example of a convergent synthesis [19]. Reprinted with permission from Zhang,* Chem. Soc. Rev. *106: 2583–2595. Copyright (2006) American Chemical Society.*

synthetic chemists, Whiting *et al.* (2010) [20] selected the synthetic route of GSK269984B using the following criteria:

- Number of stages.
- Specialized equipment needed.
- Overall yield.
- Potential for genotoxic impurities.
- Number of solvents.
- Process mass intensity (PMI).
- Raw material costs.

As a result of a detailed consideration of these factors, the chosen route showed a raw material cost reduction of one-third from the original route. Moreover, waste from the new process was reduced by 20%, and an overall yield of 46% was achieved in three stages (compared to five stages for the other routes examined).

There is a clear relationship between the number of synthetic steps and the volume of API that can be produced in a defined time by a batch process [21]. This model showed that there is a non-linear relationship between the number of synthetic steps and the batch time so that the fewer steps in a chemical synthesis, the better. Furthermore, the number of steps to be developed can also affect the cost of a project in development; that is, if there are fewer steps in a synthesis, the overall cost will be reduced [22].

4.4 Registered Starting Materials (RSMs)

Section 5.1.1 of ICH Q11 addresses the issue of where the drug substance manufacturing process actually begins. It states that, in general, changes in material attributes or operating conditions that occur near the beginning of the manufacturing process have a lower potential to impact the quality of the drug substance.

The selection and development of an RSM have a critical impact on the development and launch of a drug substance process. ICH Q7A defines an API (active pharmaceutical ingredient/drug substance) starting material as ‘a raw material, intermediate, or an API that is used in the production of an API and that is incorporated as a significant structural fragment into the structure of the API. An API starting material can be an article of commerce, a material purchased from one or more suppliers under contract or commercial agreement, or a material produced in-house. API starting materials are normally of defined chemical properties and structure’. Thus, an RSM is a compound which, in agreement with the regulatory authorities, defines the formal start of regulatory control and introduction of full cGMP for the commercial synthesis of the drug substance.

From a business perspective, a good RSM position can provide scope to save costs through new suppliers/routes; it can also improve utilization of plant capacity. Furthermore, a good RSM choice identifies those steps requiring full cGMP and should underpin the rapid and efficient approval of the regulatory chemistry manufacturing controls (CMC) submission. Since delays in regulatory approval of a drug substance directly correspond to lost sales during the product lifetime, it is good company practice to carefully manage both the selection and approval of RSMs. An industry perspective on the justification of API starting materials has recently been reviewed in a number of papers [23, 24].

There is a need to provide to the regulatory authorities a justification for the selection of a RSM, for example, the ability to detect impurities in the starting material, the fate and purge of those impurities and their derivatives in subsequent process steps. The general guidance indicates that there should be at least three synthetic steps (covalent bond forming) from a proposed RSM to the drug substance (propinquity or similarity/closeness with the API). Indeed, in a review of data of 50 API starting materials, [24] found that, regardless of the complexity of the API, the average industry propinquity was 2–3 steps from the drug substance: non-isolated intermediates are usually not considered appropriate RSMs. Therefore, isolation and purification of intermediates by crystallization is important since it not only helps to define yields but, more importantly, helps to purge impurities upstream of the API. On the other hand, one benefit of *not* isolating intermediates is an increase in overall yield of the preparation process, since there is no loss on crystallization and isolation of the product. Furthermore, overall process cycle times are often decreased, since there are no intermediate filtrations or drying procedures to be undertaken. Thus, the selection of RSMs is a critical and important commercial decision exercise for most companies.

As noted by Gavin *et al.* (2006) [25], setting impurity specifications for the RSMs is an essential part of the scale-up and regulatory consideration of API development and commercialization. As a consequence, the effect of RSMs obtained from different sources (which may have different impurity profiles) on the synthesis of API will need to undergo a rigorous assessment. Moreover, as Illing *et al.* (2008) [26] pointed out, any changes to a starting material will necessarily lead to a detailed assessment for the presence of new impurities. Industry and regulatory agencies commonly use a 0.1% threshold for the presence of new impurities in the drug substance to determine equivalence of batches made before and after the change.

4.5 Definition of an Appropriate Manufacturing Process

4.5.1 Crystallization, Isolation and Drying of APIs

The solid-state properties of compounds are of vital importance in many industries, not just pharmaceuticals [27, 28], and the final step in the synthetic process production of an API is usually crystallization. However, crystallization is often problematic since there are many competing issues to be addressed, for example, purity, yield, average size and shape, and polymorphism. These solid-state properties can affect subsequent primary unit operations such as filtration and drying and also secondary manufacturing procedures (powder flow, milling, compression, etc.). Therefore, the challenge is to understand crystallization and other processes and scientifically define a set of conditions that reproducibly produces the compound according to the above criteria on progressively larger scales.

Worldwide, API manufacture (synthesis/crystallization) is typically performed, batch-wise, in glass-lined stirred tank reactors (STRs) [29]. However, it is well known that STRs have, for example, limited heat transfer characteristics, which can lead to long processing times, and when combined with relatively non-uniform mixing, it can lead to the formation of undesired by-products (impurities) [30]. Although there is a slow move towards adopting continuous crystallization processes for API manufacture [31], at present batch manufacture still dominates API production.

4.5.2 Types of Crystallization

There are a number of types of crystallization.

4.5.2.1 *Reaction Crystallization*

This is where two components react together and the product crystallizes (or precipitates) from solution. Usually this requires a low solubility of the compound in the solvent, which can often give rise to very small crystals, increased impurity levels or the undesired polymorph if not carefully controlled [32]. Salt formation is also a form of reaction crystallization, and here some level of control is also essential for delivering quality material [33].

4.5.2.2 *pH Switch*

Many compounds used as pharmaceuticals are either weak acids or bases, which show increases and decreases in solubility when the pH of the solution is altered. Therefore, if a weak acid or base is in its ionized form and a strong acid or base is added to these solutions, they will be protonated or deprotonated, respectively, and precipitate due to their reduced solubility in the solvent. Obviously, the compound requires a useful pK_a (dissociation constant) and stability towards hydrolysis to be a candidate for this type of crystallization. Clearly, this is also a type of reactive crystallization, and unless controlled, it can lead to difficulties in the production of drug substance of the requisite quality [34].

4.5.2.3 *Cooling Crystallization*

In a cooling crystallization, the compound is usually not very soluble in a solvent at ambient temperature, but it is much more soluble at elevated temperatures. Therefore, for best results, a relatively steep solubility curve is required with respect to temperature. This will be discussed in more detail later in the text.

4.5.2.4 *Evaporative Crystallization*

In this situation, the compound usually has good solubility in the solvent, which is then removed by, for example, distillation, to increase the supersaturation. Alternatively, some of the good solvent can be removed by distillation and replaced by a solvent in which the compound has inferior solubility ('put and take'), causing it to crystallize [35]. However, using this method, impurities can build up and may give rise to problems such as poor crystal habit, reduced yield and indeed impurities themselves may crystallize. In addition, the API may be deposited on the wall of the reactor (encrustation) as the solvent is removed, which can lead to decreased purity and adventitious seeding; the deposit will need to be cleaned off the reactor wall after use. Because of these potential drawbacks, evaporative crystallizations can be difficult to control and scale-up and are generally not the preferred way of manufacturing APIs. Nevertheless, examples of the scale-up of a distillation crystallization at pilot scale have been reported [36].

4.5.2.5 *Crystallization using a Miscible Non-Solvent (Anti-Solvent Crystallization)*

Anti-solvent crystallizations are often used because the temperature–solubility profile of a compound is unsuitable for a cooling crystallization. In this case, a miscible non-solvent is added in a controlled manner to a solution of the compound to reduce its solubility to a point where it crystallizes. However, it is important to note that adding large amounts of anti-solvent may result in precipitation of impurities or lead to an increased risk of oil formation, especially when performed quickly [37]. They are therefore somewhat more difficult to develop and scale-up. Nevertheless, anti-solvent crystallizations are second in popularity to cooling crystallizations. In addition, they are also often used in conjunction with cooling as a means to create supersaturation [38]. The order of addition can be very important in these types of crystallization, and normally the anti-solvent is added to a solution of the API. If the solution of API is added to the anti-solvent, then high levels of supersaturation will immediately be generated at the point of addition, leading to rapid precipitation of the compound. This can, of course, lead to small particles, the wrong polymorph (or mixtures of polymorphs), and so on, being generated, with severe consequences for any downstream processes. Thus, controlled anti-solvent addition, combined with seeding, can be used to control the properties of the final API [39].

4.5.3 Design of Robust Cooling Crystallization

Before commencing a scale-up of a crystallization process for the delivery of an API, there should be a good understanding of the solid-state properties of the compound. In particular, it is usually preferable to use the thermodynamically stable polymorphic form of the compound, although in some cases a metastable or even amorphous forms are utilized in the formulation. On the other hand, the consequence of the unintended transformation of a metastable to stable polymorphic transformation in a formulation is amply illustrated by the AIDS compound ritonavir [40, 41]. It is also preferable to develop relatively simple processes to obtain the desired API properties.

Since a cooling crystallization is a common method with regard to the production and purification of compounds, knowledge of the solubility of a compound in solvents with respect to temperature is essential. It should be noted that the solution that remains after a re-crystallization is known as the mother liquor, which contains the dissolved impurities and some of the compound that has not crystallized (loss to liquors); that is, the yield is always less than 100%. As a rule of thumb, those solvents in which the solute shows good solubility at elevated temperature and low solubility at lower temperatures should be considered as candidate solvents for crystallization. In this respect, recent work has shown that if a compound has solubility of around 5–20 mg/mL at room temperature and if the solubility doubles for every 20 °C increase in temperature, this justifies attempting a cooling crystallization [42]. In industrial API manufacture, the desired output is a process that gives reproducible chemical purity (usually >99%), the desired polymorph and particle properties in 10 relative volumes of solvent and 90% yield. As an example, Table 4.1 shows the purification of dirithromycin afforded by a range of solvents [43]. Note that in this case, the acetone:water system gave the biggest reduction in impurity levels and also the largest yield.

Figure 4.2 shows the temperature–solubility curve of AZD3342 (an MMP12 inhibitor that was proposed for the oral treatment for chronic obstructive pulmonary disease (COPD) [44]) with respect to temperature in 50:50 v/v 1-propanol:water and 67:33 v/v industrial methylated spirits (IMS, ethanol denatured with methanol):water. Note the rapid increase in solubility with respect to temperature, making these two solvents ideal for a cooling

Table 4.1 Impurity reduction for dirithromycin by re-crystallization from various solvents [43].

Re-crystallization solvent	Recovery (%)	Total impurity reduction (%)	Dirithromycin B reduction (%)
Tert-butyl methyl ether	79	1.5	0.04
Tetrachloroethene	76	−0.27	0.1
Heptane–toluene	84	0.86	0.04
Amyl acetate	31	1.49	0.02
Methanol–water	80	0.73	−0.06
Ethanol–water	63	1.44	−0.12
Acetone–water	90	2.15	0.22
THF–water	83	1.9	0.3
Acetonitrile	83	1.38	0.07
EtOAc-Et_3N	60	1.53	−0.27
IPA	63	2.06	0.22

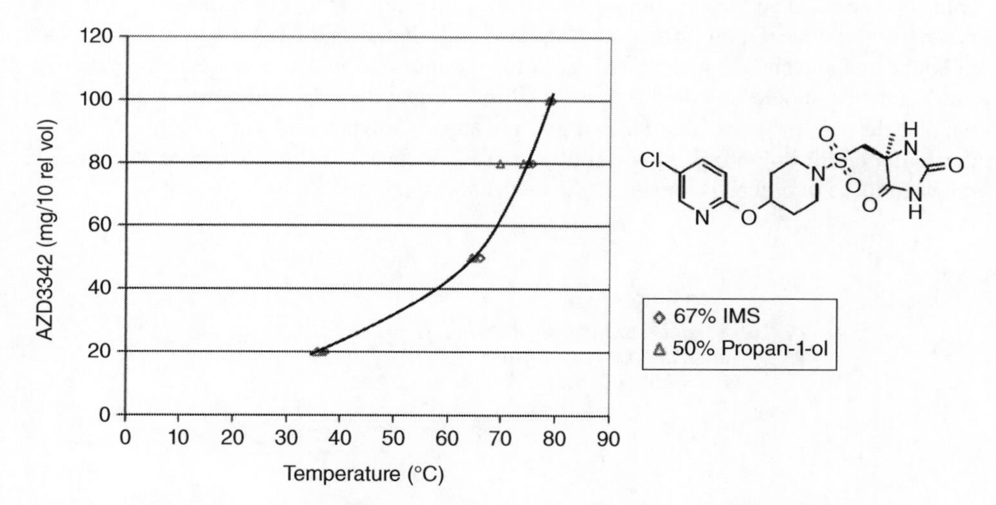

Figure 4.2 Temperature–solubility curve for AZD3342 in 1-propanol:water and IMS:water.

crystallization: the IMS:water system was used in early manufacturing campaigns, but was changed to the 1-propanol:water system for operational reasons.

In the situation where a compound shows good solubility in a solvent, but the solubility does not vary much with respect to temperature, then an evaporative or anti-solvent crystallization should be considered. Self-evidently, the compound should be stable under the conditions it is likely to experience in the laboratory and plant and must not react with the solvent. In some circumstances, mixtures of two or more solvents can also be used as a cooling crystallization medium; however, a single solvent is preferred from recycling and sustainability standpoints [45, 46]. Furthermore, ICH Class III solvents (e.g. 2-propanol, ethyl acetate, ethanol, and isopropyl acetate) should be utilized wherever possible as the crystallization medium as these are considered less deleterious to health than Class II solvents (e.g. methanol and acetonitrile) and Class I solvents. As such, higher residual levels of ICH Class III solvents are tolerated in the final API compared to Class II solvents. Note that Class I solvents would never be used in practice, since these include known carcinogens such as benzene and carbon tetrachloride. Thus, the choice of solvents used in chemical development processes is an important consideration, and a number of publications have appeared from various pharmaceutical companies outlining their preferred solvents for process development [47–49]. A summary of these guides is given by Prat *et al.* (2014) [50].

The driving force in solution crystallizations is supersaturation. A saturated solution is defined as one in which a solution of compound is in thermodynamic equilibrium with excess solid at a specified temperature. It is possible, however, to cool a hot solution to give solutions that contain more dissolved solid than that represented by equilibrium saturation. These solutions are known as supersaturated solutions.

In 1897, Ostwald introduced the terms *labile* and *metastable* supersaturation to classify supersaturated solutions in which spontaneous nucleation would and would not occur. This leads to an area of solution which upon cooling, although supersaturated, does not spontaneously crystallize. Upon further cooling, the solution passes the point where the amounts of solute and solvent are such that the solution is supersaturated and labile, and crystallization occurs. It should be noted that supersaturation may be achieved not only by cooling, but also through evaporation, or the addition of an anti-solvent, and so on. Figure 4.3 shows the temperature–solubility curve (solid) and the metastable limit (dashed): the area in between these two lines is known as the metastable zone (MSZ).

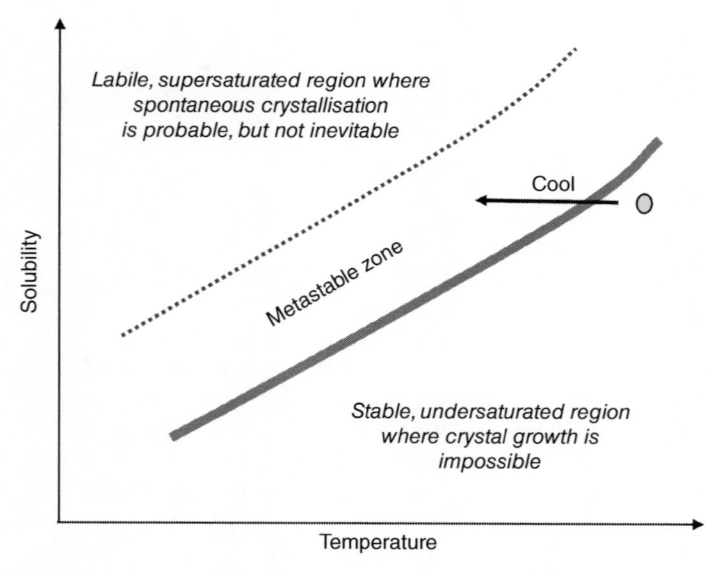

Figure 4.3 *Solubility–temperature and metastable zone limit curves.*

In the MSZ region, the delay in the appearance of nucleation, known as the induction time, τ, is the time that elapses as the molecules organize themselves into a three-dimensional arrangement prior to crystallization of the bulk of material. In practice, the determination of the width of the metastable zone is very important in the design of crystallization processes. Clearly, inside the MSZ, the supersaturation increases as the temperature decreases, and at some point nucleation will occur. However, the wider the metastable zone (the metastable zone width, MSZW), the higher the supersaturation generated and the faster the rate of nucleation. Thus, high levels of supersaturation often lead to the production of small crystals with the possibility of agglomeration, polymorphism and reduced purity. Unlike the temperature–solubility curve, which is a thermodynamic quantity, the limit of the metastable zone is not well defined and depends on the temperature, impurities, stirring and scale; that is, it is kinetic in nature [51]. In general, the faster the cooling rate or more impure the sample to be crystallized, the wider the MSZW. In turn, this can lead to a wide particle size distribution and as a consequence long filtration and drying times [52]. Figure 4.4 shows the MSZW as a function of the cooling rate for sibenadet hydrochloride (AR-C68397AA). As can be seen, the metastable zone for this compound is relatively wide (~30 °C), and diverges as the cooling rate increases.

Black *et al.* (2013) [53] have summarized current thinking on the measurement of solubility of organic compounds. Although there are a number of ways to measure solubility, the determination of the solubility of a compound and MSZW can be conveniently made using a Crystal16 instrument (Technobis Group, the Netherlands), which uses turbidity to measure clear (solubility) and cloud points (crystallization) on a small scale of 1–2 mL. Although the Crystal16 is a very useful instrument, it should be remembered that the MSZW can be a function of scale, and results may be different when using larger-scale equipment [54]. Figure 4.5 shows the data obtained from Crystal16 runs for two

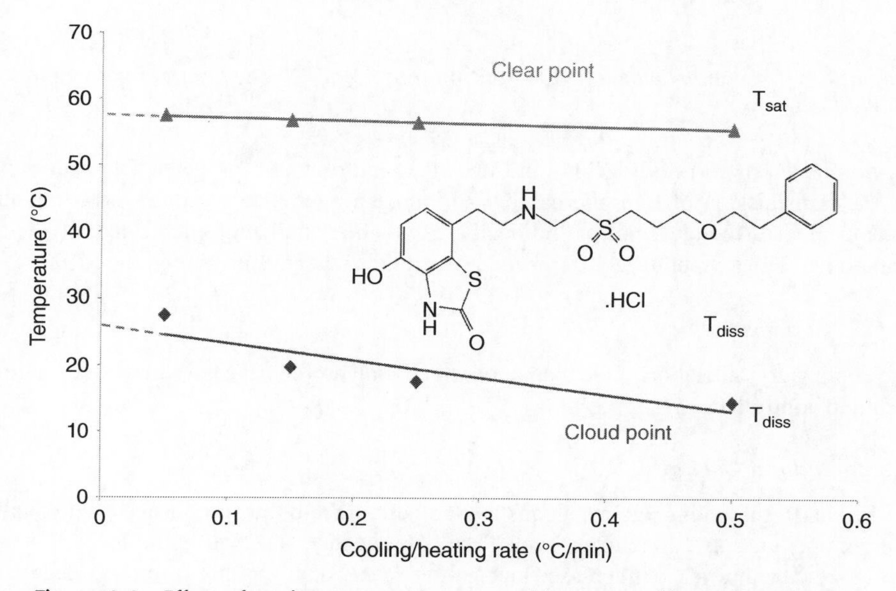

Figure 4.4 *Effect of cooling rate on the metastable zone width of sibenadet HCl.*

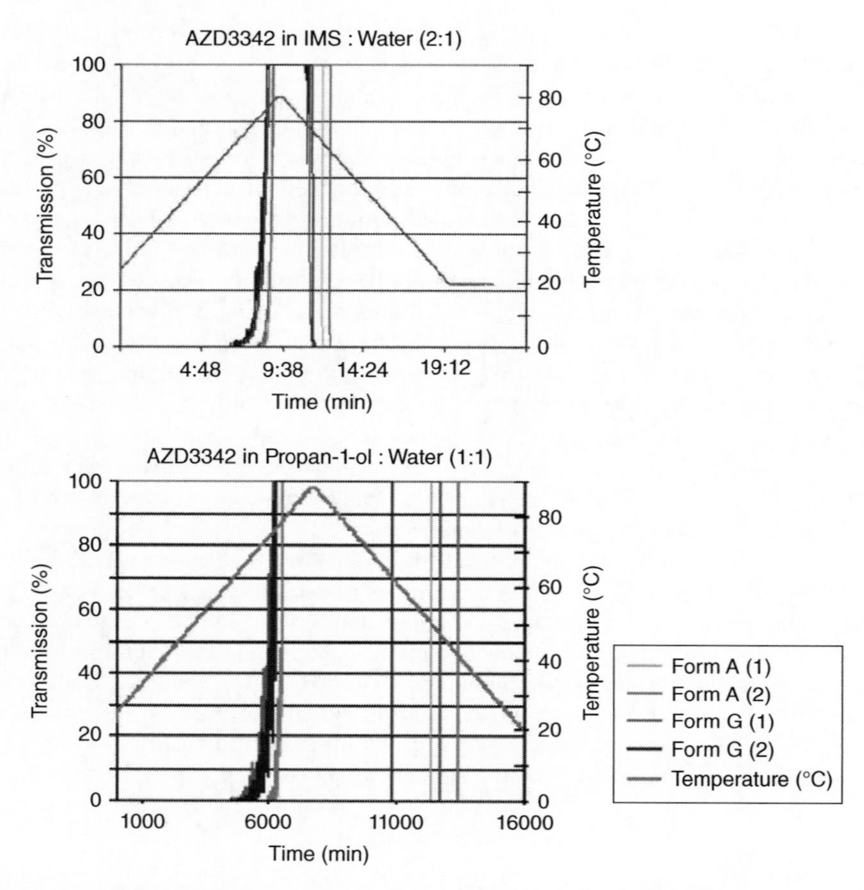

Figure 4.5 *Crystal16 data for AZD3342 polymorphs A and G. (See insert for color representation of the figure.)*

polymorphs (A and G) of AZD3342 in IMS:water and propan-1-ol:water. The data shows how the transmission of light increases with increasing temperature until the compound dissolves (100%, the clear point) and decreases when crystallization is detected (0%, the cloud point). The temperature difference between these two events in then the MSZW.

4.5.3.1 Cooling Profiles

For a cooling crystallization, three main types of cooling profile have been identified: natural, linear and controlled.

4.5.3.2 Natural Cooling

This results from cooling against a constant temperature and thus produces an exponential fall in temperature as a function of time. The consequence is a high initial level of supersaturation with the risk of increased nucleation rates, small crystal sizes and decreased purities. Since natural cooling is neither reproducible nor scalable, it is best avoided in industrial processes.

4.5.3.3 Linear Cooling

This is the most often used cooling profile used in industrial practice because it is straightforward to program, for example, a 0.1 °C–0.3 °C min^{-1} decrease in temperature with respect to time is a typical cooling program used on scale-up in large-batch STRs. Fast cooling rates ('crash' cooling) should be avoided since this can lead to a high supersaturation being generated before crystallization takes place.

4.5.3.4 Controlled Cooling

Controlled cooling, or 'cubic cooling', aims to maintain a constant supersaturation level throughout the run, which should lead to a more constant mass deposition throughout crystallization. Thus, after seeding in the MSZ, a temperature reduction program is initiated following the profile generated by Equation (4.1)

$$T = T_o - \left[(T_o - Tf) \left(\frac{t}{\Delta t} \right)^3 \right] \tag{4.1}$$

where T_0 is the start temperature, and T_f is the final temperature of the cooling profile. The start time of the temperature profile and its duration are given by t and Δt, respectively. If this profile cannot be programmed into a cooling controller, then it can be approximated with a number of linear steps. Figure 4.6 shows the form of the cooling curves for natural, linear and controlled cooling with respect to temperature and time and the respective supersaturation profile and time profiles.

Clearly, using controlled cooling, the solution concentration and hence supersaturation is maintained close to a constant level, and hence any large-scale nucleation should be avoided. In this situation, crystal growth is favoured if the solution is seeded.

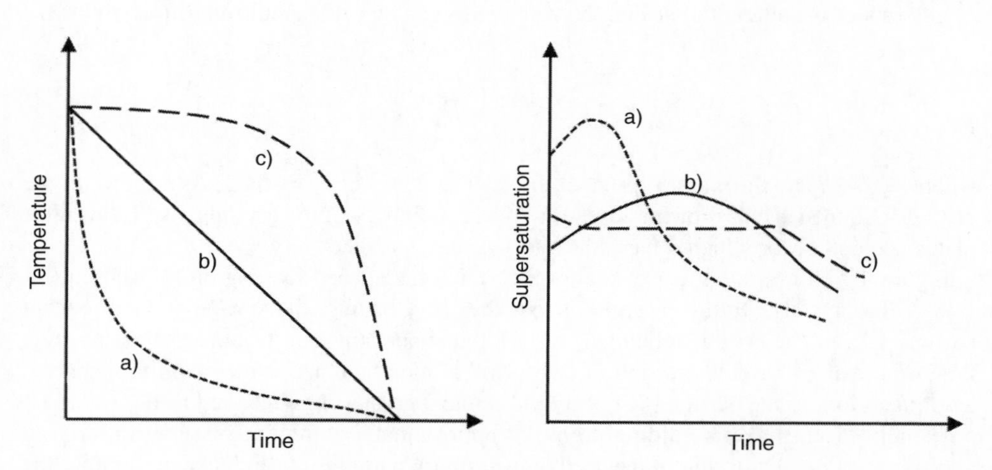

Figure 4.6 *Cooling crystallization scenarios: (a) natural cooling, (b) linear cooling, (c) controlled cooling.*

4.5.3.5 Seeding

The primary purpose of seeding is to suppress nucleation and make crystal growth the dominant process. It is often performed to control the particle size or polymorphic form of a compound [55–58] and generally improves the quality of the crystals produced [59]. Adding seeds should also be considered if the system exhibits difficulty with nucleation or if it oils (separation of a second liquid phase of the compound at high supersaturation) [58, 60].

The usefulness of seeding is a function of the amount and activity of the seed, the supersaturation conditions at which the seed is added and the cooling or anti-solvent addition program used [61]. It may be necessary to activate or decontaminate the seeds, which can be achieved by washing or wet grinding in the crystallization solvent before they are added to the solution to be crystallized. It is important that during the crystallization process the solution does not cross from the metastable zone into the labile region before seeding the solution, as this will result in nucleation and hence make any subsequent seeding redundant. Seeding has the effect of reducing the MSZW compared to an unseeded crystallization [62].

In terms of where to seed in the MSZ, it is usually best to seed at low supersaturations, since crystal growth will be more controlled here, and larger crystals may be obtained. As noted by Beckman (2000) [58], the best place to seed is close to the solubility line and no more than a quarter to a half way into the metastable zone. With regard to the amount of seed, if it is too low, then there will not be enough surface area to take up the supersaturation. As a consequence, there will not be enough crystal growth to reduce the supersaturation, and hence the solution may show a nucleation event due to 'unconsumed' supersaturation leading to a bimodal particle size distribution [61]. Beckmann (2000) [58] has indicated that the amount of seed added to a crystallization is typically ≤10% w/v and is usually added close to the stirrer for good dispersion. According to Heffles and Kind (1999) [63], the mass of seeds used should be of the order of 0.1 to 3% w/w. Of course, the seed needs to be of cGMP quality if these relatively high levels are used. If non-cGMP seed is to be used, then the amount used is typically reduced to, for example, 0.1% w/w.

In terms of the amount of seed to be added, this can be calculated from Equation (4.2):

$$m_s = m_c \left(\frac{d_s}{d_c} \right)^i \tag{4.2}$$

where m_s and d_s are the mass and size of the seed, and m_c and d_c are the mass and size of the crystals obtained. The exponent i relates to the morphology of the crystals, that is, needles, plates and cubes, for which it has values of 1, 2 and 3, respectively.

Figure 4.7 shows the effect of the level of micronized seed loading on the subsequent size of the crystals. In this case, the higher the seed loading, the smaller the subsequent particle size of the crystals obtained; that is, the supersaturation has been taken up by a larger number of particles, which in turn grow to a certain size. Note also the increased agglomeration, as the particle size of the subsequent crystals has been reduced.

Usually there is also a holding period of approximately a half to one hour after seed addition to allow dispersion of the seed to its primary particle size and take up some of the supersaturation before the cooling program is initiated.

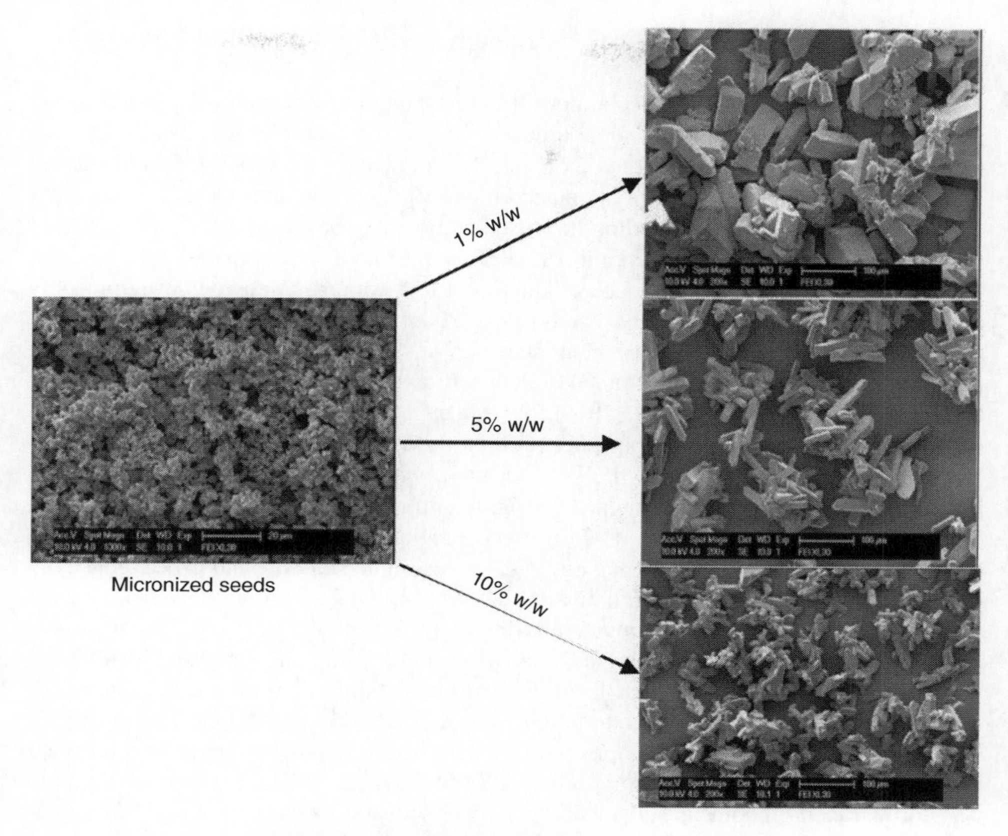

Figure 4.7 *Effect of seed loading on particle size.*

Warstat and Ulrich (2010) [64] have developed some heuristic rules regarding batch cooling crystallization. Using citric acid, adipic acid, potassium sulphate and a heterocyclic organic salt as test substances, they arrived at the following conclusions:

- The seeds should not be added more than 30%–40% into the MSZ.
- The supersaturation should be maintained as low as possible.
- For substances with a wide metastable zone (≥15 K), a seed mass calculated using Equation 4.2 can be used.
- Controlled cooling and increasing the seed mass helps; however, at very high seed loadings, the mean size of the product decreases. In addition, it can lead to agglomeration.
- For substances with a narrow MSZW, the amount of seed used should be higher than calculated from Equation 4.2, and controlled cooling should always be used.
- Milled seeds lead to breeding effects and uncontrolled crystallization and are not preferred.
- A suspension of seed crystals leads to better results compared to addition of dry seeds.

4.6 In-Line Process Analytical Technology and Crystallization Processes

The use of process analytical technology (PAT) to monitor and control crystallization processes has been available for some time. As well as the examples given here, see also the chapter in this book on PAT and its applications (Chapter 9). In 1997, Dunuwila and Berglund [65] reported the use of attenuated total reflectance Fourier transform infra-red (ATR FTIR) spectroscopy to monitor the crystallization of maleic acid in the process vessel in real time; that is, it did not need removal of a sample and a phase separation to measure the solution concentration of the solute. The use of PAT for crystallization development is now considered to be routine, and Birch *et al.* (2005) [66] have published a PAT-based strategy for crystallization development: Barret *et al.* (2005) [67] and Byrn *et al.* (2006) [68] have also reviewed the use of PAT for batch crystallization process understanding. More recently, Bordewaker *et al.* (2015) [69], Chanda *et al.* (2015) [70] and Simon *et al.* (2015) [71] have given industry perspectives on the use of various PAT tools in API manufacturing processes. In addition, the FDA is interested in PAT applied to crystallization processes, since it allows real-time processing, monitoring and control [72, 73].

Some of the benefits of using PAT in crystallization operations are that they can aid (along with appropriate experimental designs) process understanding and hence reduce the occurrence of batch failures in production. In addition, PAT measurements can enhance product quality through continuous monitoring and feedback control [74]. PAT measurements can also help reduce cycle times, improve manufacturing efficiency and identify the root causes of process deviations. There are various definitions of how PAT is used, for example, on-line or in-line, which are often confused. Therefore, to be clear, on-line PAT techniques are those where a sample is withdrawn from the process vessel and returned after analysis. For example, Dharmayat *et al.* (2006) [75] monitored the $\alpha \rightarrow \beta$ polymorphic transformation of glutamic acid using a specially designed flow through X-ray diffractometer. On the other hand, in-line techniques, which are preferred nowadays, monitor the crystallization in real time *in the process vessel*. Some in-line PAT techniques used to monitor crystallization processes are the following:

- Turbidity [76]
- Focussed beam reflectance measurements (FBRM), Lasentec, various sizes available in lab and plant [77]
- In situ visualization, for example, particle vision monitor (PVM) [60] or endoscope-based systems [78]
- Attenuated total reflectance ATR UV-Vis spectroscopy [79]
- ATR FTIR spectroscopy [80]
- Near-IR (NIR) spectroscopy [81]
- Raman spectroscopy (useful for polymorphs) [82]

Most of the in-line techniques have models that can be used on a very small laboratory scale as well as those that can be introduced into large plant reactors [77, 83]. During development of crystallization processes, it is often beneficial to use a number of these techniques in the same laboratory process vessel to aid data interpretation. For example, Deneau and Steele (2005) [60] used a combination of ATR UV/visible spectroscopy, PVM and FBRM to deduce the mechanism of oiling out and crystallization of a compound and redesign it as a seeded, cooling crystallization that did not oil out. Since the crystallization

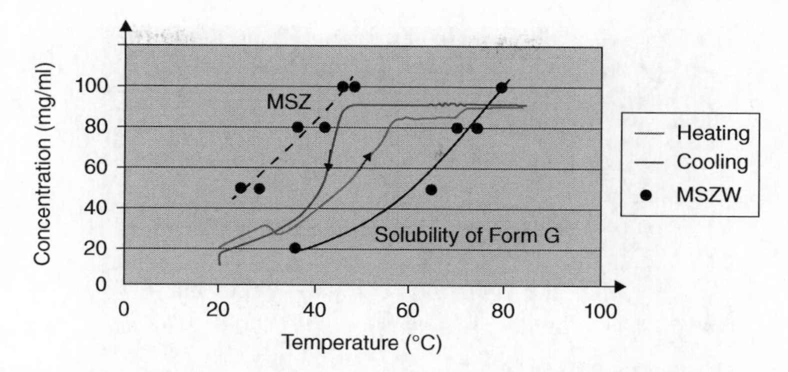

Figure 4.8 *UV Data for an unseeded linear cooling crystallization of AZD3342.*

process is being monitored in real time using PAT, the possibility exists to employ feedback control strategies to ensure that crystallizations remain under control by monitoring the supersaturation or employing a technique known as direct nucleation control [74, 80].

Figure 4.8 shows the concentration data (measured by an in-line ATR UV/visible spectroscopy probe) obtained using a linear cooling profile versus a cubic cool after seeding a solution of AZD3342. As can be seen, this compound had a relatively wide MSZW of approximately 40 °C (Crystal16 data), and when allowed to spontaneously crystallize, it led to significant agglomeration of the crystals. Agglomeration (non-reversible binding of crystals together) is often undesirable in API production since it can, for example, entrap mother liquor and reduce the purity of a compound [84].

For the unseeded crystallization, the dissolution (the pink line) of the crude material (which was an isopropyl acetate solvate) showed a number of inflexions on heating and indicated that it probably underwent a number of solid-state transformations on heating. The solution was then cooled using a linear cooling profile (blue line), and the solution concentration measurements showed a relatively fast de-supersaturation after the metastable zone boundary was reached. The corresponding FBRM data (Figure 4.9) showed that after an initial increase in counts the smaller size ranges decreased with time. On the other hand, the larger size ranges increased in number, which taken together indicated that agglomeration was taking place in the crystal slurry. A photomicrograph showed the presence of a large agglomerates of the compound.

This experiment was repeated using seeding and a cubic cooling profile, and Figure 4.10 shows the corresponding ATR UV/visible spectroscopy data. The solution was seeded at 68 °C and cooled using the cubic cooling profile described by Equation 4.1. The UV data shows that, after seeding, the de-supersaturation profile was above and close to the solubility curve where, of course, the supersaturation is low. As a consequence, crystal growth becomes the dominant process. When compared to the unseeded crystallization with linear cooling, the FBRM data showed only growth with no agglomeration, which was confirmed by optical microscopy (Figure 4.11). The photomicrograph showed large single crystals of the desired Form G (confirmed by X-ray powder diffraction (XRPD)).

Although the polymorphic form of AZD3342 was not monitored during these crystallizations, it is possible to do so by using in-line Raman spectroscopy. For example, Thirunahari *et al.* (2011) [85] used Raman spectroscopy to define the conditions (and

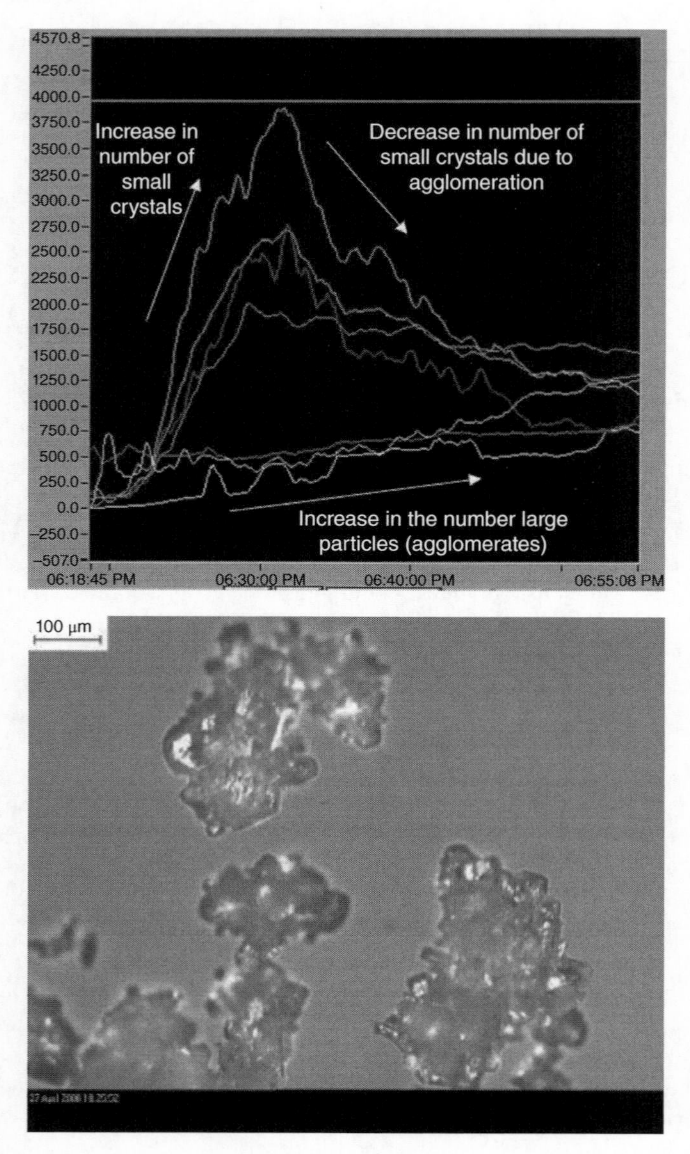

Figure 4.9 *FBRM Lasentec data and Photomicrograph for an unseeded, linear cool crystallization of AZD3342. (See insert for color representation of the figure.)*

generate a design space) for the production of a metastable form (Form I) of tolbutamide that had a better morphology than the thermodynamically stable Form II.

4.6.1 Other Unit Operations

Crystallization is not the only unit operation that the drug substance undergoes before being formulated. After crystallization, the drug substance has to be separated from its

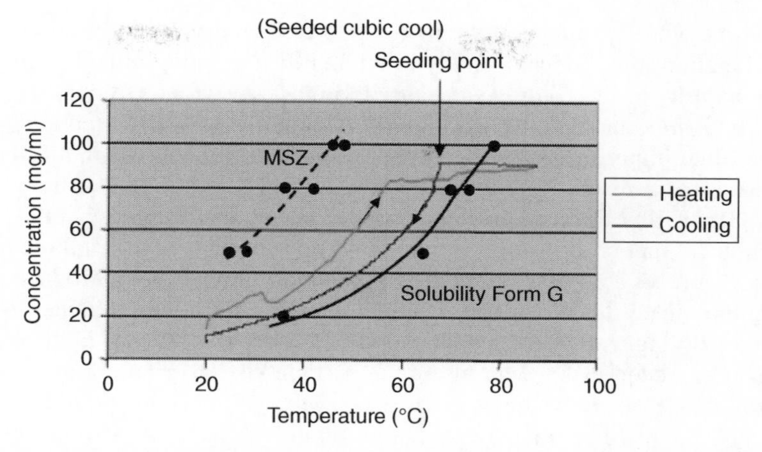

Figure 4.10 *UV Data for the dissolution and crystallization of AZD3342 using seeding and a cubic cooling profile (controlled cooling).*

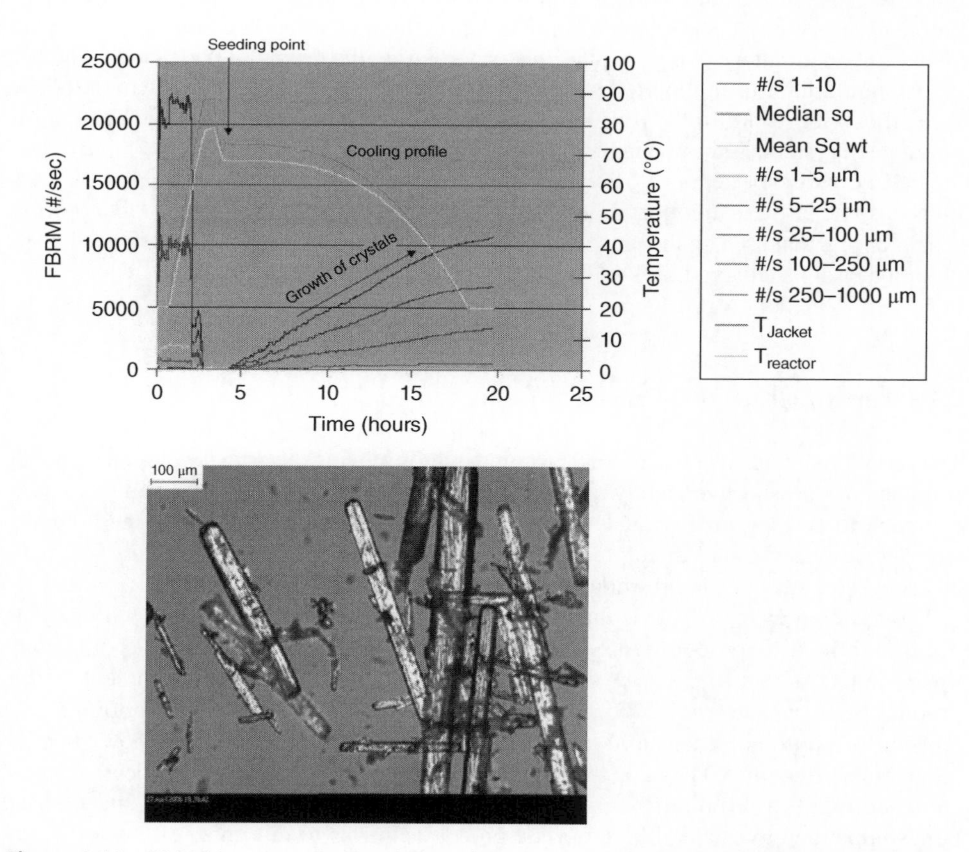

Figure 4.11 *FBRM Lasentec data and optical microscopy of AZD3342 crystals using seeding and a cubic cooling profile (controlled cooling). (See insert for color representation of the figure.)*

mother liquor (which contains the rejected impurities), usually by pressure filtration (or centrifugation), washed and finally dried and then possibly milled or micronized. Since the impurity level of a drug substance is almost always a CQA, washing the filter cake is a very important step in a compound's manufacture. This step is designed to remove residual impurities from the mother liquor left in the voids in a filter cake by the application of a miscible, pure solvent. For example, Estime *et al.* (2011) [86] studied the effect of filter cake washing on the size and shape of API crystals and found that an impurity that co-crystallized with the drug substance required three filter cake washes to ensure its complete removal. Likewise, drying can also introduce problems post filtration, since the filter cake in a pressure filter is often ploughed to enhance efficiency of drying. As shown by MacLeod and Muller (2012) [87], changes due to filtration and agitated drying can affect the particle size distribution of crystals post crystallization; that is, they showed that there was a decrease in the size of the large particles and an increase in the number of small particles with time due to cake ploughing.

On the other hand, agglomeration and particle size enlargement through granulation type processes have also been reported to occur during drying [88, 89]. Drying can also cause problems, especially if the material being dried is a hydrate [90]. Of course, PAT can play a vital role in monitoring any changes that occur during the drying of APIs, and a number of techniques have been reported in this respect [91–94]. Similarly, milling can introduce variability with regard to the production of the API for formulation [95, 96]. Thus, all of these post-crystallization manipulations require careful control to deliver the material of the desired particle size, purity, polymorphic form, and so on, from a QbD point of view. Indeed, some of these unit operations may also need to be examined using a QbD process separate from that used on the preceding crystallization step to ensure that the desired quality of material is delivered.

4.7 Applying the QbD Process

The adoption of QbD by pharmaceutical companies is moving the drug development model from one of retrospective quality by testing (QbT) to a prospective science and risk-based approach to develop process understanding. The problem with QbT is that the products have already been made; therefore, if any defect is detected, it cannot be easily rectified, and so a batch may need to reworked or even rejected [97].

The basic premise of QbD is that the drug substance should be designed with patient needs and the drug product process design in mind. Therefore, it is necessary to determine the CQAs that impact, for example, safety and efficacy (impurity content), drug product performance (dissolution, stability) and drug product manufacturability (content uniformity). For each unit operation, it is necessary to understand how process parameters affect the CQAs and then determine the operating ranges for process parameters and input material attributes that achieve the desired quality (design space). The next step is then to establish appropriate process controls to minimize effects of variability on the CQAs. It should also be emphasized that continuous process improvement is a vital element of QbD [98].

The creation of a design space begins with the definition of the QTTP (quality target product profile). From a QbD viewpoint, the first requirement is to understand what factors affect the quality of the drug substance. These can be established from prior knowledge, a QRA and early experiments. Typical CQAs for the drug substance include, for example, moisture content, polymorphic form, crystal morphology, particle size distribution, and so on, that can affect downstream processing and manufacture of the drug product. A list of primary and secondary QAs can then be drawn up and matched with how much regulatory flexibility is desired. The team then prepares scientific evidence to support the design space. If a wide-ranging design space is desired, it is probable that a larger amount of resources and experimental work will be needed.

4.7.1 Quality Risk Assessment (QRA)

For a more comprehensive review of QRA, refer to Chapter 2 of this book by Noel Baker. Briefly, a QRA is an assessment of the risk of variation of material attributes (starting materials, etc.) and processes parameters to the CQAs of the associated API. It is a quantitative model often based on a failure mode, effects and criticality analysis (FMECA), which is used to facilitate the identification of those factors affecting the quality of the drug substance. Once the risks are identified, they can then provide clearer direction and attention for defining the design space and any subsequent control strategies that will be needed.

The first step is the definition of the CQAs or properties of the drug substance that have the potential to affect the safety and efficacy of the drug product. The second step is identification of the unit operations with the potential to affect a CQA. Assessment of the CQAs can be done in two ways: the top down approach or the bottom up approach [99]. The aim of the top down approach, which is typically used early in the project's life, is to identify those process steps that can affect the CQAs. Alternatively, the bottom up approach is usually applied when the process is better understood, since it should be easier to identify possible failure modes in each of the process steps and operations, and hence the consequences of the failure on the CQA can be more accurately evaluated.

A cause-and-effect diagram, also known as a Fishbone or Ishikawa diagram, is a useful tool to capture and document the brainstorming of all the potential factors that may influence the process under consideration. Figure 4.12 shows an example of an Ishikawa diagram for an anti-solvent crystallization. The Ishikawa diagram provides the linkage between those unit operations that affect the CQAs and the parameters in the unit operation that might have an impact.

Next, a prioritization matrix or cause-and-effect table can be drawn up. A list of each of the unit operations and a list of all the process parameters that could affect them is made and evaluated with respect to failure mode, failure effect and its impact on quality of the API. Table 4.2 shows a process parameter that might vary during manufacture, in this case, the re-crystallization dissolution temperature for the crude AZD3342, which was an isopropyl acetate solvate. This is then assessed with regard to its probability, severity and detectability in the manufacture of the API (on a scale of 1 to 5, i.e. not likely to highly likely). The process parameter is then scored with respect to probability and severity, which are then multiplied together to yield a risk number. The detectability of such an event is then estimated,

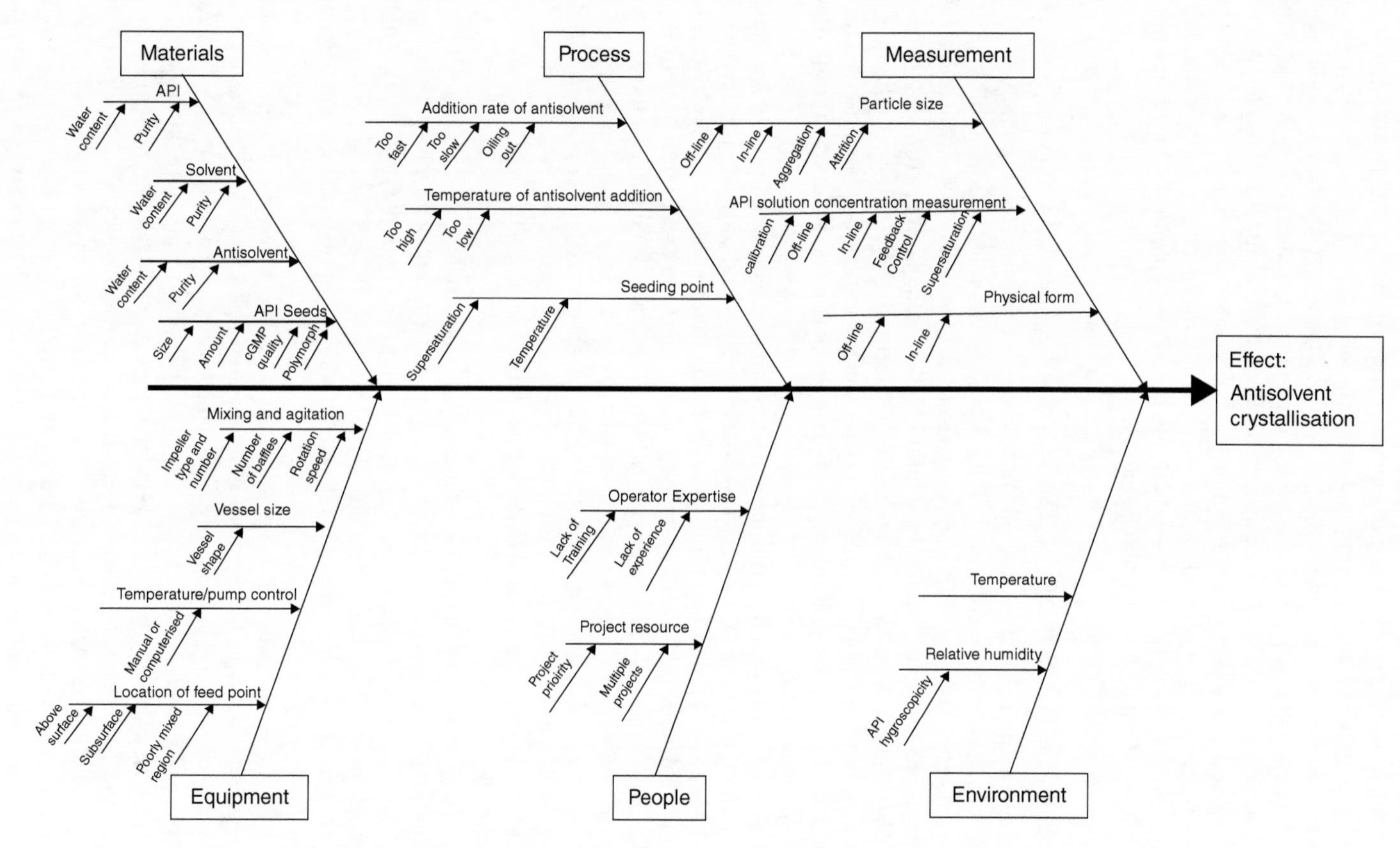

Figure 4.12 *Typical Ishikawa diagram for an anti-solvent crystallization process.*

Table 4.2 *Potential critical process parameter.*

Process parameter	Failure mode	Failure effect	Quality impact
Re-crystallization dissolution too low	Not all AZD3342 isopropyl acetate solvate dissolved	Mixture of forms produced	Levels of isopropyl acetate in solid after crystallization too high

Table 4.3 *Scoring table.*

Probability	Severity	Risk number	Detectability	Risk priority number
2	4	8	4	32

and this is multiplied by the risk number to generate a risk priority number (RPN). An example of an entry in a cause-and-effect matrix for this system is shown in Table 4.3.

As shown in Table 4.3, the scoring for this process parameter indicated that it was not likely to occur, thus resulting in a low probability estimate of 2. However, if it did occur, then the severity would be high (4), resulting in a halt in production. Fortunately, production of the wrong polymorph or mixtures of solid forms would be easily detected by, for example, XRPD, and hence it is given a high score of 4. Taken together, the RPN is then calculated to be 32. This is not too high, but still large enough to give some concern. Of course, the RPN could be reduced if an in-line PAT technique like Raman or NIR spectroscopy was used to monitor the crystal form during the crystallization process. If the wrong form was detected during the crystallization, then it might be possible to introduce a rescue procedure, for example, reheating the slurry to the correct dissolution temperature and restarting the crystallization.

For convenience, Tables 4.2 and 4.3 can be combined to give a complete scoring table (Table 4.4).

Other processes parameters, for example, cooling rate (too high/too slow), stirring rate (too fast/too slow) and seeding level (too much/too little) can then be added and scored accordingly so that a comprehensive picture can built up of how all of the process parameters could affect the crystallization process.

As a visual aid, the RPN can be colour-coded to emphasize those factors that need special attention. In this case, the re-crystallization dissolution temperature, which might be too low prior to crystallization, would yield a relatively high RPN (32) and could be colour-coded yellow. However, if the probability of the re-crystallization dissolution temperature being too low was even higher, say 3, then the RPN would become 64 and would thus be coloured-coded red.

There needs to be justification for including or omitting particular parameters to be investigated; therefore, presenting output from the FMECA investigation and cause-and-effect matrix all help to strengthen the case for the selection/omission of parameters. It is important to note that the ranges and levels of parameters to be investigated should be in line with the scale of the operation and intended manufacturing facility.

Table 4.4 *Scoring table for a crystallization process.*

Process parameter	Failure mode	Failure effect	Quality impact	Probability	Severity	Risk number	Detectability	Risk priority number
Pre-crystallization dissolution too low	Not all AZD3342 isopropyl acetate solvate dissolved	Mixture of forms produced	Levels of isopropyl acetate in solid after crystallization too high	2	4	8	4	32

4.8 Design of Experiments (DoE)

In order to fully understand the nature of the complex interactions between process conditions and the control of, for example, the final particle size distribution, many experiments at bench scale would be required to vary the necessary parameters. However, a DoE approach is generally recognized as being a more efficient way of investigating the experimental variables that can affect a process rather than examining the effect of one variable at a time (OVAT) [100]. Thus, once key variables have been established, it will usually be necessary to perform an experimental study incorporating the identified variables in a fractional factorial DoE. A fuller review of DoE is given in Chapter 7 of this book by Martin Owen and Ian Cox, but basically, a DoE involves varying all the relevant factors simultaneously over a set of planned experiments, which connect the results via a mathematical model. Using this approach, it should be possible to gain the maximum information from the minimum number of experiments. In other words, using statistical experimental design and multivariate methods allows simultaneous investigation of non-independent crystallization factors. Interactions between the crystallization conditions can then be modelled in unbiased screening experiments designed to obtain the maximum amount of data from a minimum number of experimental runs.

It is necessary to characterize acceptable ranges of key and critical process parameters contributing to the identification of a design space, which helps provide an 'assurance of quality'. Commercial software is available, for example, Modde or Design Expert, to assist with experimental design. However, it is important that the DoE used for the drug substance includes an appropriate resolution to assess interactions if a fractional factorial is used [101] (Marco 2013).

In a relatively early example of DoE use in the crystallization arena, Braatz *et al.* [102] performed multivariate screening experiments to create a main-effects model for the investigation of conditions during a pharmaceutical crystallization process. More recently, Sato *et al.* (2015) [103] used DoE to control the residual solvent and particle size distribution for orantinib, which was produced by the conversion of its potassium salt to the free acid by neutralization with HCl in a 2-propanol/water mixture (a reactive pH-switch type of crystallization). In this work, they performed a total of 36 experiments designed to screen and optimize the crystallization process parameters. They found that temperature had a large effect on the particle size and also that as the particle size increased, the residual solvent level decreased. Following on from these measurements, they designated the temperature and solvent ratio as critical process parameters (CPPs) and constructed a verified design space. Of course, the use of DoE is a general approach to experimentation that can be applied throughout primary and secondary manufacture [104].

Clearly, much work is initially done in the laboratory with the goal of understanding the nature of the synthetic and crystallization processes. From a QbD perspective, some of the issues related to understanding scale-up and equipment problems associated with small molecule manufacturing process have been discussed by am Ende *et al.* [105]. They make the point that when performing risk assessments and designing experiments, the data generated should be able to predict how the scale and equipment used in the commercial manufacture impact the design space. In this respect, Brueggemeier *et al.* (2012) [36] used the pilot scale manufacture (175 kg) of ibipnabant to successfully test their laboratory predictions and the scale independence of the process.

Forthcoming regulatory requirements will include such guidance for when the drug substance has been developed at a commercial scale if the process parameters have not been demonstrated to be scale independent, that is, the drug substance has to be verified at commercial scale [103]. am Ende *et al.* (2010) [105] have developed a risk mitigation tree for this purpose.

By their nature, many DoEs are repetitive and relatively routine in nature, and hence data collection can be logistically challenging, since it may tie up a particular piece of equipment or experimental rig for a considerable period of time. Therefore, to maximize laboratory productivity, automation should be encouraged, wherever possible, to collect the data from the instruments. In this way, equipment and vessels can be used to perform experiments during the working day, at nights and/or over weekends (in conjunction with suitable risk assessments of course!). Furthermore, the accuracy and degree of control that programmed computerized systems can achieve make them very suitable for repetitive runs [106].

4.9 Critical Process Parameters (CPPs)

During the development of an API, it is essential that a control strategy is established to ensure process performance. As part of the control strategy employed, it is vital that the CPPs are identified such that their effect on the CQAs of the API are understood [107].

By definition, any process parameter that impacts a CQA is a CPP; that is, it is a significant risk to the API quality. Usually, all process parameters are assessed and classified as either critical or non-critical for the quality of both APIs and the isolated intermediates. Nosal and Schultz (2008) [98] have discussed in detail the ISPE Product Quality Lifecycle Implementation (PQLI) definition of criticality. Here, criticality is defined by Equation (4.3).

$$\textbf{\textit{Criticality}} = \textbf{\textit{Severity}} \times \textbf{\textit{Occurrence}} \times \textbf{\textit{Detection}} \tag{4.3}$$

The effect of process parameters can be determined in a number of ways, for example, statistically designed experiments, subjective scientific opinion and historical precedent [107]. From a regulatory perspective, it has been stated by Marco (2013) [101] that the criticality of process parameters must be based on impact (severity) and *not* on residual risk after the implementation of the control strategy. Furthermore, any results that are not consistent with existing scientific knowledge should be justified. Siebert *et al.* (2008) [107] have devised a flowchart to determining whether a parameter is critical or not.

4.10 Design Space

ICH Q8 (R2) defines the design space as:

> The multidimensional combination and interaction of input variables (e.g. material attributes) and process parameters that have been demonstrated to provide assurance of quality.

Working within the design space is therefore *not* considered a change. However, working outside the design space *is* considered a change and would therefore need regulatory approval. A design space might be determined per unit operation (e.g. reaction, distillation and crystallization), or a combination of unit operations and a design space that spans multiple unit operations can provide additional flexibility (ICH Q11). As noted by Lepore and Spavins (2008) [108], the design space is arrived at by an iterative process such that is not a true design space until it has been demonstrated that an 'appropriate understanding of attributes needed to assure the Quality Target Product Profile has been achieved'.

ICH Q8 (R2) states that the design space can be arrived at by a first-principles approach whereby a combination of experimental data and mechanistic knowledge of chemistry, physics and engineering can be used to model and predict performance. Alternatively, using non-mechanistic or empirical approaches that utilize DoEs, combined with linear and multiple-linear regression analyses, can also be used. In addition, scale-up correlations, which correlate operating conditions between different scales or pieces of equipment, need to be examined. Furthermore, the design space can be described in a number of ways, for example, linear ranges of parameters, mathematical relationships, time-dependent functions or, combinations of variables. Visually, the design space can be represented as a contour plot or a surface plot. A design space verification protocol for a small molecule drug substance has been presented by Watson *et al.* (2013) [109].

As an example of a first-principles approach that avoided the crystallization of a hydrate in a water-based crystallization medium, Black *et al.* [110] were able to define a design space, without the need for multivariate DoE investigations, which routinely afforded the desired anhydrated form of the compound. In practice, however, more complex DoE-based investigations are typically required to generate a design space. For example, an investigation into the crystallization of casopitant mesylate that utilized DoEs has been reported by Castagnoli *et al.* (2010) [111]. In this paper, some API salt crystals were produced via a reactive crystallization of casopitant free base with methanesulphonic acid in acetone/ethyl acetate followed by the addition of an anti-solvent (isooctane) to complete the crystallization. Potential CPPs such as the impurity profile of the free base, the agitation rate, the ethyl acetate, acetone and isooctane amounts, the quantity of seed used and temperature and isooctane addition rate were identified and investigated using a suitable DoE. The experiments were monitored using in-line FBRM, PVM, FTIR and the temperature during the crystallizations and off-line using HPLC, NMR, GC, XRPD, and so on. From the data generated, it was found that the impurities present in the API were mainly a function of the impurity profile of the free base and, to a smaller extent, the ethyl acetate volume. The particle size of crystals was strongly correlated with the volume of acetone used, seeding temperature and, again, the purity of the casopitant free base. A further iteration suggested that the API CQAs were determined by the free base impurity profile, the relative amounts of ethyl acetate and acetone present during the crystallization and the seeding temperature. In terms of the design space, these workers used a Bayesian predictive approach (as described in Reference [112]) rather than those more commonly used in these types of studies. By using this Bayesian methodology, they found a smaller region (compared to the other methods) that resulted in drug substance that better met its specifications.

4.11 Control Strategy

The control strategy is defined as 'a planned set of controls, derived from current product and process understanding that assures process performance and product quality' (ICH Q10). Garcia *et al.* (2008) [113] and Davis *et al.* (2008) [114] describe an approach and technical process for developing and implementing a control strategy. According to Davis *et al.* [114], the control strategy model for pharmaceutical development operates on three levels.

Level 1 relates to the CQAs of the product and is focussed on patient safety, efficacy and quality; any business considerations may also be identified at this stage too, for example, cost and manufacturability.
Level 2 outlines those controls necessary to enable product CQAs to be met, for example, the CPPs and material attributes.
Level 3 includes analytical, engineering and other control methods.

More recently, Thomson *et al.* (2015) [115] have summarized ICH Q11 control strategy in the general development of drug substance as a series of bullet points:

- Controls on material attributes (including raw materials, starting materials, reagents and primary packaging for drug substance).
- Controls implicit in the design of the manufacturing process (e.g. order of addition of reagents).
- IPCs, including in process tests and process parameters.
- Controls on the drug substance (e.g. release testing).

There are a number of publications dealing with the control strategies employed for the production of drug substance via a cooling crystallization. For example, Cimarosti *et al.* (2012) [116] describes the different control features and methodologies that were employed for the production of the preferred Form I of the compound (an n-propanolate was also possible, which they wanted to avoid). The control elements were described in terms of (a) attribute controls, (b) parametric controls, that is, working within proven acceptable ranges for CPPs and (c) procedural controls relating to the equipment and the 'how' by which the operations were carried out, that is, equipment, sequence of operation. By carrying out a risk assessment, the following CPPs were identified: seeding temperature, cooling and aging time post crystallization as well as the filtration temperature. In practice, even though the risk of conversion of Form I to the n-propanolate was low, an NIR method was developed and used in support of their control strategy to produce Form I. Thus, their control strategy to produce Form I was that the solution should be seeded with Form I (attribute control) at a particular temperature with a known amount of seed (parametric control). The solution was then cooled at a set rate from 70 °C to 20 °C (parametric control). On cessation of cooling, the ageing suspension (parametric control) was monitored with the NIR probe (process control) to ensure there was no conversion from Form I to the n-propanolate. After crystallization, the cake was washed (procedure control) at a set temperature (parametric control). A second example is the paper by Lobben *et al.* (2015) [117] that describes the control strategies employed in the synthesis and crystallization of bravanib alaninate. The API was produced by an anti-solvent process, which was monitored using in-line Raman spectroscopy. They used an eight-variable, two-level DoE to resolve the main effects in the crystallization and develop a robust design space. Particle size was identified as a CQA from a formulation perspective, and hence wet milling, using a rotor stator instrument, was carried out before filtration.

4.12 References

[1] Federsel, H.-J. (2010) Process R&D Under the Magnifying Glass: Organization, Business Model, Challenges, and Scientific Context, *Bioorg Med Chem*, **18**, 5775–5794.

[2] O'Brien, M.K., Kolb, M., Connolly, T.J. *et al.* (2011) Early Chemical Development at Legacy Wyeth Research, *Drug Discovery Today*, **16**, 81–88.

[3] Mehta, B. (2013) Understanding ICH Q11-FDA's Guidance on the Development and manufacturing of Drug Substances, *Pharm Tech*, **35**, 64–68.

[4] Broadhead, J., Schlindwein, W., and Potter, D. (2013) Quality by Design: Perspectives on the Current Status of Implementation and Challenges for the Future, *GMP Rev*, **11**, 15–16.

[5] Ye, Q., Huang, Y., Rusowicz, A. *et al.* (2008) Understanding and Controlling the Formation of an Impurity During the Development of Muraglitazar, a PPAR Dual Agonist, *Org Process Res Dev*, **14**, 238–241.

[6] Scott, C. and Black, S.N. (2005) In-Line Analysis of Impurity Effects on Crystallization, *Org Process Res Dev*, **9**, 890–893.

[7] Okamoto, M., Hamano, M., Igarashi, K. *et al.* (2004) The Effect of Impurities on Crystallization of Polymorphs of a Drug Substance, *J Chem Eng Jpn*, **37**, 1224–1231.

[8] Huang, Y., Ye, Q., Zhenrong, G. *et al.* (2008) Identification of Critical Process Impurities and their Impact of Process Research and Development, *Org Process Res Dev*, **12**, 628–636.

[9] Raman, N.V.V.S.S., Prasad, A.V.S.S., and Ratkanar, R.K. (2010) Strategies for the Identification, *Control and Determination of Genotoxic Impurities in Drug Substances: A Pharmaceutical Industry Perspective, J Pharm Biomed Anal*, **55**, 662–667.

[10] Robinson, D.I. (2010) Control of Genotoxic Impurities in Active Pharmaceutical Ingredients: A Review and Perspective, *Org Proc Res Dev*, **14**, 946–959.

[11] Gordiano, A., Kobel, A., and Gally, H.U. (2011) Overall Impact of the Regulatory Requirements for Genotoxic Impurities on the Drug Development Process, *Eur J Pharm Sci*, **43**, 1–15.

[12] Looker, A.R., Ryan, M.P., Neubert-Langille, B.J. *et al.* (2010) Control of Genotoxic Impurities in Active Pharmaceutical Ingredients: A Review and Perspective, *Org Process Res Dev*, **14**, 946–959.

[13] Cimarosti, Z., Bravo, F., Stonestreet, P. *et al.* (2010) Application of Quality by Design Principles to Support Development of a Control Strategy for the Control of Genotoxic Impurities in the Manufacturing Process of a Drug Substance, *Org Proc Res Dev*, **14**, 993–998.

[14] Burt, J.L., Braem, A.D., Ramirez, A., Mudryk, B. *et al.* (2011) Model-Guided Design Space Development for a Drug Substance Manufacturing Process, *J Pharm Innov*, **6**, 181–192.

[15] Cui, Y., Song, X., Reynolds, M. *et al.* (2012) Interdependence of Drug Substance Physical Properties and Corresponding Quality Control Strategy, *J Pharm Sci*, **101**, 312–321.

[16] Morrison, H.G., Tao, W., Trieu, W. *et al.* (2015) Correlation of Drug Substance Particle Size Distribution with Other Bulk Properties to Predict Critical Quality Attributes, *Org Process Res Dev*, **19**, 1076–1081.

[17] Carey, J.S., Laffan, D., Thomson, C. *et al.* (2006) Analysis of the Reactions Used for the Preparation of Drug Candidate Molecules, *Org Biomol Chem*, **4**, 2337–2347.

[18] Anson, M.S., Graham, J.P., and Roberts, A.J. (2011) Development of a Fully Telescoped Synthesis of the S1P1 Agonist GSK1842799, *Org Process Res Dev*, **15**, 649–658.

[19] Zhang, T.Y. (2006) Process Chemistry: The Science, Business, Logic, and Logistics, *Chem. Rev*, **106**, 2583–2595.

[20] Whiting, M., Harwood, K., Hossner, F., Turner, P.G., and Wilkinson, M.G., (2010) Selection and Development of the Manufacturing Route for EP_1 Antagonist GSK269984B, *Org Process Res Dev*, **14**, 820–831.

[21] Butters, M., Catterick, D., Craig, A. *et al.* (2006) Critical Assessment of Pharmaceutical Processes – A Rationale for Changing the Synthetic Route, *Chem. Rev*, **106**, 3002–3027.

[22] Lathbury, D. (2010) Develop an Early Lead with CROs, *Manuf. Chem.*, **39–40**.

[23] Faul, M.M., Kiesman, W.F., and Smulkowski, M., (2014a) Part 1: A Review and Perspective of the Regulatory Guidance to Support Designation and Justification of API Starting Material, *Org Process Res Dev*, **18**, 587–593.

[24] Faul, M.M., Argentine, M.D., Egan, M. *et al.* (2015) Designation and Justification of API Starting Materials: Proposed Framework for Alignment from an Industry Perspective, *Org Process Res Dev*, **19**, 915–924.
[25] Gavin, P.F., Olsen, B.A., Wirth, D.D. *et al.* (2006) A Quality Evaluation Strategy for Multi-Sourced Active Pharmaceutical Ingredient (API) Starting Materials, *J Pharm Biomed Anal*, **41**, 1251–1259.
[26] Illing, G.T., Timko, R.J., and Billet, L. (2008) Drug Substance Starting Material Selection, *Pharm Tech*, **32**, 52–57.
[27] Paul, E.L., Tung, H.-H., and Midler, M. (2005) Organic Crystallization Processes, *Powder Tech*, **150**, 133–143.
[28] Variankaval, N., Cote, A.S., and Doherty, M.F. (2008) From Form to Function: Crystallization of Active Pharmaceutical Ingredients, *AIChE Journal*, **54**, 1682–1688.
[29] Carpenter, K.J. and Wood, W.M.L. (2004) Industrial Crystallization for Fine Chemicals, *Adv Powder Tech*, **15**, 657–672.
[30] Stitt, E.H. (2002) Alternative Multiphase Reactors for Fine Chemicals: A World Beyond Stirred Tanks? *Chem Eng Journal*, **90**, 47–60.
[31] Lawton, S., Steele, G., Shering, P. *et al.* (2009) Continuous Crystallization of a Pharmaceutical Using a Continuous Oscillatory Baffled Crystalliser, *Org Process Res Dev*, **13**, 1357–1363.
[32] Lee, H.L., Lin, H.Y., and Lee, T. (2014) Large-Scale Crystallization of a Pure Metastable Polymorph by Reaction Coupling, *Org Process Res Dev*, **18**, 539–545.
[33] Kim, S., Lotz, B., Lindrund, M. *et al.* (2005) Control of the Particle Properties of a Drug Substance by Crystallization Engineering and the Effect on Drug Product Formulation, *Org Process Res Dev*, **9**, 894–901.
[34] Alatalo, H., Hatakka, H., Kohonen, J. *et al.* (2010) Process Control and Monitoring of Reactive Crystallization of L-Glutamic Acid, *AIChE Journal*, **56**, 2063–2076.
[35] Harris, A. (2005) Drug Substance Development and Technology Transfer in *Technology Transfer: An International Good Practice Guide for Pharmaceuticals and Allied Industries* (ed. M. Gibson), DHI Publishing LLC, pp. 57–97.
[36] Brueggemeier, S.B., Reiff, E.A., Lynberg, O.K. *et al.* (2012) Modelling-Based Approach Towards Quality by Design for the Ibipinabant API Step, *Org Process Res Dev*, **16**, 567–576.
[37] Kim, S., Wei, C., and Kiang, S. (2003) Crystallization Process Development of an Active Pharmaceutical Ingredient and Particle Engineering via the Use of Ultrasonics and Temperature Cycling, *Org Process Res Dev*, **7**, 997–1001.
[38] Nagy, Z.K., Fujiwara, M., and Braatz, R. (2008) Modelling and Control of Combined Cooling and Antisolvent Crystallization Processes, *J Proc Control*, **18**, 856–864.
[39] Nonoyama, N., Hanaki, K., and Yabuki, Y. (2006) Constant Supersaturation Control of Antisolvent-Addition Batch Crystallization, *Org Process Res Dev*, **10**, 727–732.
[40] Chemburkar, S.J., Bauer, J., Deming, K. *et al.* (2000) Dealing with the Impact of Ritonavir Polymorphs on the Late Stages of Bulk Drug Process Development, *Org Process Res Dev*, **4**, 413–417.
[41] Bauer, J., Spanton, S., Henry, J. *et al.* (2001) Ritonavir: An Extraordinary Example of Conformational Polymorphism, *Pharm Res*, **18**, 859–866.
[42] Muller, F., Fielding M., and Black, S. (2009) A Practical Approach for Using Solubility to Design Cooling Crystallizations, *Org Process Res Dev*, **13**, 1315–1325 (2009).
[43] Wirth, D.D. and Stephenson, G.A. (1997) Purification of Dirithromycin. *Impurity Reduction and Polymorph Manipulation, Org Process Res Dev*, **1**, 55–60.
[44] Fitzgerald, M.F. and Cox, J.C. (2007) Emerging Trends in the Therapy of COPD: Novel Anti-Inflammatory Agents in Clinical Development, *Drug Disc Today*, **12**, 479–486.
[45] Capello, C., Fischer, U., and Hungerbühler, K. (2007) What is a Green Solvent? *A Comprehensive Framework for the Environmental Assessment of Solvents, Green Chem*, **9**, 927–934.
[46] Perez-Vega, S., Ortega-Rivas, E., and Salmeron-Ochoa, I. (2013) A System View of Solvent Selection in the Pharmaceutical Industry: Towards a Sustainable Choice, *Environ Dev Sustain*, **15**, 1–21.
[47] Henderson, R.K., Jiménez-González, C., Constable, D.J.C. *et al.* (2011) Expanding GSK's Solvent Selection Guide – Embedding Sustainability into Solvent Selection Starting at Medicinal Chemistry, *Green Chem*, **13**, 854–862.

[48] Dunn, P.J. (2012) The Importance of Green Chemistry in Process Research and Development, *Chem Rev*, **41**, 1452–1461.
[49] Prat, D., Pardigon, O., Flemming, H.-W. *et al.* (2013) Sanofi's Solvent Selection Guide: A Step Toward More Sustainable Processes, *Org Process Res Dev*, **17**, 1517–1525.
[50] Prat, D., Hayler, J., and Wells, A. (2014) A Survey of Solvent Selection Guides, *Green Chem*, **16**, 4546–4551.
[51] Ulrich, J. and Strenge, C. (2002) Some Aspects of the Importance of Metastable Zone Width and Nucleation in Industrial Crystallizers, *J Cryst Growth*, **237–239**, 2130–2135.
[52] Moore, W.P. (1994) Optimize Batch Crystallization, *Chem Eng Prog*, 73–79.
[53] Black, S., Dang, L., Liu, C. *et al.* (2013) On the Measurement of Solubility, *Org Process Res Dev*, **17**, 486–492.
[54] Kadam, S.S., Kramer, H.J.M., and ter Horst, J.H. (2011) Combination of a Single Primary Nucleation Event and Secondary Nucleation in Crystallization Processes, *Cryst Growth Des*, **11**, 1271–1277.
[55] Sudo, S., Sato, K., and Harano, Y. (1991) Solubilities and Crystallization Behaviours of Cimetidine Polymorphs A and B, *J Chem Eng Japan*, **24**, 237–242.
[56] Beckmann, W., Nickisch, K., and Budde, U. (1998) Development of a Seeding Technique for the Crystallization of the Metastable A Modification of Abecarnil, *Org Process Res Dev*, **2**, 298–304.
[57] Beckmann, W., Otto, W., and Buddle, U. (2001) Crystallization of the Stable Polymorph of Hydroxytriendione: Seeding Process and Effects of Purity, *Org Process Res Dev*, **5**, 387–392.
[58] Beckmann, W. (2000) Seeding the Desired Polymorph: Background, *Possibilities, Limitations and Case Studies, Org Process Res Dev*, **4**, 372–383.
[59] Stelzer, T. and Ulrich, J. (2010) No Product Design without Process Design (Control)? *Chem Eng Tech*, **33**, 723–729.
[60] Deneau, E. and Steele, G. (2005) An In-Line Study of Oiling Out and Crystallization, *Org Process Res Dev*, **9**, 943–950.
[61] Lee, S. and Hoff, C. (2008) Large Scale Aspects of Salt Formation: Processing of Intermediates and Final Products in *Pharmaceutical Salts* (ed. P.H. Stahl and C.G. Wermouth), Zürich: Wiley VCH, pp. 191.
[62] Mersmann, A. (1996) Supersaturation and Nucleation, *Trans I Chem E*, **74**, 812–820.
[63] Heffles, S.K. and Kind, M. (1999) *Seeding Technology: An Underestimated Critical Success Factor for Crystallization*, 14th International Symposium on Industrial Crystallization, IChemE, UK, pp. 2234–2246.
[64] Warstart, A. and Ulrich, J. (2010) Seeding During Batch Cooling Crystallization – An Initial Approach to Heuristic Rules, *Chem Eng Tech*, **29**, 187–190.
[65] Dunuwila, D.D. and Berglund, K.A. (1997) ATR FTIR Spectroscopy for In Situ Measurement of Supersaturation, *J Cryst Growth*, **179**, 185–193.
[66] Birch, M., Fussel, S.J., Higginson, P.J. *et al.* (2005) Towards a PAT-Based Strategy for Crystallization Development, *Org Proc Res Dev*, **9**, 360–364.
[67] Barrett, P., Smith, B., Worlitschek, J. *et al.* (2005) A Review of the Use of Process Analytical Technology for the Understanding and Optimization of Production Batch Crystallization Processes, *Org Process Res Dev*, **9**, 348–355.
[68] Byrn, S.R., Liang, J.K., Bates, S. *et al.* (2006) PAT-Process Understanding and Control of Active Pharmaceutical Ingredients, *J Proc Anal Tech*, **3**, 14–19.
[69] Bordawaker, S., Chandra, A., Daly, A.M. *et al.* (2015) Industry Perspectives on Process Analytical Technology: Tools and Applications in API Manufacturing, *Org Process Res Dev*, **19**, 1174–1185.
[70] Chanda, A., Daly, A.M., Garrett, A.W. *et al.* (2015) Industry Perspectives on Process Analytical Technology: Tools and Applications in API Development, *Org Process Res Dev*, **19**, 63–83.
[71] Simon, L. L., Hajnalka, P., Marosi, G. *et al.* (2015) Assessment of Recent Process Analytical Technology (PAT) Trends: A Multi-Author Review, *Org Proc Res Dev*, **19**, 3–62.
[72] Yu, L.X., Lionberger, R.A., Rawa, A.S. *et al.* (2004) Applications of Process Analytical Technology to Crystallization Processes, *Adv Drug Del Rev*, **56**, 349–369.

[73] Wu, H., Dong, Z., Li, H., and Khan, M. (2015) An Integrated Process Analytical Technology (PAT) Approach for Pharmaceutical Crystallization Process Understanding to Ensure Product Quality and Safety: FDA Scientists' Perspective, *Org Process Res Dev*, **19**, 89–101.
[74] Saleemi, A.N., Steele, G., Pedge, N.I. *et al.* (2012) Enhancing Crystalline Properties of a Cardiovascular Active Pharmaceutical Ingredient using a Process Analytical Technology Based Crystallization Feedback Control Strategy, *Int J Pharm*, **430**, 56–64.
[75] Dharmayat, D., Calderon De Anda, J. *et al.* (2006) Polymorphic Transformation of L-Glutamic Acid Monitored Using Combined On-Line Video Microscopy and X-Ray Diffraction, *J Cryst Growth*, **294**, 35–40.
[76] Harner, R.S., Ressler, R.J., Briggs, R.L. *et al.* (2009) Use of a Fiber-Optic Turbidity Probe to Monitor and Control Commercial-Scale Unseeded Batch Crystallizations, *Org Process Res Dev*, **13**, 114–124.
[77] Adlington, N.K., Black, S.N., and Adshead, D.L., (2013) How To Use the Lasentec FBRM Probe on Manufacturing Scale, *Org Process Res Dev*, **17**, 557–567.
[78] Levente, S.L., Nagy, Z.K., and Hungerbuhler, K. (2009) Endoscopy-Based in Situ Bulk Video Imaging of Batch Crystallization Processes, *Org Process Res Dev*, **13**, 1254–1261.
[79] Billot, P., Couty, M., and Hosek, P. (2010) Application of ATR-UV Spectroscopy for Monitoring the Crystallization of UV Absorbing and Non-Absorbing Molecules, *Org Process Res Dev*, **14**, 511–523.
[80] Khan, S., Ma, C.Y., Mamud, T. *et al.* (2011) In-Process Monitoring and Control of Supersaturation in Seeded Batch Cooling Crystallization of l-Glutamic Acid: From Laboratory to Industrial Pilot Plant, *Org Process Res Dev*, **15**, 540–555.
[81] Barnes, S.E., Thurston, T., and Coleman, J.O., (2010) NIR Diffuse Reflectance for On-Scale Monitoring of the Polymorphic Form Transformation of Pazopanib Hydrochloride (GW786034); *Model Development and Method Transfer, Anal Methods*, **2**, 1890–1899.
[82] Févotte, G. (2007) In Situ Raman Spectroscopy for In-Line Control of Pharmaceutical Crystallization and Solids Elaboration Process: A Review, *Trans IChemE*, **85**, 906–920.
[83] Hart, R., Herring, A., Howell, G.P. *et al.* (2015) Real-Time Monitoring and Control of Critical Process Impurities During the Manufacture of Fostamatinib Disodium, *Org Process Res Dev*, **19**, 357–361.
[84] Brunsteiner, M., Jones, A.G., Pratola, F. *et al.* (2005) Toward a Molecular Understanding of Crystal Agglomeration, *Cryst Growth Des*, **5**, 3–16.
[85] Thirunahari, S., Chow, P.S., and Tan, R.B.H. (2011) Quality by Design (QbD)-Based Crystallization Process Development for the Polymorphic Drug Tolbutamide, *Cryst Growth Des*, **11**, 3027–3038.
[86] Estime, N., Teychené S., Autret J.M., and Biscans B. (2011) Impact of Downstream Processing on Crystal Quality During the Precipitation of a Pharmaceutical Product, *Powder Tech*, **208**, 337–342.
[87] MacLeod, C.S. and Muller, F.L. (2012) On the Fracture of Pharmaceutical Needle-Shaped Crystals During Pressure Filtration: Case Studies and Mechanistic Understanding, *Org Process Res Dev*, **16**, 425–435.
[88] am Ende, D.J., Birch, M., Breneke, S.J., and Malony, M.T. (2013) Development and Application of Laboratory Tools to Predict Particle Properties Upon Scale-Up in Agitated Filter-Dryers, *Org Process Res Dev*, **17**, 1345–1358.
[89] Birch, M. and Marziano, I. (2013) Understanding and Avoidance of Agglomeration During Drying Processes: A Case Study, *Org Process Res Dev*, **17**, 1359–1366.
[90] Laurent, S., Couture, F., and Roques, M. (1999) Vacuum Drying of a Multicomponent Pharmaceutical Product Having Different Pseudo-Polymorphic Forms, *Chem Eng Proc: Proc Intensification*, **38**, 157–165.
[91] Zhou, G., Ge, Z., Dorwant, J., Izzo, B. *et al.* (2005) Determination and Differentiation of Surface and Bound Water in Drug Substances by Near Infrared Spectroscopy, *J Pharm Sci*, **92**, 1058–1065.
[92] Parris, J., Airiau, C., Escott, R. *et al.* (2005) Monitoring API Drying Operations with NIR, *Spectroscopy*, **20**, 34–41.
[93] Burgbacher, J. and Wiss, J. (2008) Industrial Applications of On-line Monitoring of Drying Processes of Drug Substances Using NIR, *Org Process Res Dev*, **12**, 235–242.
[94] Hamilton, P., Littlejohn, D., Nordon, A. *et al.* (2013) Investigation of Factors Affecting Isolation of Needle-Shaped Particles in a Vacuum Agitated Filter Drier Through Non-Invasive Measurements by Raman Spectrometry, *Chem Eng Sci*, **101**, 878–885.

[95] Hu, Y., Erxleben, A., Hodnett, B.K *et al.* (2013) Solid-State Transformations of Sulfathiazole Polymorphs: The Effects of Milling and Humidity, *Cryst Growth Des*, **13**, 3404–3413.
[96] Mohan, A.E., Lamberto, D.J., and Nakagawa, H., (2015) Effects of Comilling on Final API Physical Attributes, *Org Process Res Dev*, **19**, 1148–1158.
[97] Yu, L.X. (2008) Pharmaceutical Quality by Design: Product and Process Development, *Understanding and Control, Pharm Res*, **25**, 781–791.
[98] Nosal, R. and Schultz, T. (2008) PQLI Definition of Criticality, *J Pharm Innov*, **3**, 69–78.
[99] Bohlin, M., Black, S., and Wheatcroft, H. (2009) New Developments in Scale-Up and QbD to Ensure Control Over Product Quality, *Am Pharm Rev*, 52–57.
[100] Armstrong, N.A. (2006) *Pharmaceutical Experimental Design and Interpretation*, 2nd ed. Boca Raton, Florida: Taylor & Francis.
[101] Marco, G. (2013) QbD-Experience to Date on Assessment of Applications and Forthcoming Regulatory Developments Part 1, *Quality by Design Symposium*, De Montfort University, Leicester 11/4/2013.
[102] Braatz, R.D., Togkalidou, T., Johnson, B.K. *et al.* (2001) Experimental Design and Inferential Modelling in Pharmaceutical Crystallization, *AIChE J*, **47**, 160–168.
[103] Sato, H., Watanabe, S., Takeda, D. *et al.* (2015) Optimization of a Crystallization Process for Orantinib Active Pharmaceutical Ingredient by Design of Experiment to Control Residual Solvent Amount and Particle Size Distribution, *Org Process Res Dev*, **19**, 1655–1661.
[104] Weissman, S.A. and Anderson, N.G. (2015) Design of Experiments (DoE) and Process Optimization. *A Review of Recent Publications, Org Process Res Dev*, **19**, 1605–1633.
[105] am Ende, D.J., Seymour, C.B., and Watson, T.J.N. (2010) A Science and Risk Based Proposal for Understanding Scale and Equipment Dependencies of Small Molecule Drug Substance Manufacturing Processes, *J Pharm Innov*, **5**, 72–78.
[106] Bernlind, C. and Urbaniczky, C. (2009) An Efficient Laboratory Automation Concept for Process Chemistry, *Org Process Res Dev*, **13**, 1059–1067.
[107] Seibert, K.D., Sethuraman, S., Mitchell, J.D. *et al.* (2008) The Use of Routine Process Capability for the Determination of Process Parameter Criticality in Small-Molecule API Synthesis, *J Pharm Innov*, **3** 105–112.
[108] Lepore, J. and Spavins, J. (2008) PQLI Design Space, *J Pharm Innov*, **3**, 79–87.
[109] Watson, T., Bonsignore, H., and Callaghan-Manning, E.A., (2013) A Design Space Verification Protocol for a Small Molecule Drug Substance, *J Pharm Innov*, **8**, 67–71.
[110] Black, S.N., Phillips, A., and Scott, C.I. (2013) Crystallization Design Space: Avoiding a Hydrate in a Water-Based System, *Org Process Res Dev*, **13**, 78–83.
[111] Castagnoli, C., Yahyah, M., Cimarosti, Z., and Peterson, J.J. (2010) Application of Quality by Design Principles for the Definition of a Robust Crystallization Process for Casopitant Mesylate, *Org Process Res Dev*, **14**, 1407–1419.
[112] Peterson, J.J. (2010) What Your ICH Q8 Design Space Needs: a Multivariate Predictive Distribution, *Pharma Manufacturing*, **8**, 23–28.
[113] Garcia, T., Cook, G., and Nosal, R. (2008) PQLI Key Topics – Criticality, Design Space, and Control Strategy, *J Pharm Innov*, **3**, 60–68.
[114] Davis, B., Lundsberg, L., and Cook, G. (2008) PQLI Control Strategy Model and Concepts, *J Pharm Innov*, **3**, 95–104.
[115] Thomson N.M., Siebert K.D., Tummala S. *et al.* (2015) Case Studies in the Applicability of Drug Substance Design Spaces Developed on a Laboratory Scale to Commercial Manufacturing, *Org Process Res Dev*, **19**, 925–934.
[116] Cimarosti, Z., Rossi, S., Tramarin, D. *et al.* (2012) Application of the Quality by Design Principles for the Development of the Crystallization Process for Piperazinyl-Quinoline and Development of the Control Strategy for Form I, *Org Process Res Dev*, **16**, 1598–1606.
[117] Lobben, P.C., Barlow, E., Bergum, J.S. *et al.* (2015) Control Strategy for the Manufacture of Brivanib Alainate, a Novel Pyrrotriazine VEGFR/FGFR Inhibitor, *Org Process Res Dev*, **19**, 900–907.

5

The Role of Excipients in Quality by Design (QbD)

Brian Carlin

De Montfort University, United Kingdom

5.1 Introduction

Excipients are substances, other than the active drug or pro-drug, which are included in the manufacturing process, or are contained in a finished pharmaceutical dosage form [1]. The excipients, which often constitute the major proportion, enable the finished pharmaceutical dosage form. There must be a clear justification for inclusion of any pharmaceutical excipient in the formulation. It should provide one or more functionalities in the presence of the active drug and any other components. Excipients may be included in the formulation to

- Aid processing of the system during manufacture.
- Protect, support, or enhance stability, bioavailability, or patient acceptability.
- Assist in product identification.
- Enhance any other attribute of the overall safety and effectiveness of the drug product during storage or use.

There are clearly safety and quality benefits of using well-established excipients that have already been administered to humans by the intended route, and in similar dosage forms, and obtained from reputable quality suppliers. However, even commonly used excipients can no longer be regarded as totally inert/inactive substances, and their behavior in the formulation can only be understood by careful evaluation during product development.

Pharmaceutical Quality by Design: A Practical Approach, First Edition.
Edited by Walkiria S. Schlindwein and Mark Gibson.

Understanding the role of excipients in QbD can be divided into three stages:

- Development pharmaceutics (the design).
- Design of experiments.
- Control strategy.

The design of pharmaceutical dosage forms requires full understanding of all design inputs, without which the quality of the design will be compromised:

- Raw material properties (excipients and active pharmaceutical ingredient (API)).
- Formulation.
- Process and equipment.

It is beyond the scope of this chapter to teach the requisite raw material and process understanding for basic pharmaceutical product design. Some degree of development pharmaceutics knowledge is assumed on the part of the reader [2–4]. The focus is on those excipient-related aspects of QbD relatively unaddressed prior to the twenty-first century cGMP initiative [5], but increasingly known to be critical to finished product quality. Primarily, these relate to

- Complexity (of both raw materials and finished products).
- Criticalities or latent conditions.
- Direct or indirect impact of excipient variability. Indirect effects are less predictable and may be experimentally inaccessible.

Complexity has been defined as the repeated application of simple rules in systems with many degrees of freedom that gives rise to emergent behavior not encoded in the rules themselves [6]

Excipients have been described as the Cinderella of the pharmaceutical world. They do all the work but don't get to go to the ball! [7]. The era of the inactive (relatively ignored) ingredient is over.

5.2 Quality of Design (QbD)

The quality of a design is measured by product performance against predictions based on the design inputs. In QbD, this is formalized by the product critical quality attributes (CQAs) selected to meet the patient-centric quality target product profile (QTPP), the "prospective summary of the quality characteristics of a drug product that ideally will be achieved to ensure the desired quality, taking into account safety and efficacy of the drug product" [8]. The desired quality includes manufacturing robustness (right first time) as stated in the proposed quality metrics initiative [9].

The performance of a good design is therefore a function of the (known) input design parameters. If the design cannot perform with respect to known inputs, it is not a good design. Common excipient examples of bad design are direct compression mixes not using direct compression grades of excipients, or suspensions designed using viscosity. An immediate release tablet or capsule without disintegrant is at greater risk of excipient-related effects at some stage in the product lifecycle. Risk also increases if an immediate release formulation is not validated as non-rate-limiting on release. The choice of excipients is often linked to the choice of process: for example, wet granulation requires a wet binder. Several examples can be found in the literature [10–12].

Poor designs should be self-limiting but, unfortunately, prototypes developed with limited pharmaceutics expertise can accumulate sufficient stability and clinical data to inhibit redesign, running the risk of eventual regulatory or manufacturing failure. If identified limitations are not addressed early in the design stage, it becomes increasingly expensive, perhaps prohibitively so, to fix them. The pharmaceutical industry is not unique in this respect (e.g., bug fixes in software development) but is particularly susceptible to filing deadlines, which often precede manufacture at full scale.

Pharmaceutical formulations have multiple competing technological objectives. Compromises or trade-offs increase the risk of susceptibility to raw material variability. Can the competing priorities be uncoupled? Disintegrants uncouple release from tablet robustness. Structured vehicle formers uncouple viscosity from suspending power. Any product associated with the terms "over" or "under" needs only minor excipient variability to push it over the edge (e.g., overgranulation, underlubrication).

Good designs emphasize robust processes to cope with raw material variability, rather than dependence on a particular critical material attribute (CMA). However, not all CMAs can be avoided, such as particle size of poorly soluble API (dissolution), and excipient chemical compatibility with API. Surface area, particle size, morphology, composition, and degree of hydration are all possible CMAs for magnesium stearate, the classic example of multiple competing effects in a single excipient. Some CMAs will only be discovered in commercial production after scale-up and greater raw material experience.

Problems arise when unknown factors affect product performance. No matter how good the product in terms of the original design parameters, product quality may suffer if impacted by unforeseen factors, unknown at the time of design. *RMS Titanic* is the classic example of a robust design impacted by an unknown failure mode. The ship was not designed for tangential scraping along obstacles (neither was *Costa Concordia*). However, the first of the line, the sister ship *Olympic*, survived three collisions, and was retired after 25 years due to obsolescence, with the nickname of "Old Reliable." Good robust designs are not immune from unknown failure modes.

Recognized quality standards for pharmaceutical excipients are defined in various pharmacopoeiae. Suppliers should provide materials that comply with the specified pharmacopoeial standards. Excipients are a rich (but not exclusive) source of unknowns due to their complexity, which is often underestimated by overreliance on pharmacopoeial specifications. Pharmacopoeial or supplier specifications do not address fitness for purpose in a specific application. The user may have to perform additional tests to demonstrate that the excipient is suitable for use in a particular product or drug delivery system. In practice, each excipient must be shown to be compatible with the formulation and packaging components and effectively perform its desired function in the product. It is therefore important that during product and process development, manufacturing, and testing, account is taken of excipient performance in the formulation. The problem is exacerbated by the misguided tendency in pharmaceutical design to rely on fixed formulas and processes. This allows raw material variability to feed forward and impact finished product quality with no compensatory mechanisms. The combination of fixed processes and formulas with variable raw materials is not good design practice.

Good design addresses the knowns but cannot cover the unknowns. It is not possible to assess the robustness of a design with respect to excipient impact if the design is subject to unknown excipient attributes beyond the Certificate of Analysis (CofA) and the limited excipient batch/source experience at the time of marketing authorization. These unknowns, which represent potential failure modes, are not unknowable, but require early discussion with the excipient manufacturers and their authorized distributors. Unknown excipient attributes and variability are "known unknowns" in the sense that the excipient manufacturer should know, and what was unknown to the user becomes known on discussion. These unknowns are knowable, preferably before they cause problems.

Finished product performance will be satisfactory until impacted by an unknown, a so-called black swan event [13]. Past satisfactory performance is no predictor of future performance. Or, to put it another way, absence of evidence of a problem is not evidence of absence of the problem! Scale-up will often confound understanding, and much of the excipient-related impact will not be understood at time of filing. Good designs should take manufacturability into account, well before scale-up and commercialization. Reviewers are always concerned about the impact of scale-up if applications only contain data from small-scale or pilot batches. Manufacturing risk can be mitigated by involving production and quality groups (as can excipient-related risk, by involving suppliers) early in the development program.

A certain number of good designs will eventually fail, the problem being that there is no way of predicting which ones will fail or when they will fail. The Concorde crash in 2000 ended a 24-year perfect safety record. One designs for Olympic performance but the control strategy should be Titanic, covering as many potential failure modes as possible.

5.3 Design of Experiments (DoE)

It is necessary to decide on the level of excipient-related experiments to support QbD, over and above the minimum dictated by filing requirements, such as stability studies. A formalized DoE is not in itself QbD, and the reader is referred to Chapter 7 of this book for the theories of experimental design and practical examples. Herwig complains of DoEs being treated like recipes, with no real reflection on the actual design and strategic meaning. A lot of data are generated in a systematic way, and results are presented in colorful plots with intuitive software, but why this experimental plan and how are the boundaries of the DoE chosen? [14].

Once a design, or prototype, is available, risk assessment guides what is to be done experimentally in development and what is *not* to be done in development (DoE vs. control strategy in QbD terms). What is seen as critical gets tested and what is thought to be non-critical does not, the risks being deemed insignificant, improbable, or detectable. There is some risk in choosing what is critical, as the experimental design is less able to assess the potential impact of excipients deemed noncritical. Sighting studies cover more variables than a targeted experiment but have less power to identify subtle effects. It should be remembered that a null finding in an experiment is not evidence of absence of a specific problem, just as absence of accidents does not imply the presence of safety!

With increasing insight and regulatory scrutiny on excipients, Orloff [15] highlights the "potential for QRM to degenerate into a non-value added exercise of identifying noncritical, improbable, low risk scenarios indefinitely." This is true if it triggers unnecessary experimentation for the sake of perceived regulatory compliance but, arguably, the designated noncritical, improbable, low-risk scenarios are also important. If the applicant identifies something as critical, and institutes controls, that risk is mitigated. However, as many other potential failure modes as possible should be identified, and integrated into the control strategy if necessary. Regulatory authorities currently tend to focus on critical items, but as they gain cumulative experience of originally noncritical excipients turning critical post approval, the focus will shift. Including only critical items in a submission could be counterproductive for this reason.

Excipients are often divided into critical and noncritical categories, the latter receiving less attention experimentally. Such arbitrary classification runs the risk of surprises if subsequent experience invalidates the assumption of non-criticality. O'Keeffe *et al.* have criticized the binary classification into critical versus non-critical and advocate ranking attribute and parameter criticalities with the relative risks of not achieving the desired quality attributes, a "Spectrum of Importance" with respect to process parameters and material attributes [16].

A better approach is to ask if a specific excipient is design critical, delivering a dominant functionality essential to finished product performance. Performance in the finished product is a function of the amount and quality of the excipient, which can be experimentally determined. Examples of design-critical excipients include rate-controlling polymers for modified release, suspending agents, and disintegrants. Lubricants also fall into this category but often more for their side effects on tablettability and release rate. Lubricant criticality during development is often mitigated during development by adding the minimum level on the basis of the ejection force only to find during production that a higher level should have been used to prevent sticking.

The epitome of a so-called noncritical excipient is the filler. The term *filler* is misleading because, even when used in the true packaging sense of the word. Rayon wadding in a bottle can release enough furfural to cross-link gelatin capsules [17]. The properties of the "filler" impact tablet and capsule performance and can prove critical if these properties and interactions in the product are not fully understood. Microcrystalline cellulose controls water distribution and wet mass rheology in wet granulation, yet is rarely specified in terms of water interactivity. Tablets relying on the disintegration of microcrystalline cellulose without adding disintegrant also increase the criticality of the filler.

The absence of incidents despite the presence of noncritical excipients in DoE batches should not be taken as confirmation of non-criticality. This is a common inductive fallacy, the medical analogy being "no evidence of disease is not evidence of no disease" (NED ≠ END)! By definition a noncritical excipient is not expected to have any experimental impact, so absence of impact on the CQAs is not predictive. The criticality of that excipient may increase later in the product lifecycle due to cumulative changes, especially if subject to univariate change control. Changing things one by one can give satisfactory results until one change too many. Even if not design critical, all excipients in the product are potentially performance critical: appropriate contingencies should be built into the control strategy. Early discussion with the excipient suppliers will strengthen the control strategy with respect to excipient risk management.

Most experiments measure the impact of the excipient level rather than variability. To measure the impact of variability of CMAs, a large number of excipient samples, statistically representative of the variability to be expected from the supplier (or multiple suppliers), is required. Access to supplier data beyond purchased CofA results is needed to ensure representative sampling. The supplier should be the manufacturer of the excipient or an authorized distributor. Other resellers are unlikely to be able to provide relevant data or samples (in addition to supply chain traceability concerns) [18].

The continuous processes used to manufacture many excipients tend to run at center specification, and material at or near the edge of the specification may not be commercially available. Targeting a new continuous process setpoint could consume several hundred tons of material, at least half of which will be out of specification if running on a specification limit. If the supplier is aware of specific requirements, it may be possible to reserve material from an excursion in a relevant direction, or drive the process through a specification limit as part of a campaign shutdown. Alternative approaches are also available to the user, such as grade bracketing, fractionation, or blending. IPEC has published a QbD sampling guide [19] to address the issues in sourcing appropriate excipient samples for QbD purposes.

Quality before quantity is important before deciding on the number of experiments. Simply running large numbers of experiments to cover potential failure modes is unscientific because null results (lack of evidence of problems) are not evidence of no problem. If multi-sourcing of an excipient is evaluated during development, any intersource differences mean either that the product design should be changed to eliminate the sensitivity, or that the excipient should be specified to distinguish between the acceptable and the unacceptable sources.

Multivariate data analysis, as described in Chapter 8, is better than the traditional "change one variable at a time" to minimize the number of experiments and identify interactions. It should be noted that correlation does not prove causation, so mechanistic understanding should be sought. Sugihara *et al.* [20] demonstrate an example of ephemeral or mirage correlation, with a simple mathematical model where the variables spontaneously correlate, anti-correlate, and decouple. Model development should be underpinned by mechanistic understanding. Empirical models can guide future experimentation but should not be used for Good Manufacturing Practices (GMP)-critical decisions. Apte [21] lists common mistakes in application of DoE, particularly reliance on "mix and match" experiments dictated by the DoE software and recording their effects on preselected CQAs.

For pharmaceutical products using batch manufacture (the majority), experimental determination of the impact of excipient variabilities is only feasible on a small scale. Unfortunately, the results may be confounded on scale-up. Confirmatory batches at full scale will be too few to confirm the impact of variability (as opposed to the already known dominant formulation effects) or to detect other subtle changes.

5.4 Excipient Complexity

Using only established and approved excipients from pharmaceutically aligned manufacturers and distributors is essential to finished product quality and patient safety. However, this does not preclude other excipient-related problems in production. Pharmaceutical quality is generally high (6σ), but this is due to inspection and rejection rather than manufacturing efficiency: 2-3σ compares unfavorably with other regulated industries. Excipients are a potential source of risk to manufacturing quality. The complexity of both the excipients and the finished drug product, and the potential for interactions, are often underestimated, even in the hands of those skilled in the art.

From the earlier definition of complexity [6], the repeated application of simple rules in systems with many degrees of freedom that gives rise to emergent behavior not encoded in the rules themselves… it can be seen that the reliance on simple rules such as pharmacopoeial compliance and GMP, coupled with the many degrees of freedom inherent to excipients and finished pharmaceutical products, gives ample scope for the emergence of quality excursions and product failures.

In addition to chemical synthesis (the norm for small molecule APIs), excipients come from many sources, including animal, vegetable, and mineral. They are often polymeric, polydisperse, and multicomponent. Their composition is often complex and uncertain, with many minor or concomitant components, which is why the IPEC Americas Composition Committee published anexcipient composition guide [22].

Excipient composition, physical properties, and performance in specific applications may be dependent on the source, manufacturing history, processes, and raw materials. Compositional complexity may not be adequately defined in the supplier or pharmacopoeial specifications[22]. For simple reagents in solution (e.g., buffers), the specification (essentially purity standards) may be relied upon. For more complex excipients, especially when relying on solid properties, it is necessary to think beyond the CofA.

Reliance on pharmacopoeial specification alone is a risk and has driven commoditization of many excipients. The term *commodity* implies fungibility or interchangeability regardless of source. A Pharmaceutical Quality Research Institute (PQRI) survey [23] demonstrated the lack of such interchangeability with 76% of drug product manufacturers performing additional tests to determine excipient suitability for their intended use. In addition to functionality and processability, concerns were related to stability and impurities. Surprise was expressed about the 25% frequency of drug product manufacturers testing excipient suitability for processing using experimental (laboratory) scale batches, or pilot scale manufacturing batches. Given the general lack of understanding of excipient composition and functionality, the assumption of interchangeability of sources is a risk in many applications. Pharmacopoeial standards define minimum purity and safety requirements but do *not* define fitness for purpose in a specific application. NSF/IPEC/ANSI 363-2014 defines quality as "the suitability of an excipient for its intended use" in addition to attributes such as identity, strength, and purity [24]. Reliance solely on pharmacopoeial specifications also facilitates economically motivated adulteration.

A 2-3σ pharmaceutical manufacturing efficiency implies a direct cost of poor quality (COPQ) of 10%–15%. Indirect COPQ, typically threefold, may be further increased if regulatory authorities start to use quality metrics to categorize applicants in terms of risk [9].

> The fundamental problem we identify is the inability of the market to observe and reward quality. This lack of reward for quality can reinforce price competition and encourage manufacturers to keep costs down by minimizing quality investments. [25]

With greater recognition of their supply chain and technological risks, the information value of excipients will increase. From a design point of view, critical process parameters (CPPs) are preferred as they are in the hands of the drug product manufacturer. CMAs, if known, are in the hands of the excipient manufacturer. Greater excipient supplier involvement will be required to minimize inadvertent dependency on CMAs, or to control unanticipated CMAs in response to product criticalities.

The International Conference on Harmonization (ICH) Q9 guidance defines quality as "the degree to which a set of inherent properties of a product … fulfils requirements" [26]. The risk of equating quality with compliance is illustrated by the melamine in milk scandal. Under the ICH definition, melamine could be described as a quality-enhancing additive, the intent being to improve compliance to specification. Inappropriate reliance on compliance actually makes compliant materials riskier than non-compliant materials. The latter should be prevented from entering the chain by the quality system, the detectability of known but non-compliant attributes eliminating the risk. In most cases of economically motivated adulteration, compliance with specification is not the issue.

In his book *The Black Swan: The Impact of the Highly Improbable*, Taleb describes the extreme impact of rare and unpredictable events (black swans), the risk being accentuated by the tendency to place too much reliance on what we know and ignore or underestimate what we donot know [13]. Pharmaceutically, there tends to be overreliance on the CofA, which focuses on pharmacopoeial parameters, but is of limited relevance to determining excipient fitness for purpose in an application. Other unspecified excipient attributes may vary in an uncontrolled manner in the background, but will be unknown to the user unless discussed with the excipient supplier. This must be addressed during development, either by designing the formulation and/or process to be robust enough to cope with the variability of such previously unknown attributes, or by appropriately specifying the excipient to limit the impact on finished product CQAs.

Black swan logic makes what you do not know far more relevant than what you do know. Any risk assessment based solely on pharmacopoeial parameters may not be valid due to the unknowns. Risk assessment must also take into account that absence of raw material impact during development is not evidence that there are no problems (it is difficult to prove a negative). Gaps in excipient knowledge devalue risk assessment. It is easier to construct a (poor) design based on known attributes (especially of limited functional relevance) than to accommodate unknowns.

Excipient unknowns fall into several categories:

- Unknown to the user, known to supplier (do not rely on CofA alone).
- Unknown to the supplier (especially if supplier is unaware of the application).
- Unknown to both.

Risk assessment requires that unknowns (not unknowable) be addressed with all stakeholders, including the excipient suppliers. Unknowns add to system complexity, undermine risk assessment, and make it difficult to develop meaningful models. It has been said that all models are wrong; the practical question is how wrong do they have to be to not be useful? [27]

Excipient unknowns include

- Composition.
- Drivers of functionality or performance.
- Limited utility of pharmacopoeial attributes.
- Other unspecified attributes.

- Variability.
- Criticalities or latent conditions in the finished product.

5.5 Composition

Excipients are typically more complex than well-characterized APIs. Non-biologic APIs are predominantly single synthetic small molecule chemical entities, manufactured in batches with well-characterized impurity profiles (unintended or unavoidable constituents which differ from the labeled chemical entity). However, this is the exception rather than the rule for excipients, which are often polymeric or multicomponent with ill-defined compositional profiles. Excipients are also often manufactured using continuous production on much larger scales than APIs or drug products. Unlike APIs, the sum of the labeled entity and defined impurities will not add up to 100% due to other concomitant components. The functionality of some excipients may actually depend on so-called "impurities." For example, pure dicalcium phosphate is poorly compactable, and non-crystallizing sorbitol solution depends on presence of oligomers to prevent cap locking.

NSF/IPEC/ANSI 363-2014 (Good Manufacturing Practices (GMP) for Pharmaceutical Excipients) [24] specifies consistent excipient composition, and, where possible, limits for excipient composition, including known impurities. NSF/IPEC/ANSI 363-2014 also states that manufacturing processes shall be adequately controlled so the excipient composition falls within established limits. It will be difficult for users to agree on meaningful (application-specific?) compositional limits with their suppliers unless they understand in some detail the manufacturing processes and raw materials used to manufacture their excipients. A more realistic definition of an excipient "impurity" is any component, other than the labeled entity, that needs to be controlled. Again, greater understanding of the excipient manufacturing history is required for such a definition of an "impurity."

Often, it is the multicomponent nature of the excipient that drives many chemical incompatibilities with APIs. For example, although one might theoretically avoid the classic amine incompatibility with reducing sugars, by using non-reducing sugar excipients, trace levels of reducing sugar impurities may thwart the avoidance strategy [28]. Even for the most commonly used excipients, it is necessary to understand the context of their manufacture in order to identify potential API interactions with trace components.

It is common practice to approve specific excipient sources, but it is better to identify the mechanism of reaction and specify the excipient with respect to the level of reactive components. Risks associated with changes in excipient-induced API degradation are particularly insidious if only detectable by long-term stability studies, when at-risk product will already be in the market. Change control and notification are less effective if the impact cannot be immediately assessed. A source-specific API-incompatibility suggests that the reactant is something other than the labeled excipient entity.

5.6 Drivers of Functionality or Performance

Excipient functionalities are qualitative classifications describing the purposes or roles of an excipient in a drug product, and are the rationale for inclusion in the formulation. Excipient performance is a more holistic term embodying the actual expression of the excipient properties, including functionalities, in a specific drug product. The current regulatory environment and the paradigm of QbD go beyond simply identifying excipient function and emphasize performance through the identification, evaluation, and control of CMAs that assure consistent performance throughout a product's lifecycle [29].

Excipient CMAs may not be identifiable or evaluable using the tests and specifications listed in compendial monographs. To minimize the risk of inappropriate overreliance solely on compendial specifications, the USP Excipient Performance chapter <1059> [29] is designed to provide an overview of material attributes and tests for many functional categories of excipients (USP). These additional tests are not typically included in excipient monographs, and are not exclusive. The appropriate tests and specifications to ensure consistent and reliable excipient performance, in terms of finished product quality, may not come from either the monograph or chapter <1059>. A thorough understanding of the formulation and manufacturing processes, the dosage form performance requirements, and the physicochemical properties of each ingredient (including the manufacturing history of each ingredient) is essential for a meaningful risk assessment. Pharmacopoeial guidance on excipient performance or functionality-related characteristics will not in itself eliminate the risk of performance- or functionality-related surprises. Finished product CQAs may be dependent on ill-defined physical properties of excipients, and such CMAs may themselves be dependent on the method of excipient manufacture and complex precursor materials.

5.7 Limited Utility of Pharmacopoeial Attributes

Many pharmacopoeial attributes are derived from measures of the original producer's consistency. Subsequent producers will comply with the monograph, but their raw material feed stocks and production processes may differ, affecting other excipient properties. Adding to the risk, many excipients have other industrial applications, pharmaceutical consumption often constituting a relatively small proportion.

Pharmacopoeial methods may not be relevant in determining fitness for purpose of an excipient in a particular application or in determining interchangeability of multiple sources. A good example is viscosity. A medium-viscosity grade of polymer could be provided by one supplier as a true medium molecular weight distribution and by another as an average of high and low molecular weight distributions. If the molecular weight distribution, not the viscosity, is the true CMA, expect surprises on finished product quality when changing sources. The difficulty of characterizing macromolecular excipients has been reviewed by Apte [30].

Reliance on pharmacopoeial viscosity methods is also risky given that most are dilute solution apparent viscosities, measured using simple viscometers. Viscosity is often the least relevant rheological parameter in many applications (especially suspensions), and the dilute concentration may not reflect the effective in-use concentration. For example, polymer matrix controlled-release tablets will have effective in-use concentrations at least an order of magnitude higher than the pharmacopoeial viscosity methods. Not only may the rank order of viscosities differ but other rheological effects, such as gelation, may intervene as well [31].

Pharmacopoeial compliance ensures neither fitness for purpose in a specific application, nor equivalence between sources. Pharmacopoeial compliance should be regarded as a minimum standard, not a guarantee of interchangeability between multiple sources. Pharmaceutically aligned suppliers will often provide additional data to demonstrate functional equivalence in support of change control, but cannot warrant interchangeability in a specific application.

A PQRI survey found that more than 70% of all respondents performed additional functionality or processability testing on excipient from a new supplier. About 25% of the time, excipient suitability testing involved laboratory or pilot scale manufacturing batches [26]. Although reported as "higher than expected," such findings are not surprising given the limited understanding of excipient composition and performance in complex systems.

Reliance solely on pharmacopoeial attributes runs the risk of failing to identify CMAs, introduction of non-interchangeable material, and consequent risk to finished product quality.

5.8 Other Unspecified Attributes

Focus only on pharmacopoeial attributes is the pharmaceutical industry example of blindness to uncertainty associated with black swan theory [13]. If other unknown attributes of an excipient vary uncontrolled in the background, there is risk to finished product quality, especially for fixed processes and formulations. Arbitrarily tightening ranges of pharmacopoeial attributes in response to such finished product quality problems does not address the problem and adds an unnecessary compliance burden. An excipient specification should never be tightened without discussion with the supplier, to ensure that the tighter specification is within their process capability. Detectability of risk from unknown CMAs is low because quality control based solely on pharmacopoeial attributes will give compliant results up to the point of product failure. Assessing the risk from unspecified attributes is made more difficult in that many unknown CMAs will be application specific. Some excipients have pharmacopoeial attributes, none of which could be regarded as CMAs.

5.9 Variability

Robustness is defined as the ability of a manufacturing process to tolerate the expected variability of raw materials, operating conditions, process equipment, environmental conditions, and human factors. Robust formulations should be "able to accommodate the typical variability seen in the API, excipients, and process without the manufacture, stability, or performance of the product being compromised" [32]. In addition to uncertainty as to composition or performance, excipient risk also arises if the true variability is underestimated. Assessment of what variability is to be expected or is typical requires knowledge of the excipient manufacturing history and process capability.

The risk of underestimating excipient variability is mainly attributable to four main causes:

- Equating excipients with reagents.
- Unilateral assessment.
- Limited excipient experience.
- Continuous manufacture of excipients.

Some excipients are simple reagents such as buffers. Add the right amount of the right purity, and performance (chemistry) is guaranteed. This also applies to APIs, where composition and performance (efficacy) are understood. Process-dependent excipient composition, performance, and variabilities are less understood. This risk of underestimating excipient complexity is reinforced if the users do not discuss the use of excipients in their products with the excipient suppliers.

During development and scale-up, the number of excipient batches (and suppliers) may be insufficient to meaningfully characterize the impact of excipient variability on that product. The risk can be mitigated by reference to similar products in commercial production using those excipients. The excipient supplier can also supply data additional to that on the limited number of purchased batches. This could be CofA data or summary statistics. In-process data or data on attributes beyond the CofA might also be available.

Many common excipients are manufactured in much larger volumes (>10,000 tons per annum) than pharmaceutical products, often by continuous production. In continuous production, the "batch" is the production campaign, extending over variable periods up to several months. In practice, the continuous production is usually time-sliced into discrete batches, but even days or weeks of production can still be many tons. Of necessity, the data for some parameters on the CofA will be a composite or an average. Reliance on CofA data alone runs the risk of underestimating the true variability if the in-process data is noisier than the CofA data. Access to the most relevant in-process data also requires traceability back from the individual packed unit, the need being dependent on how noisy the in-process data is relative to CofA data. Supplier process capability should always be determined by supplier in-process data, never by user estimates from limited CofA data. User testing is much less statistically relevant in the assessment of excipient variability compared to the much larger supplier databases [33,34].

An inadequate assessment of true excipient variability also incurs regulatory risk. A reviewer may reasonably query excipient specification limits which are much wider than the narrow range of excipient data typically presented. The applicant runs the risk of delay, either justifying their reliance on supplier limits, or being asked to explain the potential impact of operating in specification regions beyond their experience space, if not explained in the control strategy. Reviewers have been known to demand that excipient specification limits be narrowed, which is risky as they do not know the supplier process capability. Even greater risk ensues if the applicant agrees to the narrower limits without supplier agreement, as material meeting such specification may not be feasible, technically and/or commercially. Regulators are paying increasing attention to discussion of excipient impact in applications. If no information beyond the formula level and pharmacopoeial compliance is provided, then an assessor cannot judge the significance of excipient impact.

5.10 Criticalities or Latent Conditions in the Finished Product

Critical is a much-used term in QbD that relates to anything which affects the safety or efficacy of the finished product. CPPs and CMAs must be controlled to ensure conformance with the finished product CQAs, the surrogates for safety and efficacy.

MIL-STD-1629A uses the word "critical" as a severity classification, one stop short of catastrophic, and defines a criticality as a relative measure of the consequences of a failure mode and its frequency of occurrences [365The ICH defines criticality in terms of severity, probability, and detectability, but omits the common physics/mathematics usage as a transition between two states. A criticality or critical transition is defined as being in a state, or at a point, where some quality, property, or phenomenon undergoes a definite change [36].

This latter definition is rarely used in QbD but is equally important because a criticality, a point of transition from one state to another, can be critical (in the ICH/MIL-STD sense), if encountered during production. Synonyms include thresholds, nonlinearities, discontinuities, tipping points, percolation effects, and edges. Critical transitions have been described as catastrophic bifurcations, where a minor trigger can invoke a self-propagating shift to a contrasting state [37,38].

An example of a criticality is the critical micelle concentration (CMC), commonly encountered in dissolution testing. Dissolution of API (or lack thereof) in media where the surfactant is present below CMC is not predictive of the behavior above CMC, where there is a disproportionate (with respect to surfactant concentration) increase in drug solubility, proportional to the number of surfactant micelles.

Percolation effects are also examples of criticalities, or critical transitions, in pharmaceutical systems, especially in powder mixes and compaction. There are no fundamental mixing rules or prediction of cooperative properties in powder mixes. Force transmission and resultant densities within a compact are inhomogeneous and dependent on tablet geometry. The term "explosive percolation" refers to the characteristic binary step function where the system goes from one state to another with little or no warning. Analogies relevant to movement within the design space would be like falling off a cliff, or stepping on a landmine. These are latent conditions if you are unaware of the topography or location of the mines.

Conflicting technological objectives are another source of criticalities. The closer the formulation is to a performance margin, the greater the impact of excipient variability. Ranging studies during development are useful: if you can vary the level of an excipient by ±50% and maintain product performance, then the impact of variability of that excipient is generally going to be less than that associated with a more impactful ± 5% titration. However, if you are trying to balance too many multiple competing objectives, then you will have a very narrow operating margin with much greater susceptibility to excipient variabilities and unknowns. Good examples can be found with design-critical rate-controlling polymers in modified release. The higher the level of gelling-matrix-former, the less the impact from variability in the excipient attributes. If faster release is required, it might be better to maintain a high level of a "weaker" polymer rather than reduce the original polymer to a level where the impact of excipient variability is greater. Similarly, maintaining a high loading of a rate-controlling controlled release coating is preferable to reducing to a level where it is subject to the impact of both the coating precision and variability of the excipient attributes.

If the finished drug product undergoes a critical transition, there may be a severe impact (out of specification) with little warning (low probability, low detectability). Excipients may disproportionately impact CQAs if minor excipient variability interacts with a criticality in the application, and the minor excipient variability is suddenly governing the transition from one state to another. A hitherto "noncritical" excipient attribute has now become critical. The offending application-specific excipient CMA may be a known attribute, and the variability may also be within normal limits and prior experience, but the drug product is no longer robust to variability in that particular excipient attribute.

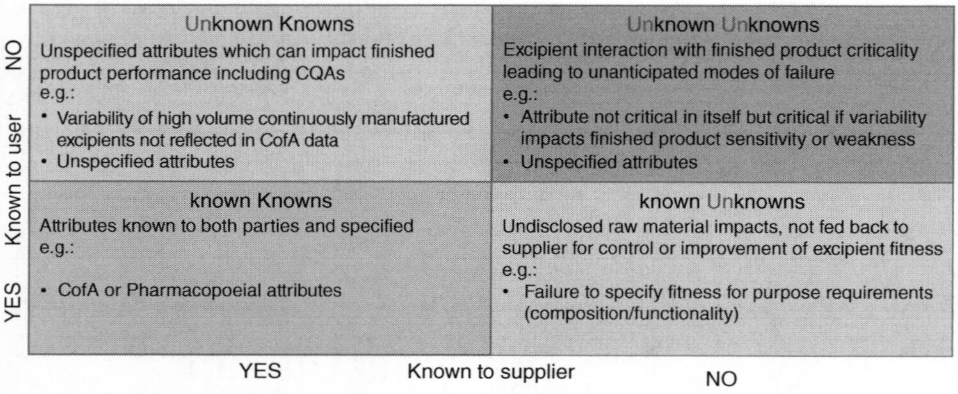

Known to user	YES (Known to supplier)	NO (Known to supplier)
NO	**Unknown Knowns** Unspecified attributes which can impact finished product performance including CQAs e.g.: • Variability of high volume continuously manufactured excipients not reflected in CofA data • Unspecified attributes	**Unknown Unknowns** Excipient interaction with finished product criticality leading to unanticipated modes of failure e.g.: • Attribute not critical in itself but critical if variability impacts finished product sensitivity or weakness • Unspecified attributes
YES	**known Knowns** Attributes known to both parties and specified e.g.: • CofA or Pharmacopoeial attributes	**known Unknowns** Undisclosed raw material impacts, not fed back to supplier for control or improvement of excipient fitness e.g.: • Failure to specify fitness for purpose requirements (composition/functionality)

Figure 5.1 *Examples of "known and unknowns" to users and suppliers.*

Even worse, a critical transition is analogous to moving into a parallel universe, where the rules are different. It is not a case of too much noise taking the existing model beyond a specified limit but rather a new model. It is difficult to demonstrate being a state of control if the model and the understanding underpinning the design have unexpectedly changed. An analogy might be the impact of the tide going out on a body of navigable water. If the design space was unknowingly based on high tide, then movement within the design space might suddenly be curtailed if low water levels had not been factored into the design. Criticalities may also be scale-dependent, but detectability drops with the decreasing number of experimental batches as the process is scaled-up.

As implied by the synonym "latent conditions," criticalities are not intentionally incorporated into the design. Criticalities are "unknown unknowns," in that neither the excipient manufacturers, nor the manufacturer of the drug product, are aware of the presence of such potential weaknesses or susceptibilities within the product (Figure 5.1). They are artifacts of the design analogous to the bugs in software systems, arising out of the complexities of the subsystems and how they interact. Excipients are complex materials which are used in complex systems, giving rise to myriad interactions. A specific "unknown unknown" is by definition unknowable before it impacts product quality, but pharmaceutically aligned suppliers of excipients should be able to provide general guidance to make designs more robust with respect to potential excipient-correlated failure modes.

A poor design will cause quality problems, but a good design does not always guarantee absence of problems. Criticalities and other unforeseen failure modes will always bedevil complex systems.

5.11 Direct or Indirect Impact of Excipient Variability

Ignoring interactions, the performance or impact of an excipient in a finished product will be a function of its concentration (c) and expression (strength, potency, efficacy) of the relevant excipient attribute (x)

$$Performance = f\left(c \times x\right) \tag{5.1}$$

$$Performance = f\left(x\right) for\ fixed\ formulae \tag{5.2}$$

Interactions will result in dependency on other formulation or process variables. Design-critical excipients by definition deliver a specific functionality to the product and require titration in the formula. A rate-controlling polymer in a sustained release matrix or coating is a good example: the lower the concentration the faster the release, and vice versa. In this case, "x" would be considered a CMA.

The ability of the design to accommodate an excipient variability can be modeled in terms of capability indices. The smaller the range of excipient performance variability relative to the relevant product specification limits, the lower the chance of out-of-specification finished product.

For a process under control (with normally distributed data), the process capability index (C_{pk}) is

$$C_{pk} = \min\left[\frac{USL - mean}{3\sigma}, \frac{mean - LSL}{3\sigma}\right] \tag{5.3}$$

where the USL and LSL are the upper and lower specification limits. A C_{pk} for a single-sided specification can be similarly calculated:

$$C_{pk} = \left[\frac{Max - mean}{3\sigma}\right], or = \left[\frac{Min - mean}{3\sigma}\right] \tag{5.4}$$

If $C_{pk} = 1$, the actual range of variability matches the allowable range, and there is no margin for drift. For this reason, a $C_{pk} > 1.33$ is preferred.

A single-sided specification C_{pk} can be used to illustrate the indirect effect of an originally noncritical excipient going critical, as a result of product drift, leading to an interaction with a critical transition. The variability in the noncritical excipient, which may involve a known attribute within its norms of variability, has hitherto had no detectable impact on finished product quality. If there is a criticality or latent condition in the finished product, one can postulate a virtual de facto limit beyond which product quality suffers. Operating away from the criticality is equivalent to a high C_{pk}. However, if the product drifts toward the criticality, then the previously innocuous variability of the noncritical excipient may start to govern the transition. As the product drifts toward the criticality, C_{pk} (with respect to the virtual limit) decreases. Once C_{pk}=1, the noncritical excipient has become critical.

5.12 Control Strategy

Excipient risk management must address three sources of excipient-related risk:

- Safety.
- Supply chain integrity.
- Technological risks associated with complexity and uncertainty.

The intrinsic safety of an excipient relates to the level of human exposure via a specified route of administration. A regulatory safety assessment as part of a finished product marketing application is a prerequisite to pharmacopoeial listing, but compliance with a pharmacopoeial monograph alone should not be relied upon. The source of the excipient, and the integrity of the supply chain, is also safety critical. If not subject to Good Manufacturing [27] & Distribution [39] Practices and change control [40], there is the potential for contamination, adulteration, substitution, undeclared additives, and/or degradation. Compliance with a pharmacopoeial specification is not in itself an adequate guarantee of quality for human use.

Excipient unknowns compromise risk assessment but are not unknowable. Pharmaceutically aligned suppliers can identify excipient aspects unknown to the user and pre-empt criticalities in the finished product by identifying potential failure modes (if they are aware of the application). NSF/IPEC/ANSI 363 – 2014 excipient GMP requires suppliers to consider "requirements not stated by the customer but necessary for the specified or intended use, where known," when determining excipient quality [24].

ICH Q9 recognizes that the manufacturing of a drug product, including its components, necessarily entails some degree of risk, including risk to the drug product quality throughout the product lifecycle[26]. Quality Risk Management (QRM) ensures patient safety by providing a proactive means to identify and control potential quality issues during development and manufacturing. QRM facilitates better and more informed decisions if quality problems arise, provides regulators with greater assurance of a company's ability to deal with potential risks, and can beneficially affect the extent and level of direct regulatory oversight [26]. Raw material compliance and GMP do not eliminate the impact of variability. QbD attempts to minimize the risk that raw material variability will adversely affect the finished product CQAs. Formulation and process design must accommodate the raw material variability (robustness). The API typically gets most attention, but simply defining an excipient as noncritical does not guarantee absence of impact on product quality.

The steps outlined below are iterative and not strictly sequential. Risk assessment is an ongoing process throughout the product lifecycle, not just during development. QbD often refers to continuous improvement, but continual monitoring is also essential to better understand the limitations of the product in commercial reality, with old assumptions, models, and analyses under constant revision [41]:

- Communication with suppliers.
- Build in compensatory flexibility.
- Risk assessment.
- Contingencies.

5.13 Communication with Suppliers

Early discussion with excipient suppliers is recommended to ensure fitness for purpose, especially for new applications. Proceeding in the absence of, or contrary to supplier advice is risky. If the application raises safety issues the supplier should refuse to supply. If a supplier cannot provide application support, historical data, or is unwilling to comply with pharmacopoeial requirements, the option of selecting another supplier or an alternative material should be pursued. Even for conventional applications, supplier insight into failure modes will strengthen risk assessment. Many grades of excipients are formulated off the shelf with no understanding, or specification, of attributes that govern performance in a specific application. If the supplier changes or stops supplying that specific grade there is risk to finished product quality and risk of shortage.

Development personnel should visit their suppliers for insight into the manufacturing background and properties of the excipients. Compliance audits do not count in this respect. Design Review Based on Failure Mode (DRBFM), states that good discussions during preliminary design can achieve the same result as validation testing in identifying design weaknesses [42]. Communication with suppliers should continue throughout the product lifecycle.

5.14 Build in Compensatory Flexibility

Fixed processes together with fixed formulations are poor designs where excipient variability can feed forward to the detriment of finished product quality. If the excipient variability cannot be reduced, the control strategy becomes meaningless. QbD requires that flexibility be built into the system to compensate for the variability inherent in the raw materials.

The overwhelming emphasis in QbD has been on process controls, with a near complete retention of fixed formulations. Building flexibility into the formulation itself also provides compensation against the impact of raw material variability. Quantitative variation of an excipient level in accordance with a validated algorithm could counter the incoming variability of the API or other excipients. Excipients with functional concentration optima, such as glidants and lubricants, are obvious candidates to deliver fixed performance with variable composition versus the traditional fixed composition and variable performance.

Whether the formulation is fixed or not, excipient risk assessment also benefits from excipient ranging studies during development. If the prototype contains a certain level of excipient, what happens to functionality and performance as the level is titrated downward? If increased variability of the finished product (or failure) occurs close to the target level, then there is a greater risk of a criticality within the formulation and/or susceptibility to variability of that excipient. On the other hand, if the level has to be halved in order to see effects, it suggests that the excipient level is not near a criticality and that the target level offers a reserve of performance. Sensitivity to raw material variability is generally greater nearer the margin. Demonstrating understanding of the impact of such changes also facilitates quantitative formula changes during the product lifecycle. If a fixed formula is not critical, why maintain it at the expense of product consistency?

The traditional focus has been on excipient consistency (variable performance), but under QbD the logic inverts. How can excipient variability be turned to competitive advantage by pharmaceutical manufacturers in pursuit of consistent performance and finished product quality? Many excipients when used as food ingredients are specified by a functionality which is standardized by addition of varying amounts of agreed food-grade diluent, thus eliminating batch-to-batch performance differences. It is to be hoped that the pharmaceutical fixation on composition (often ill-defined), at the expense of performance and finished product quality, is eliminated by QbD.

5.15 Risk Assessment

ICH Q9 states that it is neither always appropriate, nor always necessary, to implement a formal risk management process. The use of informal risk management processes can also be considered acceptable. "The evaluation of the risk to quality should be based on scientific knowledge and ultimately link to the protection of the patient; and the level of effort, formality and documentation of the quality risk management process should be commensurate with the level of risk" [26]. More importantly, the risk assessment ethos should apply throughout the product lifecycle, regardless of risk assessment method.

Excipient risk assessment starts with risk identification, defined in ICH Q9 as a "systematic use of information to identify hazards referring to the risk question or problem description. Information can include historical data, theoretical analysis, informed opinions, and the concerns of stakeholders." Are excipient suppliers not stakeholders with informed opinions? They have the knowledge of variability and unspecified attributes without which excipient risk assessment is flawed. Their application knowledge can identify potential excipient-related modes of failure, which the user might not have unilaterally identified.

5.16 Contingencies

Contingencies for foreseeable events, such as scale-up and expansion of raw material experience space, should be included in the control strategy. Evaluating new excipient sources would be expected under conventional change control, but if the full variability of a single source was not evaluated during development, similar additional testing can be specified in the control strategy.

Unforeseen events also need to be anticipated by continual monitoring. Unexplained increases in variability or new correlations could be indicators of impending criticalities or transitions. A design which is too resistant to change could undergo a critical transition unless redesigned for more gradual adaptive response or to strengthen the preferred state [38].

Any models developed during development need to be continually updated to confirm ongoing relevance and validate their predictive power.

Excipient sourcing must be under control of the R&D and/or quality departments. Commodity buying is not QbD and is the highest risk sourcing strategy. Joint due diligence between the users and their excipient suppliers is the lowest-risk sourcing strategy.

5.17 References

[1] IPEC (n.d.). *FAQs: What are Pharmaceutical Excipients?* http://ipecamericas.org/about/faqs#question1 (accessed August 30, 2017).
[2] Rowe, R.C., Sheskey, P.J., and Cook W.G.. (Eds) (2012) *Handbook of Pharmaceutical Excipients* (7th ed.), Pharmaceutical Press, London.
[3] Aulton, M.E. and Taylor, K.M.G. (Eds) (2013) *Aulton's Pharmaceutics: The Design and Manufacture of Medicines* (4th ed.), Churchill Livingstone Elsevier, London.
[4] Gibson, M. (Ed.) (2009) *Pharmaceutical Preformulation and Formulation* (2nd ed.), Informa Healthcare, New York.
[5] FDA, (2004) *Pharmaceutical cGMPs for the 21st Century – A Risk-Based Approach*, http://www.fda.gov/downloads/drugs/developmentapprovalprocess/manufacturing/questionsandanswersoncurrentgoodmanufacturingpracticescgmpfordrugs/ucm176374.pdf (accessed August 30, 2017).
[6] Christensen, K. and Moloney, N.R. (Eds) (2005). *Complexity & Criticality* (1st ed.), Imperial College Press, London.

[7] Moreton, R.C. (1996) Tablet excipients to the year 2001: A look into the crystal ball, *Drug Development and Industrial Pharmacy*, **22**, 11–23.
[8] Yu, L.X., Amidon, G., Khan, M.A. et al. (2014) Understanding pharmaceutical quality by design review article, *The AAPS Journal*, **16**(4), 771–783.
[9] FDA (2015) *Request for Quality Metrics: Guidance for Industry* (draft guidance), http://www.fda.gov/downloads/drugs/guidancecomplianceregulatoryinformation/guidances/ucm455957.pdf (accessed August 30, 2017).
[10] Dave, V.S., Saoji, S.D., Raut, N.A., and Haware, R.V. (2015) Excipient variability and its impact on dosage form functionality, *Journal of Pharmaceutical Sciences*, **104**(3), 906–915.
[11] Kushner, J., Langdon, B.A., Hiller, J.I., and Carlson, G.T. (2011) Examining the impact of excipient material property variation on drug product quality attributes: A quality-by-design study for a roller compacted, immediate release tablet, *Journal of Pharmaceutical Sciences*, **100**(6), 2222–2339.
[12] Siew, A. (2015) Excipient selection and solid-dosage drug performance, *Pharmaceutical Technology*, **39**(1), 32–39.
[13] Taleb, N.N. (2010) *The Black Swan: The Impact of the Highly Improbable* (2nd ed.), Random House Publishing Group, New York, ISBN 978-0-6796-0418-1.
[14] Herwig, C. (2015) Can we fix the negative perception of quality by design? *The Medicine Maker*, **11**, 18–19.
[15] Orloff, J. (2011) The promise and threat of quality risk management, *Pharmaceutical Technology*, **35**(2), 38–40.
[16] O'Keeffe, D., Campbell, C., and O'Donnell, K. (Feb 2016) A spectrum of importance—challenging the concept of critical/non-critical in qualification and validation activities | IVT, *Journal of Validation Technology*, **22**(1).
[17] Hartauer, K.E. (1993) The effects of rayon coiler on the dissolution stability of hard shelled gelatin capsules, *Pharmaceutical Technology*,**17**, 76–83.
[18] IPEC (2013) *Certificate of Analysis Guide for Pharmaceutical Excipients*, The International Pharmaceutical Excipient Council publisher.
[19] IPEC (2015). *The Excipient QbD Sampling Guide*, The International Pharmaceutical Excipient Council publisher.
[20] Sugihara, G., May, R., Ye, H. et al. (2012) Detecting causality in complex ecosystems, *Science*, **338**, 496–500.
[21] Apte, S. (2011) QbD not QED, *Journal of Excipients and Food Chemicals*, **2**(3), 50–52.
[22] IPEC (2009) The IPEC Composition Guide, http://ipecamericas.org/system/files/IPEC_CompositionGuide-Final_7-9-2013.pdf (accessed August 30, 2017).
[23] Pharmaceutical Quality Research Institute (2006) *PQRI Survey Findings of Pharmaceutical Excipient Testing and Control Strategies, Used by Excipient Manufacturers, Excipient Distributors and Drug Product Manufacturers*, http://pqri.org/wp-content/uploads/2015/08/pdf/XWG_Survey_Report_FINAL14Sep06.pdf (accessed August 30, 2017).
[24] NSF (2014) *NSF/IPEC/ANSI 363 – 2014 Good Manufacturing Practices (GMP) for Pharmaceutical Excipients*, http://standards.nsf.org/apps/group_public/download.php/26765/NSF%20363-14%20-%20watermarked.pdf (accessed August 30, 2017).
[25] Woodcock, J. and Wosinska, M. (2013) Economic and technological drivers of generic sterile injectable drug shortages, *Clinical Pharmacology & Therapeutics*, **93**(2), 170–176.
[26] ICH Q9 (2005) *Quality Risk Management*, http://www.ich.org/fileadmin/Public_Web_Site/ICH_Products/Guidelines/Quality/Q9/Step4/Q9_Guideline.pdf (accessed August 30, 2017).
[27] Box, G.E.P., Hunter, J.S., and Hunter, W.G. (2005), *Statistics for Experimenters* (2nd ed.), John Wiley & Sons, New York.
[28] Dubost, D.C., Kaufman, M.J., Zimmerman, J.A. et al. (1996) Characterization of a solid state reaction product from a lyophilized formulation of a cyclic heptapeptide. *A novel example of an excipient-induced oxidation, Pharmaceutical Research*, **13**(12), 1811–1814.
[29] Sheehan, C. and Amidon, G. (2011) Compendial standards and excipient performance in the QbD era: USP excipient performance Chapter <1059>, *American Pharmaceutical Review*, http://www.americanpharmaceuticalreview.com/Featured-Articles/37322-Compendial-Standards-and-Excipient-Performance-in-the-QbD-Era-USP-Excipient-Performance-Chapter-1059/ (accessed August 30, 2017).

[30] Apte, S. (2010) Uncharacterized and uncharacterizable attributes of macromolecular excipients, *Journal of Excipients and Food Chemicals*, **1**(3), 1–2.

[31] Fu, S.B. (2014) Inter-grade and inter-batch variability of sodium alginate used in alginate-based matrix tablets, *AAPS PharmSciTech*, **15**(5), 1228–1237.

[32] Moreton, R. (2006) *Functionality and Performance of Excipients*, http://www.phexcom.com/admin/UploadFiles/Technology/Functionality%20and%20Performance%20of%20Excipients%20.pdf (accessed August 30, 2017).

[33] Salunke, S., Giacoia, G., and Tuleu, C. (2012) The STEP (Safety and Toxicity of Excipients for Paediatrics) database. Part 1 – A need assessment study, *International Journal of Pharmaceutics*, **435**, 101–111.

[34] Salunke, S. and Tuleu, C. (2015) The STEP database through the end-users eyes – USABILITY STUDY, *International Journal of Pharmaceutics*, **492**, 316–331.

[35] US Dept Defense (1980) *MIL-STD-1629 RevA*, http://src.alionscience.com/pdf/MIL-STD-1629RevA.pdf (accessed August 30, 2017).

[36] Webster, M. (n.d.) Definition of Critical, http://www.merriam-webster.com/dictionary/critical (accessed August 30, 2017).

[37] Kuznetsov, Y. (Ed.) (1998) *Elements of Applied Bifurcation Theory*, 2nd ed., Springer, New York.

[38] Scheffer, M., Carpenter, S.R., Lenton, T.M. et al. (2012). Anticipating critical transitions, *Science*, **338**, 344–348.

[39] IPEC (2006) *The IPEC Good Distribution Practices Guide*, http://ipecamericas.org/system/files/IPEC_GDP_Guide_final.pdf (accessed August 30, 2017).

[40] IPEC (2014) *The IPEC Significant Change Guide for Pharmaceutical Excipients*, http://ipecamericas.org/system/files/IPEC_Significant_Change%20_Final_2014.pdf (accessed August 30, 2017).

[41] NASA (2004) *Risk: It is Part of the NASA Mission*, http://rmc.nasa.gov/archive/rmc_v/presentations/oconnor%20osma%20risk%20is%20part%20of%20nasa%20mission.pdf (accessed August 30, 2017).

[42] Tatsuhiko, Y. (Ed.) (2002) *Mizenboushi (Preventative Measures) Method GD3, How to Prevent a Problem Before It Occurs*, JUSE Press Ltd, Tokyo.

6

Development and Manufacture of Drug Product

Mark Gibson,[1] Alan Carmody,[2] and Roger Weaver[3]

[1] AM Pharma Services Ltd, Congleton, United Kingdom
[2] Pfizer, Canterbury, United Kingdom
[3] Weaver Pharma Consulting, Canterbury, United Kingdom

6.1 Introduction

The current regulatory requirements allow for a diverse approach to pharmaceutical drug product development. The minimum requirement is to demonstrate acceptable patient safety and efficacy. The minimum requirement is more likely to limit the extent of product and process understanding gained and less likely to result in a robust end product and manufacturing process.

A more structured and systematic approach to product development can be achieved by designing quality into a product [1, 2]. The Quality by Design (QbD) concept was actually developed mainly by the quality pioneer, Dr Joseph M. Juran, and others, in various publications [3–5]. Dr Juran believed that quality could be planned, and that most quality crises and problems could be averted with improved quality planning and the way in which a product was designed in the first place. The Juran definition of "quality" has two meanings that are still relevant to modern day QbD:

- The presence of attributes that create customer satisfaction.
- The reliability of those attributes.

Pharmaceutical Quality by Design: A Practical Approach, First Edition.
Edited by Walkiria S. Schlindwein and Mark Gibson.

Thus, creating attributes is the path taken by the application of QbD. Also, it reduces failures of the attributes and customer dissatisfactions [6].

The pharmaceutical adoption of QbD was triggered by a realisation of typical problems of low productivity, efficiency, and quality, leading to manufacturing failures and shortages of supply of medicines to patients, when compared to other industries that had already applied QbD to their processes.

Another issue at this time was the dramatic increase in the number of manufacturing supplements to regulatory applications as companies made changes to drug manufacturing processes post approval. This was required in order to improve their poor existing process and to rectify shortfalls. Not surprisingly, there was a real burden on the quality review process in regulatory agencies.

In response to all of these factors, the US FDA launched its initiative 'Pharmaceutical Quality for the 21st Century: A Risk-Based Approach' [7], recognising that increased testing does not necessarily improve product quality. Quality should be built into a product with an understanding of the product and manufacturing process along with a knowledge of the risks involved in manufacturing the product and how best to mitigate those risks.

A pilot programme was established to allow pharmaceutical companies to submit information for a new drug application demonstrating the use of QbD principles, product knowledge, and process understanding. In recent years, the US FDA has worked closely with other regulators from both the EU and Japan to further QbD objectives through the International Conference on Harmonisation of Technical Requirements for Registration of Pharmaceuticals for Human Use (ICH). This resulted in the formal implementation of 'Quality by Design' by the pharmaceutical industry, assisted by the publication of several associated ICH guidance documents including

- **ICH Q8 (R2)** [8] 'Pharmaceutical Development' (2009) – describing the expectations for the drug product pharmaceutical development section of the Common Technical Document (CTD).
- **ICH Q9** [9] 'Quality Risk Management' (2005) – describes approaches to producing quality pharmaceutical products using current scientific and risk-based approaches.
- **ICH Q10** [10] 'Pharmaceutical Quality Systems' (2010) – provides a model for an effective quality management system for the pharmaceutical and biotech industries.

However, many implementation details are not covered in these documents, and it is left to pharmaceutical companies to interpret them. This has led to a wide range of diverse approaches to QbD. This chapter is intended to explore the requirements as defined in the ICH guidances for drug product development aided with practical examples and current experience.

Although QbD is still not currently a mandatory regulatory requirement, there is an increasing expectation by the regulatory authorities that it should be applied. It can provide useful information that gives regulators more confidence that developers have established the knowledge and understanding to ensure the production of a quality pharmaceutical product. Some of the main potential benefits of applying a QbD approach to product development are as follows:

- Use of a systematic approach to product design and development, thus enhancing development capability, speed, and formulation design.
- Focus of resource and effort on upstream proactive mode rather than on downstream corrective mode.

- Improved product and process understanding, leading to reduced risk to product quality (and the patient) and better understanding of risk areas and root causes of manufacturing failures.
- Increased product and process capability, robustness, and consistency of product quality (more reliable manufacturing and fewer rejected batches).
- Knowledge-based specification setting and reduced specification testing.
- Potentially achieving operational flexibility within a defined design space. Opportunities for process improvement without having to generate more data and a reduction in regulatory 'oversight'.

6.2 Applying QbD to Pharmaceutical Drug Product Development

QbD can be applied to any pharmaceutical dosage form or manufacturing process, including small chemical molecules and large biopharmaceuticals/biologic products. The first pharmaceutical product approved based upon a QbD application was Merck & Co.'s Januvia for the treatment of diabetes in 2006. In 2013, there was a major breakthrough, when the FDA made the first QbD approval for a biological pharmaceutical product with a proposed design space, after Genentech (Roche) had submitted a Biologic Licence Application for Gazyva (obinutuzumab) for people with previously untreated chronic lymphocytic leukaemia.

According to the ICH Q8 (R2) definition, 'Quality by Design is a systematic approach to development that begins with predefined objectives and emphasises product and process understanding and process control, based on sound science and quality risk management'. Furthermore, the QbD process applied to pharmaceutical product development includes the following steps:

- Define the quality target product profile (QTPP) as it relates to quality, safety, and efficacy, considering, for example, the route of administration, dosage form, bioavailability, strength, and stability.
- Identify the approach to drug product formulation/manufacturing process.
- Identify potential critical quality attributes (CQAs) of the drug substance/raw materials/drug product, so that those characteristics having an impact on product quality can be studied and controlled.
- Identify potential critical process parameters.
- Using risk assessment and experimental approaches, determine the functional relationships that link raw material CQAs and unit operations critical process parameters (CPPs) to drug product CQAs.
- Optimise the formulation and manufacturing process in an iterative fashion to meet the QTPP defined in step 1 of this list.
- Establish the design space and control strategy.

Thus, a typical QbD process flow is shown schematically in Chapter 1, Figure 1.1. It can be seen that this is a much more structured, systematic and professional product development approach compared to the traditional approach; which was very empirical, often involving reactive issues driven formulation and process fixes with the use of testing to check if quality is present.

As demonstrated with examples in this chapter, QbD is about designing in quality then confirming by assessment. The initial QbD step begins with a pharmaceutical developer establishing a QTPP for the intended product to be developed during the product design phase. This is followed by product and process development studies conducted to establish a thorough scientific understanding of the product and manufacturing design and to achieve a robust product and manufacturing process.

For QbD, this also entails an understanding of variability in the raw materials and identifying critical material attributes, identification of the product's and process CQAs, and the association of these with the CPPs. Enablers, such as risk assessment, experimental design (DoE), and multivariate analysis, are employed to prioritise what needs to be done, avoiding what does not need to be done and collecting the relevant information in an efficient way. This subsequently leads to a proposed design space, the mathematical expression and graphical representation of the link between material attributes, CPPs, and CQAs, where assurance of quality is provided. QbD requires that a control strategy is applied to ensure that the process remains within the design space. Finally, continuous improvement is applied throughout the entire product lifecycle to manage the impact of changes and continuously improve.

The key elements of the QbD process flow are discussed in more detail in this chapter, along with practical examples given for drug products to show how QbD 'principles' can be applied to drive the design and development programme.

6.3 Product Design Intent and the Target Product Profile (TPP)

When QbD is applied to the design phase of a new pharmaceutical product, the following sequential steps can be followed for an effective design process:

- Agree upon fundamental business drivers.
- Establish the target product profile/QTPP and CQAs.
- Endorsement/agreement by the development team.
- Formulation/manufacturing process design scouting.
- Design review by the development team.
- Formulation and manufacturing approach selected.
- Formulation and process design.
- Design review by the development team.
- Formulation and process development and optimisation.
- Design review.

From both business and quality perspectives, it may seem obvious that a new product should be adequately defined and agreed upon before any significant product development is undertaken. However, with the immense pressure to start development and get products to the market by the shortest route, it is not surprising that many companies avoid investing sufficient time and resource into the initial design phase. If the product design is poor, even though the product may be well understood and well manufactured, then it is still likely to result in a poor-quality product. For example, if a tablet is friable and not very robust, it will be a poor-quality product, although the way it is made may be well understood and the manufacturing process runs smoothly, albeit with cautious handling procedures, and highly protective packaging.

Starting development too quickly can result in wasted time and money, lack of product robustness, unnecessarily complex products, missed opportunities for elegant solutions, and unhappy customers. There have been costly mistakes made in the past where a product is developed that is not wanted or the product design concept has changed during development in an uncontrolled fashion. For example, there is a need to change the dosage form because it does not meet some important customer needs, but the impact of this change is not considered. A real and very costly example of a product that was not wanted was the first inhaled insulin product, which was commercially launched but was shortly withdrawn due to poor uptake and sales [11]. A clear design intent is therefore essential prior to commencing development activities to ensure that customer needs are met. Studies have demonstrated that a relatively small investment is required during the product design stage, which can greatly influence the quality and nature of the product developed and its ultimate commercial success [12].

QbD, when applied effectively, can provide a clear understanding of the business needs and a framework to enhance a shared understanding of the options considered, risks, trade-offs, and design decisions taken. There are clear benefits for the regulators too, as they will be able to see that the design intent is supported by good science and that quality has been built into the design phase of the product. Successful product design and QbD implementation requires cooperation, input, and agreement from all the key functions across the developer company organisation such as R&D (pharmaceutical development, safety, clinical), manufacturing operations, QC/QA, regulatory, and commercial/marketing. This is to ensure that they all have input into the initial product design activities to gain 'global' agreement about the nature of the product to be developed. An effective product design phase should achieve input and buy-in from all the above- mentioned functions at the start of development to provide clear direction and objectives for the project team. A high-level description of the requirements captured in the TPP will be the major project driver (see below). Further, there will be the benefits that any risks should be identified early and can be managed. Importantly, the wasting of valuable resources should be avoided on developing a product that is not needed or does not meet the quality criteria.

As issues are identified during the design phase, they should be considered, and the product design adapted to 'engineer out' any sensitivities, thus keeping the design simple, fit for purpose, and aligned with customer needs. The subsequent additional understanding gained from the development phase can support the product's efficient manufacture and robustness. It is therefore essential to align the design and development programmes.

A good basis for capturing the design requirements for a new pharmaceutical product is the TPP. The TPP is a prospective summary of the product attributes for the intended commercial product based on all 'customer' and 'end-user' needs. Customers and end-users are defined as anyone in the supply chain, including both internal (from within the company) and external customers such as those in manufacturing, distribution, healthcare professionals such as doctors, nurses, pharmacists and, of course, patients. It must also meet the needs of payers and government agencies. The TPP may often be expressed primarily in clinical terms, but should also include the pharmaceutical, technical, regulatory, and commercial/ marketing attributes required of the product.

Most likely, the TPP will be based on the ideal product characteristics that are considered to be desirable, but in some cases the 'minimum' product requirements are captured too. For example, the TPP may stipulate that the product should ideally have at least a three-year shelf life at ambient temperature. However, it may still be viable, as a minimum, to have a two-year shelf life if the required stability cannot be achieved. The initial TPP should provide a level of detail to provide sufficient clarity to enable the product to be developed. As the TPP evolves, the impact and options should be evaluated. Typical examples of TPPs for a solid oral dosage form, a parenteral injectable dosage form, and an inhalation dosage form are given in Tables 6.1, 6.2, and 6.3, respectively, illustrating the level of detail expected at the start of product development.

Product design scouting involves listing alternative technology options that can be evaluated against the TTP targets and product design intent criteria, with consideration of 'musts', 'needs', and 'wants'.

- *MUSTs* = what the patient requires to satisfy patient safety and efficacy. Identify what is 'non-negotiable'.
- *NEEDs* = include critical to quality attributes or business critical performance, for example, minimum of two-year drug product shelf life.

Table 6.1 *TPP for a solid oral dosage form.*

Product Attributes	Target
Disease to be treated/clinical objective	Arthritis
Patient type	Adults, particularly over 40 years old and geriatrics
Route of administration	Oral
Efficacy	Anti-inflammatory activity better than existing 'gold' standard therapy
Safety/tolerability	No GI side effects
Pharmacoeconomics	Aim to reduce overall health costs by preventing disease progression
Dosage form	Film-coated tablet No more than two strengths
Dose and dose frequency	One tablet twice daily
Pack type/design	Blister calendar pack Tamper evident Must be able to be opened by patient
Manufacturing process	Wet granulation. Ideally by continuous manufacture
Aesthetic aspects	Differentiate tablet strengths
Countries to be marketed	Europe, United States, Japan
Stability	Three years at room temperature. Minimum two years
Cost of goods	The cost of goods should be no more than 20% of commercial price to be viable
Commercial price	Commercial price must be less than current 'gold' standard therapy to be competitive

Table 6.2 TPP for an injection dosage form.

Product attributes	Target
Disease to be treated/ clinical objective	To provide an injectable sustained release delivery of the drug to treat hypertension
Patient type	Adults
Route of administration	Depot by subcutaneous injection to deliver systemic therapeutic levels
Efficacy	Limited variation in drug plasma levels over 7–14 days
Safety/tolerability	Limited and/or acceptable adverse reactions. Subcutaneous injection should be less painful and have fewer adverse events compared to intramuscular injection
Pharmacoeconomics	Improved patient compliance and wastage compared to daily oral therapy should reduce overall health costs
Dosage form	Sterile solution for injection, unit dose, non-preserved
Dose and dose frequency	100 mg in a 1.2 mL volume injected weekly
Pack type/design	Pre-filled syringe fitted with safety needle. The pack must maintain product sterility and stability over product shelf life
Manufacturing process	Aseptic manufacture/ideally terminally sterilised by autoclaving
Aesthetic aspects	Clear, colourless slightly viscous solution
Countries to be marketed	Europe and United States initially
Stability	18 months minimum shelf life at 2°C–8°C and 2 days at room temperature for use in clinic
Cost of goods	No more than 10% of commercial price to be viable
Commercial price	Commercial price needs to be competitive compared to equivalent oral therapy

- *WANTs* = include desirable product or process performance standards, for example, 3+ years shelf life, use of continuous processing technology, meets compendia standards, and so on.

A Pugh matrix or decision-grid matrix is a good way of clearly summarising the output from such product design scouting (see Table 6.4). Note that the use of traffic light colours (red, amber, and green) can improve the transparency of the matrix to clearly see the best options. In the simple example given here, the dosage form must be suitable for a paediatric patient with daily dosing. Therefore, a suitable technology is required to achieve sustained release of the active pharmaceutical ingredient over 24 hours. There are clearly restrictions on tablet size that a young child can swallow, thus reducing the formulation options. If a liquid dosage form is selected, taste may be an issue that would have to be overcome. Weightings can be used to rank the different technology options, although if a single 'must' criteria cannot be met, then that option would probably be ruled out altogether. At the subsequent design review, the development team will review all the information and recommend a technology approach to advance to the next phase of formulation and product development. Ideally, the design option should be viable, robust, and straightforward to develop. The product requirements for the endorsed design option are recorded, and the pertinent quality attributes are established in a QTPP.

Table 6.3 *TPP for an inhalation dosage form.*

Product Attributes	Target
Disease to be treated/clinical objective	Treatment of chronic obstructive pulmonary disease COPD by oral inhalation of the drug
Patient type	Adults
Route of administration	Pulmonary delivery
Efficacy	Must deliver multiple, reproducible doses and required range of aerodynamic particle size distribution
Safety /tolerability	Must be safe from repeated dose (impurities, leachables, particulates). Low oropharangeal deposition
Pharmacoeconomics	Improved patient compliance with simple, easy-to-use, breath-actuated device providing low-cost treatment
Dosage form	Orally inhaled inhalation suspension 2% w/w
Dose and dose frequency	Minimum number of 50 actuation per unit, 120 mg per actuation
Pack type/design	Adaptable delivery platform. Robust, reliable, and easy-to-use pressurised metered dose inhaler (pMDI) fitted with dose counter and breath actuation device
Manufacturing process	Device simple with the minimum number of components. Robust component supply chain. Lean and efficient product manufacturing process incorporating Process Analytical Technology (PAT)
Aesthetic aspects	Colour-coded inhaler. Attractive and functional design
Countries to be marketed	6.1 United States
Stability	18 months minimum shelf life at ambient temperature
Cost of goods	Competitive cost of goods
Commercial price	Must be affordable and low cost to be competitive

It can be a valuable exercise to conduct a risk assessment to identify, assess, and prioritise risks associated with the selected technology approach. Some of the comments in the Pugh matrix (Table 6.4) referring to taste, robustness, and tablet size are some of the preliminary risks that have been highlighted in this example. In fact, the use of risk assessment methods, including a more detailed and quantitative type of risk assessment applying failure mode effects and criticality analysis (FMECA) (see Chapter 2 for more details), can be very helpful throughout the design and development process to

- Rank and quantify quality attributes and process parameters in order of importance.
- Identify potentially high-risk formulation and process variables.
- Determine which studies are necessary to achieve product and process understanding in order to develop a control strategy.

Each risk assessment should be updated after development to document the reduced level of risk due to improved product and process understanding.

Table 6.4 *An example of a Pugh matrix for formulation technology design option evaluation. Note: red = dark grey; green = light grey and yellow = medium grey*

Product requirements 'musts' & 'wants' From TPP	Formulation Technology Options				
	Liquid suspension	Modified release granules for reconstitution	Mini-tabs	Multi-layered tablet	Matrix tablet (HPMC)
Paediatric dosage form (must)	M Taste?	M Taste?	U Size?	N Too big	N Too big
Oral, once daily dosing (must)	U	U	M	M	M
Dose: 5mg, 10mg (must)	M	M	M	M	M
Shelf life > 2y at 25°C/60% RH (need)	U	M	M	M	M
Degradants & impurities below safety threshold, or qualified (must)	U	U	M	M	M
Meets pharmacopoeial requirements for oral solid dosage forms (want)	M	M	M	M	M
Use existing in-house technology (want)	M	N	M	M	M

M = Meets requirements (green); U = Unknown/unclear or Not sufficient data (amber); N = Does not meet requirements (red).

6.4 The Quality Target Product Profile (QTPP)

The concept of the QTPP and its application is novel in the QbD paradigm, although the TPP concept described earlier has been employed for some time as a business tool by pharmaceutical companies to aid product development decisions [1]. According to ICH Q8 (R2), the QTTP is defined as a 'prospective and dynamic summary of the quality characteristics of a drug product that ideally will be achieved to ensure that the desired quality, and thus the safety and efficacy, of a drug product is realized'.

The QTPP is required to identify the critical clinical performance objectives (targets), the important commercial objectives (manufacturing, marketing) and the product design requirements so that the developers can design-in customer needs or wants. Typically, the QTPP will include the intended clinical use, dosage form and route of administration, dosage form strength(s), therapeutic moiety release or delivery and pharmacokinetic characteristics (e.g. dissolution and aerodynamic performance) appropriate to the drug product dosage form being developed. It describes prospectively the quality characteristics of the intended marketed product that ideally will be achieved to ensure the desired quality, taking into account safety and efficacy of the drug product.

In practice, the first step is to list all the potential quality attributes, preferably in the format of a table, with a discussion of how the drug substance and formulation excipient CQAs relate to the finished drug product CQAs based on prior knowledge, risk assessment, and early pre-formulation studies. The latter studies are conducted to characterise the candidate drug and to determine the physico-chemical properties considered important to support product design, the appropriate dosage form and excipients, such as solubility, excipient compatibility, and stressed stability studies. Data from these studies may also influence the selection of the manufacturing process and help establish the CQAs for the product.

There are several methods and tools available that can be employed to gather information about customer needs and attributes to establish the TPP and QTPP [2]. A typical pharmaceutical company would be expected to already have a reasonable amount of internal prior knowledge and expertise, particularly if they have already established marketed products in a particular therapeutic area. It is important that input is gained from 'experts' representing all the key functions and departments from the company, and there is a suitable process for gaining agreement in a constructive way. The TPP can be used to facilitate 'buy-in' and to ensure that the science, quality, regulatory, and commercial departments' inputs are all considered and agreed upon.

The TPP can then be used to define the QTPP. The QTPP should be a joint commitment with the business (medical, marketing, manufacturing, and quality assurance) to a set of decisions that will streamline development and enhance customer and patient satisfaction with the product. Once defined, the TPP and QTPP provides details of the label claims that the pharmaceutical company is seeking for the candidate drug, including the safety and efficacy requirements that should allow regulatory approval. The QTPP should be constantly reviewed during the development process and may evolve as more information and real data becomes available. During these QTTP reviews, the impact of any new findings on the design options needs to be considered. For example, if drug solubility is low and a liquid formulation is a must, then the impact needs to be discussed, for example, creative use of solubilising agents, an alternative salt form, or perhaps a drug suspension.

Table 6.5 A typical QTPP for an oral solid dosage form product.

Product quality attribute	Quality target
Dosage form description	Film-coated, round tablet.
Dose/strengths	No more than three strengths. Target 5 mg, 10 mg, and 20 mg active pharmaceutical ingredient (API).
Appearance	Aim for as small a tablet weight as possible and coloured to differentiate tablet strengths. 5 mg white (5–8 mm diameter), 10 mg brown (6–9 mm diameter), 20 mg orange (8–2 mm).
Identity	Positive for API.
Assay	95%–105%.
Impurities	Limits for individual and total impurities.
Content uniformity	Meets USP, JP, and Ph.Eur acceptance criteria for Uniformity of Dosage Units.
Hardness	5–12 kP.
Friability	Not more than (NMT) 1%.
Dissolution	Meets requirement for immediate release, e.g. not less than (NLT) 75% at 30 min.
Disintegration	Not more than (NMT) 15 min.
Microbiology	If testing required, meets USP criteria.
Stability	Shelf life of at least two years at 25°C/60% RH.
Manufacturing process	Use a technology that already exists in the company.
Raw materials	Use the minimal number of excipients necessary to make a pharmaceutically acceptable tablet. Select from standard company range of fillers, binders, disintegrants, and lubricants from preferred suppliers. Must comply with USP, JP, and Ph.Eur. requirements.
Pack design/type	Blister calendar pack, tamper evident. Secondary pack – carton with patient instruction leaflet.

Typical examples of QTPPs for an oral solid dosage form product and a biopharmaceutical (large molecule) parenteral product, respectively, are given in Table 6.5 and Table 6.6.

The QTPP is now a key document that the regulators expect to see in the regulatory submission for a finished pharmaceutical product. The QAs and CQAs should provide confidence that the final product requirements will be met. Even though the product has not yet been developed, a prospective product specification can be established with tests and limits that the product should meet at the time of manufacture and at the end of product shelf life. It is useful to agree on what the minimum acceptable shelf life for the drug product should be to be viable, considering that sufficient time will be required for QC testing, QA release, distribution to wholesalers, pharmacists, and doctors; acceptable time will also be required for storage until the product is prescribed and used by patients.

The regulators have worked closely with the pharmaceutical industry to establish case studies and 'mock' worked examples of regulatory submissions to illustrate the design space concept with the application of QbD, PAT, and quality risk management during the development process. Scientific, regulatory, quality, and information management

Table 6.6 *A typical QTPP for a biopharmaceutical large molecule parenteral product.*

Product quality attribute	Quality target
Dosage for description	Liquid, single-use, non-preserved
Dose	5 mg/kg
Product strength/concentration range	10–20 mg/mL in 5 mL
Protein concentration per vial	100 mg
Mode of administration	Intravenous, diluted with saline or dextrose
pH	5–7 using phosphate or citrate buffer
Viscosity	Less than 20 cP at room temperature
Manufacturing process	Aseptic manufacture. Terminal heat sterilisation not possible
Pack design/type	20R Type 1 borosilicate glass vials, fluoro-resin laminated stopper

representatives from Abbott, AstraZeneca, Eli Lilly, and GlaxoSmithKline co-authored the case study based on a fictitious drug molecule Acetriptan ('ACE') [13]. In this worked example, there is a typical QTPP for Acetriptan, and the product and process development are described. The FDA has also published QbD examples for an immediate-release dosage form [14] and a modified release tablet development and manufacturing process [15] to aid generic companies to file abbreviated new drug applications (ANDAs), whereas the CMC Biotech Working Group have published a QbD case study for a biopharmaceutical product [16]. Other typical examples of the QbD approach and of QTPPs can be found elsewhere in the literature [17].

The International Society for Pharmaceutical Engineering (ISPE) has released an ISPE Guide series: 'Product Quality Lifecycle Implementation (PQLI) from Concept to Continual Improvement' (ispi.org). These guides provide a practical approach to implementation of ICH guidelines Q8 (R2), Q9 and Q10, as well as FDA guidance produced under the Pharmaceutical CGMPs for the 21st Century Initiative. The ISPE Guide series provide more insight than is given in the ICH guidelines and are intended to assist practitioners involved in development, implementation, and application in manufacturing, including those involved in continual improvement. The explanations, how-to tools, practical examples, and case studies contained in these ISPE guides have been derived from input from a wide range of companies and 'experts' who have hands-on experience of successfully applying enhanced, QbD approaches. Part 1 and Part 2 of these guides provide a helpful practical discussion with examples of criticality, as applied to CQAs and CPPs, design space, and control strategy, which are addressed in ICH Q8 (R2) [18].

6.5 Identifying the Critical Quality Attributes (CQAs)

Once the initial TPP and QTPP have been established, the next step in the process flow is to identify the relevant potential (subject to process understanding being gathered) CQAs. According to ICH Q8 (R2), a CQA is defined as 'a physical, chemical or biological, or microbiological property or characteristic that should be within an appropriate limit, range, or distribution to ensure the desired product quality' [8].

Interpretation of criticality from the ICH guidance definition can be challenging for pharmaceutical companies to apply. If the definition is strictly interpreted, then every quality attribute is critical as they all ensure product quality. For example, if a QA is an in-process or finished-product specification limit, then these limits must be critical. However, during the course of development, test results may show little variation, leading to the conclusion that they present only a small or no risk to product quality, and are thus deemed to be non-critical. Even when defined as critical, not all CQAs will have the same impact on safety and efficacy (ICH, 2011). Therefore, in practice, risk analysis is a valuable and efficient process to use for determining criticality and also a means of estimating how much impact the attribute has on patient safety and efficacy. Companies must show how the risk analysis was applied consistently for similar products. The process of determining how a quality attribute is deemed critical is illustrated in Figure 6.1.

Current discussion suggests that if the process or the control strategy are modified, some non-critical process parameters (PPs) and quality attributes (QAs) should be reassessed as they could become critical to the safety and efficacy of the product. The term 'important non-critical' is applied to such PPs and QAs. During the initial assessment, the important non-critical should be flagged (internally) in case of subsequent process improvements.

Most of the potential product quality attributes will have been captured for the TPP/ QTPP, although it is possible that more may arise later during development. The criticality of each quality attribute can be determined through a risk assessment process, as recommended in the ICH Q9 'Quality Risk Management' guidance; also see Chapter 2 of this book. ICH Q9 refers to two primary principles of quality risk management:

- The evaluation of risk to quality should be based on scientific knowledge and ultimately link to the protection of the patient.
- The level of effort, formality, and documentation of the quality risk management should be commensurate with the level of the risk.

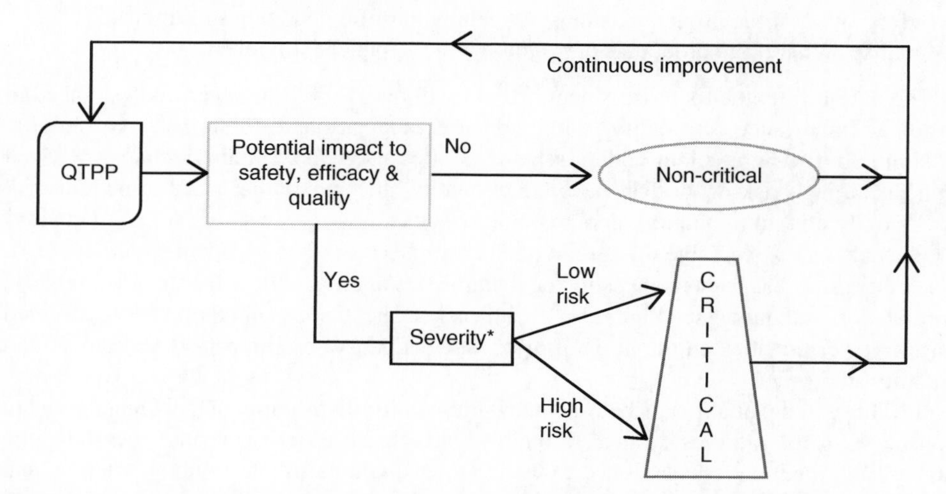

Figure 6.1 *Process of determining how a quality attribute is deemed critical. *A severity scale is used to assess relative magnitude. (Adapted from ISPE Product Quality Lifecycle Implementation (PQLI®) Guide: 'Overview of Product Design, Development and Realization: A Science- and Risk-Based Approach to Implementation', International Society for Pharmaceutical Engineering (ISPE), First Edition, October 2010, www.ispe.org)*

For a successful risk assessment process, it is essential to get input from a multi-disciplinary team of 'experts' from the various functions responsible for product development, analytical, manufacture, engineering, quality, and so on, as prior product knowledge and experience are invaluable. Sources of prior product knowledge can be obtained from laboratory, safety, and clinical experience with a specific product quality attribute, such as

- Pharmaceutical development studies.
- Use of modelling.
- Stability reports.
- Manufacturing experience:
  - Internal and vendor audits.
  - Raw material testing data.
  - Manufacturing history data.
- Trend data and technical investigations.
- Suppliers and contractors.
- Technology transfers.
- Product history and regulatory filings/feedback.
- Product quality reviews/annual product reviews.
- Complaint reports.
- Adverse event reports.
- Recalls.

The use of relevant data from similar drug molecules, formulation types, and literature references should also be applied to assess the likely risks and to relate the quality attributes to product safety and efficacy. For example, consider whether each quality attribute has an impact on the patient:

- **Efficacy**: What if the patient does not receive the complete dose?
- **Safety**: What if the product contains potentially harmful degradants or impurities?
- **Quality**: What if the product is damaged – will the patient still take it?

The output from the risk assessment process will be a list of quality attributes ranked in order of importance (criticality) with a documented rationale to support the ranking. A simple traffic light colour coding scheme is often used on a summary chart where red indicates a high risk or 'critical' to product quality attribute, amber is for a medium risk (potentially critical to quality), and green represents a low-risk or 'non-critical to quality' product attribute. See Table 6.7 for a typical example of a risk assessment summary chart. The assessment has been made using the manufacturing process flow for an oral solid dose formulation as a guide (see Figure 6.2), with each sequential unit operation being assessed in turn for criticalities on the basis of prior knowledge and preliminary development and stability studies.

In Table 6.7, the quality attributes, content uniformity, dissolution, and disintegration are considered to be the most critical to quality, safety, and efficacy because they have the highest and medium risks assigned to them. Other quality attributes with an intermediate ranking such as appearance, assay, and hardness, for example, will need to be monitored and should be part of the eventual control strategy.

Alternatively, a more formal risk management tool such as failure mode effects analysis (FMEA) or FMECA can be applied to provide a structured semi-quantitative summary of

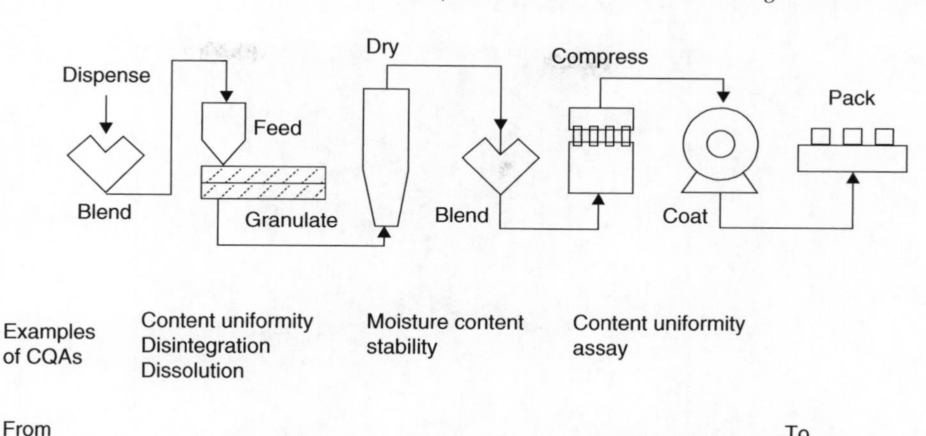

Figure 6.2 Examples of where CQAs may be impacted in the manufacturing process for an oral solid dosage form.

risk and rankings for the criticality of quality attributes. The multi-disciplinary team of 'experts' identify the potential ways in which a product may fail (failure modes) and the effect of these failures (failure effects). The impact of the failure (severity), the likelihood of it occurring (probability, likelihood or occurrence), and the ability to detect the failure (detectability) are then assigned numerical scores on the basis of predefined criteria. The individual scores are multiplied to provide an overall risk priority number (RPN), which is used to rank the risks and determine the criticality of each attribute. This type of risk assessment may also be applied to identify CPPs, discussed later in this chapter.

A slightly modified scoring approach is used in the A-Mab Case Study [16] and A-Vax Case Study applying QbD principles to vaccines [19], where

The impact ranking of a quality attribute assesses either the known or potential consequences on the product's safety and efficacy; whereas the uncertainty ranking is based on the relevance and source of the information used to assign the impact ranking. Examples of scoring for impact and uncertainty are illustrated in Table 6.8 and Table 6.9 respectively. Applying this system of scoring, if the overall severity (criticality) score for a quality attribute is 25 or above, it is deemed to be 'critical'. Scores less than 10 are deemed to be 'non-critical', and scores between 10 and 25 are 'potentially non-critical'. Upon completion of the CQA-scoring process, the complete list of attributes is reviewed to ensure that the scoring system and output is realistic.

After the risk assessment is performed to assess the criticality (severity) of the quality attributes, product development studies are conducted to confirm the criticality. It is usual to explore a suitably broad range for each attribute, taking into account the variability of the non-clinical test methods. It is important to have a robust analytical strategy in order to ensure identification of CQAs, criticality confirmation, and reliable/robust and up-to-date methods in specifications. The analytical target profile (ATP) should drive the analytical method development, ensuring that the test methods are selected and developed on the basis of quality attribute requirements (see Chapter 12 for more details of analytical methods and QbD). This testing can provide a basis for the rationale for the inclusion of

Table 6.7 *Typical example of a summary risk chart linking CQAs and prior knowledge to the unit operations for an oral solid dosage form.*

Quality attributes	Unit operations						
	Dispensing	Blending 1	Granulation	Drying	Blending 2	Compression	Coating
Appearance	Low	Low	Low	Medium	Low	Medium	High
Identification	Medium	Low	Low	Low	Low	Low	Low
Assay	Low	Medium	Low	Low	Low	High	Low
Impurities/Degradants	Low	Low	Medium	Medium	Low	Low	High
Content Uniformity	Low	High	High	Low	Low	High	Low
Dissolution	High	Low	High	Low	Low	Medium	Medium
Disintegration	High	Low	High	Low	Low	Medium	Medium
Hardness	Low	Low	Medium	Low	Low	Medium	Low
Stability	Low	Low	Low	Medium	Low	Low	Low
Water content	Low	Low	Low	High	Low	Low	Low
Microbiology	Medium	Low	Low	Medium	Low	Low	Low

Table 6.8 Example of 'impact' definition and scale from the CMC-Vaccines Working Group, 2012 [19].

Score		Efficacy	Safety & tolerability (adverse events)
Very High	25	Meaningful change in efficacy when the attribute varies significantly	Adverse events interfere with normal, everyday activities (e.g. inability to attend school, the need for medical attention or advice). Significant increase in severity and/or frequency
Moderate	8	Moderate change in efficacy when the attribute varies significantly	Sufficiently discomforting to interfere with normal, everyday activities. Moderate, but detectable increased severity and/or frequency compared to placebo
Minimal	2	Minor to no change in efficacy when the attribute varies significantly	Easily tolerated, causing minimal discomfort and not interfering with everyday activities. Similar to placebo

Table 6.9 Example of 'uncertainty' definition and scale (CMC-Vaccines Working Group, 2012).

Score		Uncertainty
Very High	5	No information available
High	4	External information available from literature or related products
Moderate	3	Data from internal laboratory or non-clinical studies or internal data extrapolated from related products
Low	2	Supportive data from clinical studies with the product
Minimal	1	Published limits widely accepted by regulatory and scientific community

attributes specifications. However, if there is not a correlation between the non-clinical and clinical responses, then non-clinical studies should not be employed for setting specification ranges. In these cases, specification ranges need to be supported by clinical studies with batches reflecting product/process variability.

An updated and finalised list of CQAs and ranges will be based on non-clinical testing; clinical results from Phases I, II, and III; and analytical capability. Meanwhile, periodic review of the initial risk assessment should reduce risk scores as increased product knowledge is gained during the product lifecycle.

6.6 Product Design and Identifying the Critical Material Attributes (CMAs)

The guidance from ICH Q8 (R2) focusses on process design, understanding, and control, but product design, understanding, and control are just as important. The key objective of product design and understanding is to identify and develop a robust product that can deliver the desired QTPP over the product shelf life [20]. Product design studies will typically include

- Physical, chemical, and biological characterisation of the drug substance.
- Identification and selection of excipient type and grade, and knowledge of intrinsic excipient variability.
- Compatibility and interactions of drug substance with excipients.
- Identification of the critical material attributes (CMAs) of both the excipients and drug substance.

A 'material', as defined in the context of pharmaceutical development by ICH, includes a wide scope such as raw materials, starting materials, reagents, solvents, processing aids, intermediates, active pharmaceutical ingredients (drug substances), and packaging and labelling materials (ICH Q7A). However, the drug substance (API) and formulation excipients will be the most important materials to investigate and understand during product design. Note that 'material attributes (MAs)'could refer to a particular physical, chemical, biological or microbiological property or characteristic. Examples of physical properties include physical description (particle size distribution and particle morphology), polymorphism, aqueous solubility as a function of pH, intrinsic dissolution rate, moisture level and hygroscopicity, and melting point. Chemical properties include purity, pKa, and chemical stability in the solid state and in solution; biological properties include partition coefficient, membrane permeability, and bioavailability. Although the vast majority of formulation excipients are pharmacologically inert, they can have a profound effect on physical or physiological properties of the final dosage form. Therefore, understanding the MAs and the potential variability of the selected formulation excipients is very important.

An input material can be quantified to be within an appropriate limit, range, or distribution, to ensure the desired quality and they are typically fixed, but since they may be changed during further processing, this needs to be checked. *CMAs* are those input materials that have a direct and high impact on the CQAs. Drug substance, excipients, primary packaging, and in-process materials may have many CMAs. Hence, they must be understood, measured, and controlled (adjusted) during manufacture so that the outputs will meet the CQAs. A systematic way of gaining product understanding is to

- Identify all possible known input MAs.
- Identify high-risk and potentially critical MAs using a risk assessment process based on good scientific knowledge and formulation expertise.
- Focus on the high-risk MAs and establish levels or ranges for further investigation.
- Design and conduct experiments employing DoE, and evaluate the data to determine if the MA is critical.
- Develop a control strategy for the CMAs.

Note that where a raw material has a CMA of medium- to high-risk impact to CQAs, it should be included as a parameter in the optimisation studies (see Section 6.8).

The specific product design experimental studies will depend on the dosage form being developed (e.g. tablet, injection, or eye drops) and the respective manufacturing process, but it will always involve testing a range of options, for example, a range of excipients at different concentrations and different combinations or testing a range of pack options.

The QTTP should dictate the product design requirements and provide some direction for development work required. For example, see Table 6.10, for a parenteral dosage form. Again, prior knowledge, pre-formulation data, and formulation experience should be applied, such as the following;

- Consider formulation options that have a successful track record.
- Use proven formulation excipients that are well understood and compatible with the drug candidate (confirm by conducting drug-excipient compatibility studies).
- Consider the chemical stability risks and constraints, and whether formulation excipients and packaging design are required to manage the risks.
- Consider the physical stability risks and constraints.
- Try and avoid the issues, and select a viable formulation option, trying to keep the formulation as simple as possible; that is, only add an excipient if there is a clear rationale. For example, to aid the processing of the dosage form during its manufacture; to protect, support, or enhance stability, bioavailability, or patient acceptability; assist in product identification; or enhance any other attribute of the overall safety, effectiveness, or delivery of the drug during storage or use. QbD and the role of formulation excipients is discussed in much more detail by Brian Carlin in Chapter 5.

Formulation and process identification is likely to be an iterative process where, initially, a qualitative formulation listing all the excipients (but not the exact levels) is defined along with the likely process unit operations. In the next steps, a qualitative formulation is defined to cover the range of doses required, and the manufacturing process unit operations are confirmed. The CQAs are aligned with their potential to impact patient safety and efficacy. Screening and selection of formulation prototypes will usually involve a combination of *in vitro* testing (e.g. dissolution), *in vivo* animal pharmacokinetic (PK) and bioavailability screening and human evaluation and PK performance testing in clinical trials.

Table 6.10 *The QTPP dictates the product design and development work.*

QTPP attributes	Design & development elements
Clear, ready-to-use solution	API concentration must be less than the solubility over the pH range
Single-use IV injection	Must be sterile. No preservative
Bolus administration	Maximum injection volume 5 mL
Dose 2.5 mg	Target concentration 0.5 mg/mL
>2-year shelf life at 25°C	Formulate at pH of optimum stability that meets solubility target Buffered (if required to maintain pH) Oxygen/light control (if required) Terminal sterilisation preferred
Biocompatibility/tolerability	Isotonic, pH range 3 to 9, endotoxin control
Degradation products & impurities Below safety threshold/qualified	pH of optimum stability, oxygen/light control, leachables and extractables control, particulate control
Packaging	Pack/formulation interactions, stability, ease of use

6.7 Process Design and Identifying the Critical Process Parameters (CPPs)

Process design is the stage of process development where an outline of the clinical trial and eventual commercial manufacturing process are identified. Process design is also described in the broadest sense in the latest US FDA Process Validation guidance [21] as the first stage of the product and process lifecycle, when the commercial manufacturing process is defined on the basis of knowledge gained through development and scale-up activities. This is discussed further in Chapter 11, along with the other lifecycle stages. In particular, see Figure 11.3 in Chapter 11, which shows the sequence of activities for formulation, process design, and optimisation and how they fit with process validation activities.

Selection of the manufacturing process will be strongly influenced by the TPP/QTPP, as this will define the dosage form (e.g. immediate-release tablet), route of administration (e.g. oral), and perhaps other manufacturing objectives, for example, to use existing technology within the company (e.g. roller compaction, direct compression, or wet granulation). If there are different technology options available, then the development team should engage in product design scouting to identify and assess the options. An important consideration is whether the manufacturing technology option will be able to deliver the CQAs of the product (refer to Table 6.7), and this may limit the options that can be used.

The selected pharmaceutical manufacturing process will usually consist of a series of unit operations to produce the desired quality product. A unit operation is a discrete activity that involves physical or chemical changes, such as those listed in Figure 6.7 for an oral solid dosage form: dispensing, blending, granulation, drying, mixing, compression, and coating. Potential manufacturing processes will be identified as part of the 'design intent'. The identification of potential unit operations and process steps will be based on prior knowledge and risk assessment.

Process parameters are referred to as the input operating parameters (e.g. mixing speed and flow rate) or process state variables (e.g. temperature and pressure) of a process step or unit operation. Each unit operation will have both input and output process variables. Sometimes the output of a one-unit operation can also be the input of the next unit operation. Process parameters are process inputs that are directly controllable, whereas the process outputs that are not directly controllable are attributes. If the attribute ensures product quality, it is deemed to be a CQA. Therefore, a process parameter is deemed to be critical when its variability has an impact on a CQA and, therefore should be monitored or controlled to ensure that the process produces product of the desired quality [8].

Process parameters may be further categorised on the basis of impact on the process, so process parameters that have been shown experimentally to impact process performance may be classified as key process parameters (KPP). The PDA Technical Report No.60 [22] includes a decision tree developed to guide the assignment of parameter definitions in conjunction with the aid of quality risk assessments (Figure 6.3).

The decision tree can be used to differentiate critical from non-critical parameters strictly according to the ICH Q8(R2) guidance definition, that is, whether there is an impact on a CQA or not. On the left-hand side of the decision tree, the process of categorisation for a

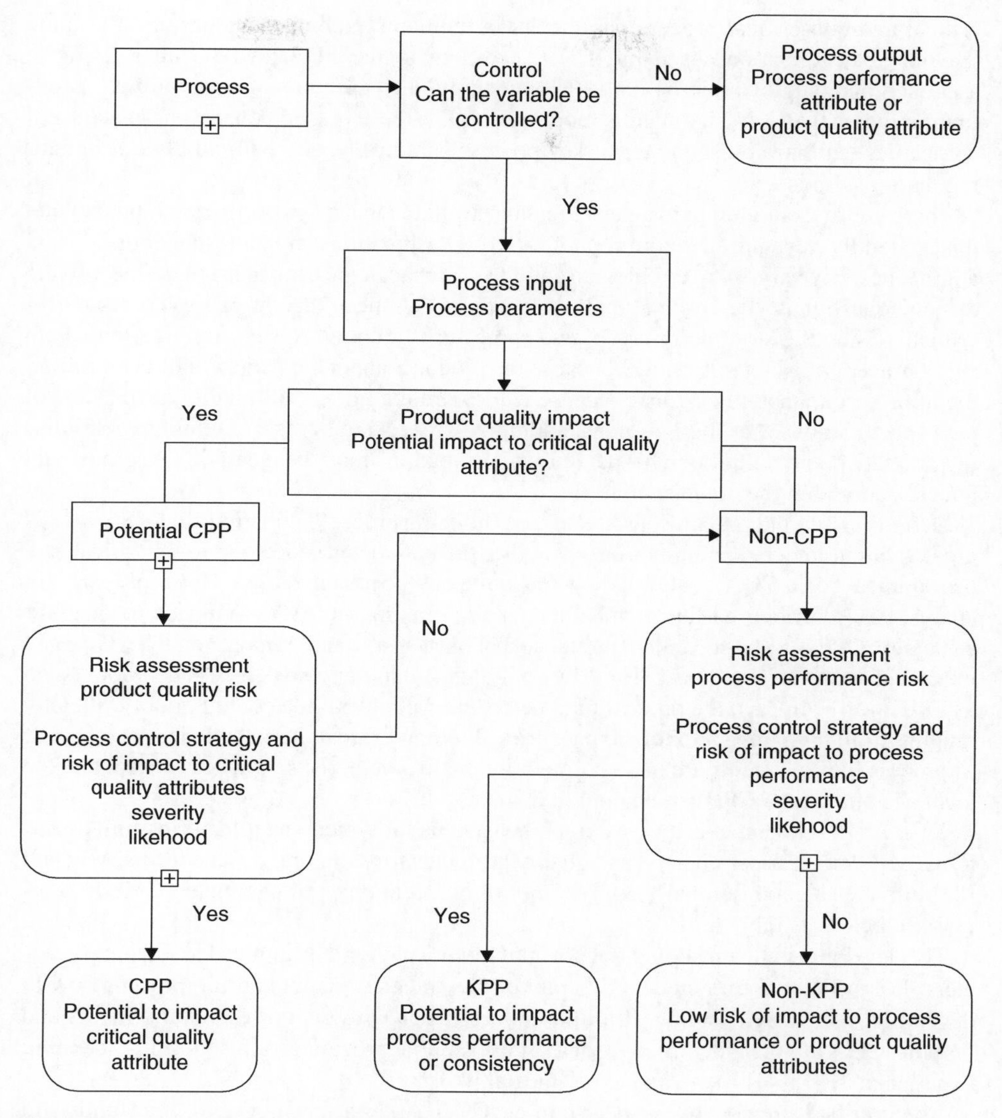

Figure 6.3 *PDA decision tree for designating parameter criticality (from PDA TR No. 60 [22]).*

potential CPP is to conduct a risk assessment to estimate the risk of impact to CQAs. On the left-hand side of the decision tree, the process of categorisation for a non-CPP is also to conduct a risk assessment, but this time to assess the risk of impact to process performance and to differentiate a KPP from a non-KPP.

However, a word of caution about using the term 'key process parameter'! The regulatory authorities (e.g. FDA and EMA) do not recognise this term and will only consider critical and non-critical process parameters in regulatory submissions because these are the only categorisations used by ICH. According to the FDA/EMA, with a lifecycle approach that employs risk-based decision making, the perception of criticality as a continuum rather than a binary state is more useful. For these reasons, many pharmaceutical companies tend not to use the term 'key process parameter', and particularly not in their regulatory submissions.

There are several other guidance documents available that discuss criticality, but beyond this generally recognised definition of a CPP, designations are not standardised, and approaches may vary. For example, the amount of impact to be critical is not defined, so the difficult question is, does even a small impact to a CQA mean that the process parameter is critical? Using the strict definition, some companies find themselves in a situation where every process parameter is critical because the product cannot be made without controlling them, or no parameter is critical because if they are controlled, all quality attributes will pass specifications. For these reasons, additional criteria are necessary to aid in determining criticality, and definitions for parameter designations must be clearly documented and understood within the organisation.

Definitions should remain consistent throughout the process validation cycle. So there is great value in understanding not only whether the parameter is deemed to be critical (i.e. has an impact on a CQA), but also, how much impact the parameter has. By employing risk analysis as described earlier in this chapter for ranking CQAs as a means to quantify criticality, CPPs can be separated into those that have a substantial impact on the CQAs and those with minor or no impact. Thus, by applying a quantitative risk assessment tool (such as FMEA or FMECA), the process parameters can be ranked in a continuum of criticality ranging from high impact to low impact critical to non-critical. As knowledge increases or as process improvements are made throughout the lifecycle, risks may be reduced, and the level of impact for a CPP can be modified.

Yu *et al.* [20] documents a very comprehensive list of typical manufacturing unit operations, MAs, process parameters, and quality attributes for solid oral dosage forms. Mitchell [23] provides a good rationale for criticality of different types of parameters, briefly summarised below in Table 6.11.

The development team should document the rationale for defining and assessing parameters. It is beneficial to include a diagram of the potential manufacturing process flow to develop a clear understanding of all the process steps with the corresponding inputs and outputs. CPPs are then identified from a list of potential process parameters using scientific experience, risk assessment, and experimental work.

As described previously for determining CQAs, risk assessment is often applied to establish potential CPPs, and DoE is recommended to establish links between various CPPs and MAs and the impact on CQAs. This should maximise opportunities to gather information and knowledge. Product design experimental work will typically start at small scale (<1 kg batch size), to evaluate potential CQAs and identify potential CPPs. It is necessary to separate out those CPPs that have a substantial impact on the CQAs and those with minor or no impact. During product and process optimisation, as knowledge increases or as improvements are made to the process, risks may reduce, and the impact of the CPP may change accordingly. This is discussed further in the next sections on product and process optimisation and control strategies.

Table 6.11 Criticality of parameter types (from Mitchell, 2013 [23]).

Parameter	Criticality assessment	Justification
Process parameters, e.g. mixing speed, flow rate	Potentially critical if impact on CQA	Need to be assessed for criticality
Fixed parameters such as equipment scale and equipment setup	Non-critical	Assess the impact and document
Parameters for sterilisation processes, cleaning processes & the preparation of process intermediates	Potentially critical	Can be assessed along with the main process or as independent processes
Formulation recipes	Non-critical	Can be considered fixed parameters. Generally have relatively tight limits that are justified during formulation development. Could vary if an operator must calculate a quantity on the basis of variable input, for example, biological activity, so this variable process parameter may lead to process variation
Holding/storage times and conditions where no processing occurs	Non-critical	Should be qualified to show little or no impact on the product
Environmental conditions during processing (room temperature, humidity), such as holding times	Non-critical	Have set limits so that they have little or no impact to the process
Process-specific environmental conditions, for example, clean rooms, cold rooms, and dry rooms	Non-critical	Included as process parameters because they are monitored to ensure product quality

6.8 Product and Process Optimisation

The objective of product and process optimisation is to confirm by assessment that a robust formulation in its primary pack and manufacturing process reproducibly complies with the QTPP. The development activities may include the following:

- *Product and process risk assessment* to identify the high risks, confirm the CPPs, and establish risk mitigations.
- *Experimentation and predictive modelling* to evaluate the product and confirm quality and robustness.
- *Final positioning and lockdown* to accept the product and position it in the design space on the basis of sound product understanding.

- *Control strategy* to define control based on product/process knowledge and any sensitivities. Establish proven acceptable ranges (PARs), design space, and response and PAT approaches.
- *Continuous improvement* to optimise manufacturing efficiency on the basis of batch data.

Use of Ishikawa fishbone (cause-and-effect) diagrams can be of great value in assisting development teams to identify potential variables that may have an impact on the desired quality attribute in a systematic way. An example of a fishbone diagram is shown in Figure 6.4 for potential material attributes (MAs) and process parameters (PPs) contributing to the quality attribute, solubility, of an API/polymer solid dispersion produced via hot melt extrusion (HME) process. The development team should reach a consensus on listing those items that are the 'most likely causes' and prioritising the development work to investigating the significance, addressing the issue, and/or controlling it.

Risk assessment is another useful tool that can be applied to identify high-risk parameters or to rank risks in terms of impact and severity and then decide what and how to study. There is an advantage of being able to objectively document and communicate the assessment and decisions made. Note that the regulators do not dictate the approach taken and methodology used, but it is up to the development company to justify these. It is recommended that all input operational parameters that may affect product quality or process performance for unit operation(s) should be risk-assessed, considering their potential impact on all relevant outputs. The development team need to discuss and agree for each process parameter as to whether there is any impact or not. If there is an impact, they

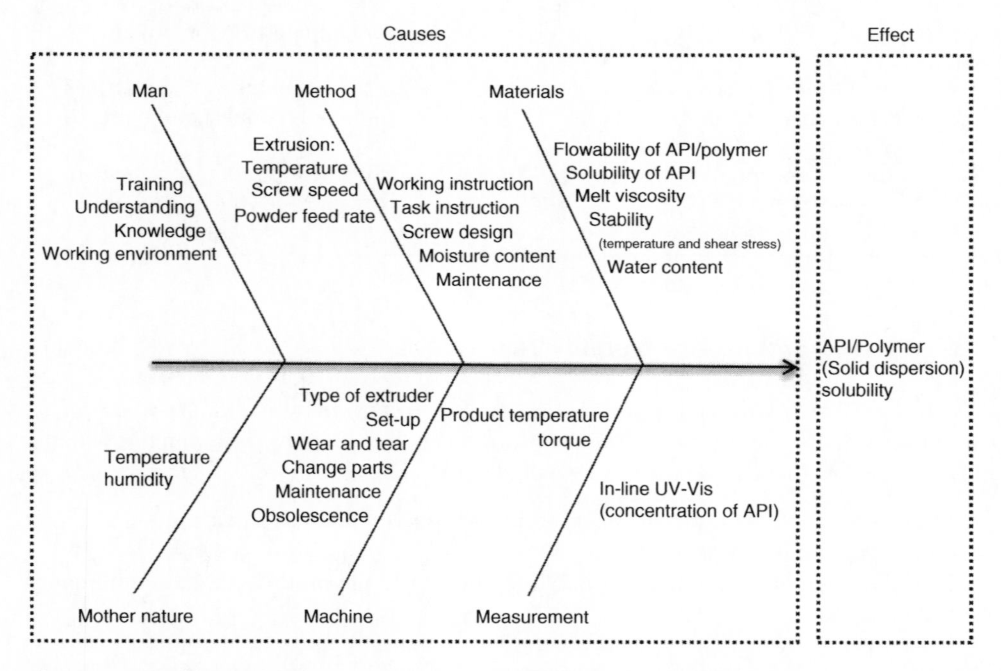

Figure 6.4 *Fishbone diagram recognising material attributes (MAs) and process parameters (PPs) contributing to the quality attribute, 'solubility', of a solid dispersion API/polymer product.*

should decide whether this is a concern. Also, they should consider whether this parameter interacts with any other parameter that can affect the output, and how this can be communicated. Each parameter is assessed individually and also in combination with other parameters. As described before, scoring can be applied according to the level of impact (overall severity score), and this can be translated into an experimental strategy, whether to perform multivariate studies, univariate studies, or no further study at all, as illustrated in Figure 6.5.

Through experimentation, the relationships between process parameters (or combination of process parameters) and quality attributes are established. This allows for criticality, process parameter ranges, and controls to be identified.

In this example, the process optimisation shows that the release target range and host cell protein are the most important. Where the CPPs are identified, they should be varied within practical constraints to establish the limits within which acceptable product can be manufactured. Combining both the drug product and process optimisation studies is usually beneficial because they are so closely linked, and there are significant efficiency savings of valuable drug substance, resources, and time. In addition, interactive (or interdependent) relationships among MAs, process parameters, and product attributes can be more easily established with combined experimental studies. Any changes in the material, formulation, or manufacturing process made to improve one attribute should be assessed to determine its potential to impact other attributes in the process. Typically, independent process variables frequently produce interactive effects. Changing one separate factor at a time (COST) is unlikely to lead to the real optimum, and yields different implications with different starting points. It also results in many experiments to produce limited information with no quantification of interactions between factors. For these reasons, design of experiments (DoE) is the recommended way forward, when the relationships between multiple process parameters is required as the entire region can be investigated in an organised way, allowing calculation of a response surface [24]. DoE applies to three main experimental objectives:

- *Screening* (fractional factorial, Plackett–Burman) to find out which factors are most influential and what are the appropriate ranges. They are used to reduce the number of process parameters by identifying the critical ones that affect the response(s).
- *Optimisation* (response surface method designs, e.g. central composite, Box–Behnken) to find the optimum, ideal set-point in the process design space, either a unique optimum or a compromise to meet conflicting demands. Also, to establish equipment settings and ranges for the parameters within which the acceptance criteria for the multiple attributes simultaneously should be met.
- *Robustness* (fractional-factorial design, Plackett–Burman) to prove that the operating ranges predicted to be acceptable following the optimisation designs will indeed result in acceptable product meeting all CQAs under worst-case conditions.

Multivariate analysis is a technique that is also recommended to characterise interdependent and synergistic effects, and reduce the resources and time to complete the experimentation and to evaluate the results. The application of predictive modelling to use exploratory data from multiple previous products and try to build a picture or model of what is really happening can also help design experiments and avoid the completion of obvious experiments and unnecessary work (refer to Chapter 7 by Martin Owen and Ian

		B			
		1	2	4	8
A	8	8	16	32	64
	4	4	8	16	32
	2	2	4	8	16
	1	1	2	4	8

A = Interaction effect ranking B = Main effect ranking

Severity	Experimental strategy
≥ 32	Multivariate Study
8–16	Multivariate, or univariate with justification
4	Univariate acceptable
≤ 2	No additional study

Inputs

Operational parameters	Historical range	Proposed range	Justification
pH load	6.8–7.2	6.5–7.5	Allow for variability coming out the bioreactor
Protein load	≤ 35 mg/ml	≤ 40 mg/ml	Allow for deviation in [Mab] concentration
no. CVs wash	≥ 3	≥ 3	De-prioritised: min added value
Elution pH	3.3–3.7	3.2–3.8	Allow for flexibility in buffer preparation and understanding on impact elution volume

Outputs

Attribute	Monomer	Host cell protein	Product Volume	Yield
Acceptable level for protein A product	≥ 96%	≤ 5000 ng/mg	≤ 52.5 CVs	≥ 95%
Release Target Range	≥ 98.5%	≤ 50 ng/mg	NA	NA
Justification	A subsequent CEX step is capable to clear 3% aggregates	A subsequent CEX step is capable to clear over 2 Logs HCP	A process fit restriction on product tank volume limits elution volume	A process fit restriction on the UFDF tetentate tank requires a minimal yield

Figure 6.5 *Use of risk assessment to determine what DoE studies to use.*

Cox and Chapter 8 by Claire Beckett *et al.*, respectively, for much more detail on these topics). The application of different types of DoE to pharmaceutical product development are covered extensively by Gibson, Ashman, and Nelson [24].

Formulation and process optimisation studies are conducted to identify the critical formulation and process parameters and to assess their impact on the CQAs of the drug product. The effects of variations in process parameters and MAs are investigated to identify the CPPs and establish limits for the CPPs (and CMAs) within which the quality of the drug product is assured. This process is illustrated in Figure 6.6.

From the experimental data, it should be possible to determine if a process parameter is critical and determine scalability. Acceptable ranges are defined for the critical parameters; otherwise, for non-critical parameters the acceptable range will be the range investigated. When more than one process parameter or material attribute is involved, these defined acceptable ranges may be termed the 'process design space' (see Section 6.9 for more details).

Due to the cost involved or availability of API, it is usual to perform most optimisation studies at laboratory or pilot scale, rather than at commercial scale. However, the manufacturing process and equipment used for product and process optimisation should be designed with commercial-scale manufacture in mind. This should eventually allow a more straightforward scale-up and technology transfer from pilot-scale at R&D to full commercial-scale manufacturing operations if the process equipment is of the same type. As the product moves through the clinical studies (Phase I to III), the manufacturing process will usually need to be scaled up to meet the increasing demand of the larger clinical trials. Scale-up may involve changes in process equipment and operation with an increase in output. For example:

- Use of larger or higher speed version of a similar type of equipment.
- Change of equipment type for a given process step.
- Increase in the batch size of 10-fold or greater on similar type of process equipment.
- Increase in output rate by more than 50% such as changing from manual to automated filling.

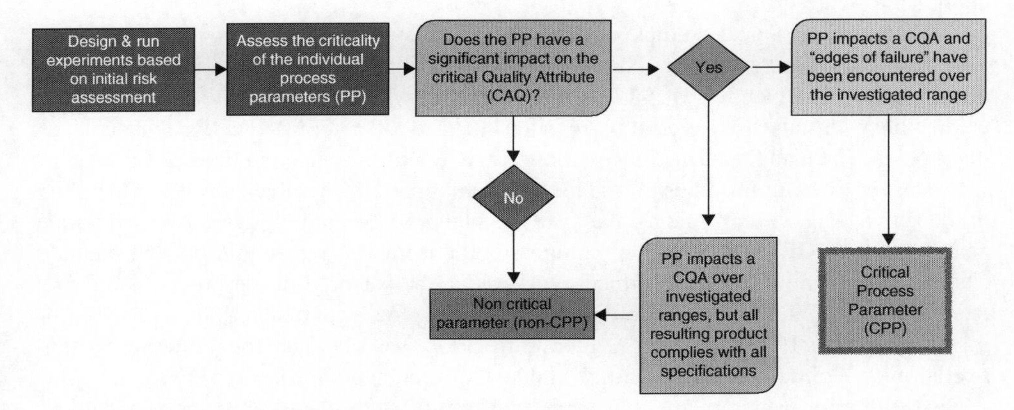

Figure 6.6 *Example of a process parameter criticality assessment decision tree.*

A number of approaches to scale-up are available depending on the dosage form and associated manufacturing process. Potential impact of scale-up can be assessed using risk assessment (considered per unit operation). Science of scale (including modelling the process at smaller scale) can be combined with DoE to optimise the approach. Scale-independent parameters can be assessed experimentally using DoE on a laboratory or pilot scale. Chemometric analysis in combination with observations and mechanistic insight is used to determine which parameter is most appropriate. Predictions are made for these parameters on a larger scale and then verified using further DoE on the larger scale. Examples of scale-independent unit operations include

- Raw material properties (reagents, excipients).
- Qualitative/quantitative composition.
- Continuously fed equipment, such as roller compactors, tablet presses, and encapsulators.

There are potential benefits of commencing an early technology transfer to supply Phase III from commercial operations to ensure that the product used in the pivotal clinical trials will be equivalent to the commercial dosage form. Alternatively, there may be opportunities to utilise continuous processing of the drug product to avoid the requirement for large-scale equipment, by simply manufacturing for longer periods instead. The advantages (and challenges) associated with continuous processing for pharmaceutical products are discussed further in Chapter 11 by Mark Gibson. To support a successful technology transfer, it is recommended that at least one experimental trial batch is made at production scale in operations to gain some assurance that the process will be acceptable at the increased scale, prior to conducting the formal scale-up and technology transfer and particularly before attempting any process validation batches.

To conclude, at the completion of product and process optimisation, there should be a clear justification to support the drug product and commercial manufacturing process steps and the CPPs, the limits for the process parameters, and confirmation that the product specifications can be met. With the successful application of QbD, there should be a good understanding of the effect of API and the functionality of the excipients, including stability, to establish a robust formulation. Similarly, for a robust process, there should be a good understanding of potential criticality of process variables and a good understanding of the approach to scale-up.

The following practical example of process optimisation was taken from the ACE Tablets Case Study [13], where a fictitious drug molecule Acetriptan (ACE) is developed as an immediate-release (IR) tablet. It is BCS Class II (low solubility/high permeability) and found to be compatible with standard excipients used in the intended roller compaction (RC) manufacturing process. The manufacturing process flow for ACE tablets is given in Figure 6.7.

Process optimisation studies for the blending unit operation involved low shear blending of the active drug with excipients in a diffusive blender. The endpoint was assessed with a near infra-red (NIR) sensor taking one measurement for every resolution of the blender. Chemometric analysis was performed by assessing the standard deviation compared to a 'standard' blend: % CV ratio, resulting in a proposed process parameter range of <5% CV for not less than 10 revolutions. A good correlation was obtained for the blending unit operation endpoint (NIR CV%) with the tablet CQA, content uniformity (Figure 6.7).

Next, the critical factors affecting blend uniformity were identified from prior knowledge and captured on an Ishikawa fishbone cause-and-effect diagram, applying risk

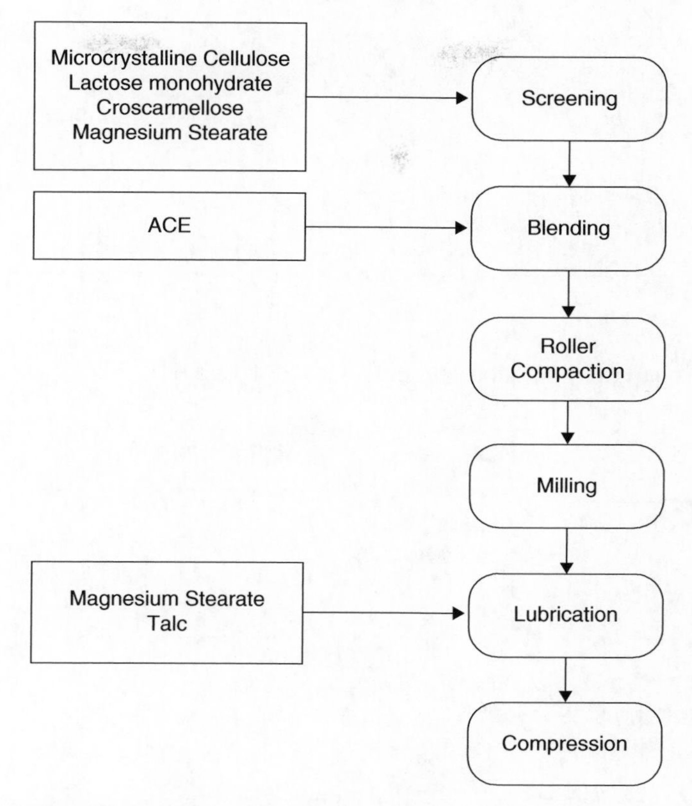

Figure 6.7 *The manufacturing process flow for ACE tablets. (From Pharmaceutical Development Case Study: 'ACE Tablets', prepared by CMC-IM Working Group (March 2008), http://www.ispe.org/pqli-resources.)*

assessment to rank each of the factors (see Figure 6.8). The proposed high-risk critical factors – relative humidity in the manufacturing plant, API particle size, and microcrystalline cellulose (MCC) particle size – were further evaluated in a blend uniformity DoE study measuring the NIR signal variation (%CV) which correlates with tablet content uniformity. The results from these DoE studies demonstrated that there was an acceptable NIR endpoint achieved for all combinations with a blend time range from 8 to 36 min (see Figure 6.8). Further, the tablet CQA of Content Uniformity met the USP criteria (specified in the QTTP) for all experimental conditions, confirming that the tablet quality was acceptable.

6.9 Design Space

The design space is a key part of an enhanced QbD process development approach, and the better understanding gained from the formulation and process optimisation of the cause-and-effect relationships between the input materials, process parameters, and the final drug

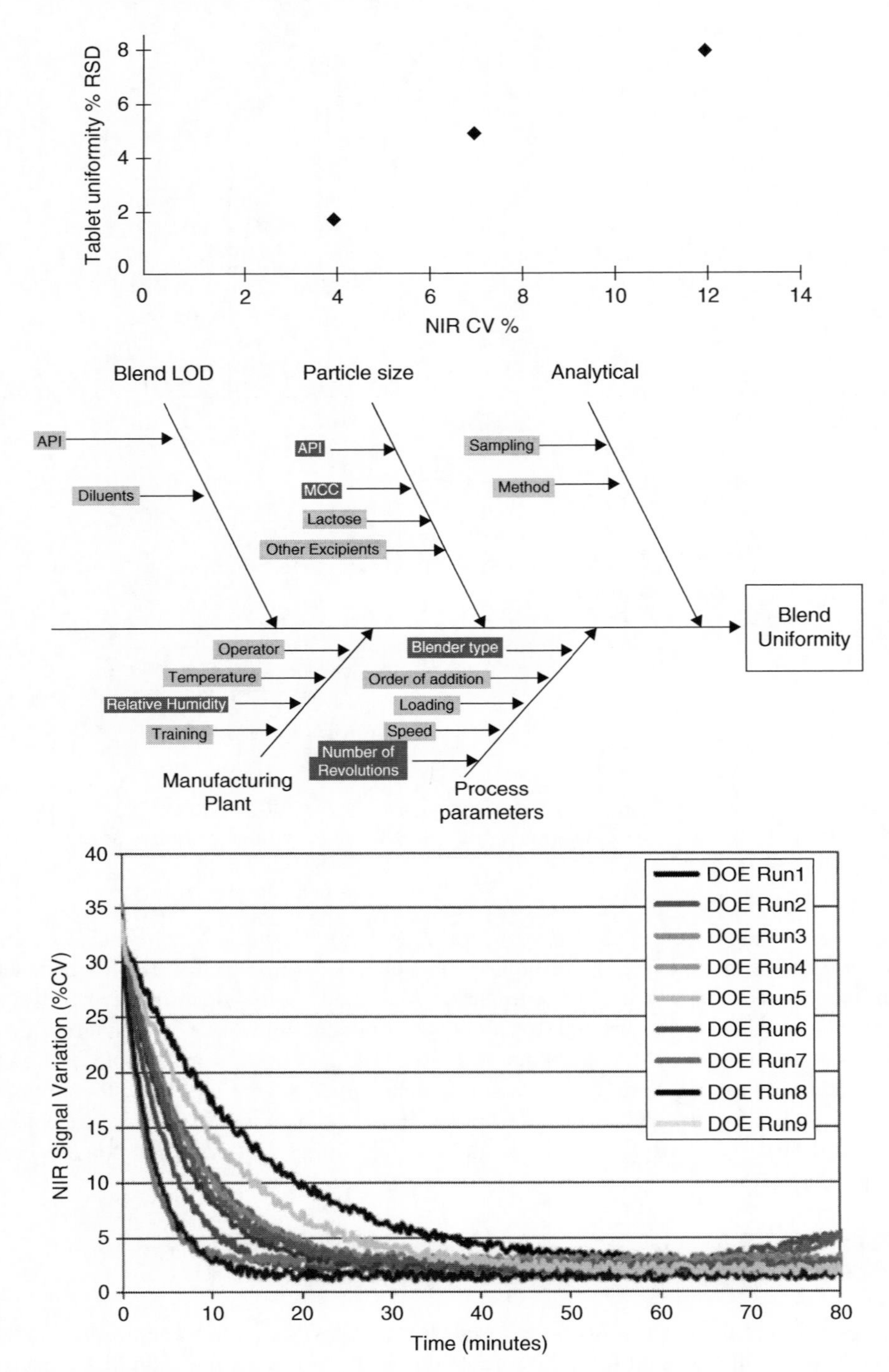

Figure 6.8 *Critical parameters affecting blend uniformity for ACE tablets. (From Pharmaceutical Development Case Study: 'ACE Tablets', prepared by CMC-IM Working Group (March 2008), http://www.ispe.org/pqli-resources.) (See insert for color representation of the figure.)*

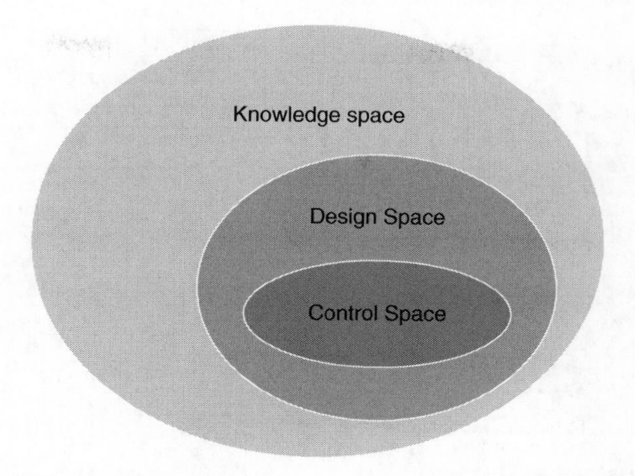

Figure 6.9 *Schematic representation of the relationship between knowledge, design, and control space.*

product quality can be used to predict a product design space. The main goal of pharmaceutical product/process development is to ensure that the patient receives a safe and efficacious dose. The product specifications should capture the range of each critical quality attribute that will consistently deliver a safe and efficacious dose, and the design space is the multidimensional combination of inputs (materials and process) that will deliver the product to meet the specifications.

The formal definition of a 'design space' according to ICH Q8 (R2) is 'the multidimensional combination and interaction of input variables (e.g. MAs) and process parameters that have been demonstrated to provide assurance of quality. Working within the design space is not considered as a change. Movement outside of the design space is considered to be a change and would normally initiate a regulatory post approval change process. Design space is proposed by the applicant and is subject to regulatory assessment and approval'. Furthermore, the ICH guidance defines three elements – knowledge space, design space, and control space – to establish a process understanding, as shown schematically in Figure 6.9. The knowledge space represents the broadest area where product and process development knowledge and understanding exists. Outside the design space, but within the knowledge space, the CQAs will not usually meet acceptance criteria. ICH Q8 (R2) guidance recommends that it is helpful to know where edges of failure could be, or to determine potential failure modes, defining 'Edge of Failure' as the boundary to a variable or parameter, beyond which the relevant quality attributes or specification cannot be met.

Even within the design space, it is possible that there could be failure of individual batches or samples within batches, if they are near the limits. For this reason, the operational limits for the CPPs are usually tighter than the design space, within the 'control space', where the normal operating range (NOR) limits for each CPP are applied. The 'proven acceptable range' is a characterised range of a process parameter for which operation within this range, while keeping other parameters constant, will result in producing a material meeting relevant quality criteria.

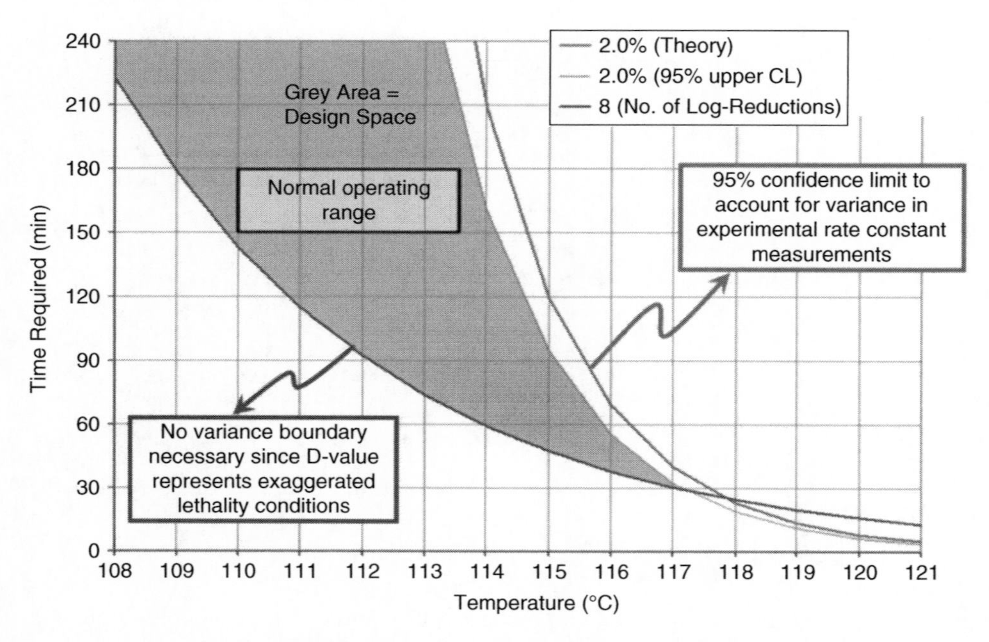

Figure 6.10 *Example of terminal sterilisation design space.*

Therefore, it can be seen that the design space is a direct outcome from the product and process optimisation DoE studies and data analysis conducted to understand the linkage and effect of MAs and process parameters on product CQAs. Those critical MAs and PPs can thus be selected for inclusion in the design space. The regulatory authorities will expect to see a rationale presented in the submission for the inclusion of any parameters in the design space. Independent design spaces can be presented for one or more unit operations, or there may be a single design space that covers multiple operations in the manufacturing process flow. The latter approach is likely to provide more operational flexibility.

The design space may be scale and equipment dependent. In this case, if the design space is determined at laboratory or pilot scale, it would need to be justified for use at commercial scale. If mechanistic understanding has been achieved or a reliable empirical predictive that is based on sound process understanding, then the design space can be translated across scale.

There are many ways of describing the design space. Examples of different potential approaches to presentation of a design space are presented in Appendix 2 of the ICH Q8 (R2) document.

Figure 6.10 shows an example of a design space for the terminal sterilisation by moist heat of a parenteral product. The process parameters that can be varied are set-point temperature and the dwell time, and the CQAs to be met are sterility assurance level (Log reduction ≥ 8) and degradation ($\leq 2\%$). By increasing the temperature and time, sterility assurance is more assured, but chemical stability becomes a problem. The design space represents the NOR whereby both CQAs can be assured.

The following practical example is shown to illustrate how a design space can be developed and presented for the CQA dissolution for an oral solid dosage tablet product. The

Table 6.12 *Structured approach to build* in vivo *understanding for dissolution CQA.*

Step	Example
1. Conduct quality risk assessment	Identify the most relevant risks (product and process variables) to *in vivo* dissolution (ICH Q9)
2. Develop appropriate CQA tests	Develop *in vitro* dissolution test(s) with physiological relevance that is(are) most likely to identify changes in the relevant mechanisms for altering *in vivo* dissolution (identified in Step 1)
3. Understand the in vivo importance of changes	Determine the impact of the most relevant risks (from Step 1) to clinical pharmacokinetics based on *in vitro* dissolution data combined with • Prior knowledge, including BCS and/or mechanistic absorption understanding • And/or clinical 'bioavailability' data
4. Establish appropriate CQA limits	Establish the *in vitro* dissolution limit that assures acceptable bioavailability
5. Use the product knowledge in subsequent QbD steps	Define a design space to deliver product CQAs, for example, to ensure *in vitro* dissolution performance within established limits Develop a control strategy to ensure that routine manufacture remains within the design space, for example, that assures dissolution limits are met during routine manufacture (ICH Q10)

structured approach taken (see Table 6.12) is to confirm a mechanistic understanding by producing product variants that incorporate the highest-risk variables and then evaluating their performance. By gaining an understanding of the *in vivo* impact and linking these to meaningful *in vitro* tests, evaluation of multiple aspects of the design space and the development of science- and risk-based specifications can be performed.

The best possible outcome is that there will be an *in vivo–in vitro* correlation (IVIVC) established, where changes in *in vitro* dissolution are directly correlated to changes in bioavailability. However, other potential outcomes are that there is no effect on bioavailability observed across a range of *in vitro* dissolution rates, or, a final option in which bioavailability is only affected for a few of the variants tested clinically. Taking this practical example further, from Step 2 in Table 6.12, the highest risk failure modes from Step 1 resulted in several dissolution retardation mechanisms being investigated, as shown in Table 6.13. An optimum dissolution test was then developed to clearly discriminate between the tablet product variants (A to D). The tablet variants were then dosed in a clinical volunteer pharmacokinetic study, together with an oral solution (control), to confirm if these *in vitro* dissolution changes would result in changed bioavailability (Step 3 in Table 6.12).

Results from the *in vivo* study indicated that tablets manufactured with a dissolution rate greater than or equal to that of the slowest dissolving formulation (Tablet Variant D) will demonstrate equivalent pharmacokinetics to the standard tablets (Tablet Variant A) and to an oral solution. Also, dissolution is not the rate determining process within the range studied. However, appropriate CQA limits for dissolution can be established (Step 4 in Table 6.12) on the basis of the premise that if a tablet product has a dissolution profile faster

Table 6.13 Developing appropriate dissolution CQA tests.

Tablet	Failure mode	Dissolution retardation mechanism
A (Standard)	N/A	N/A
B (Particle size)	Changes in API properties	Impact of API surface area on rate of dissolution
C (Processing)	Wet mass over-granulation	Impact of granule properties (size and structure) on the rate of ingress of water
D (Formulation)	Increased level of binder in the formulation Decreased level of disintegrant in the formulation	Impact of slowed tablet disintegration rate on subsequent drug dissolution

than that of the slowest profile dosed to volunteers (Variant D), then the *in vivo* performance will be comparable to pivotal clinical trials material. Finally, in Step 5, design space development is undertaken with structured, systematic investigations using DoE to investigate the impact of the variation in input materials and process parameters on the *in vitro* product dissolution (and other CQAs).

The design space boundaries were defined to ensure that batches with acceptable bioavailability would always be produced, that is, batches with a dissolution faster than that of Variant D.

6.10 Control Strategy

The knowledge gained through appropriately designed development studies to acquire product and process understanding culminates in the establishment of a design space and control strategy (see Figure 6.1). The controls can include parameters and attributes related to drug substance and drug product materials and components, facility and equipment operating conditions, in-process controls, and finished-product specifications, and the associated methods and frequency of monitoring and control (ICH Q10). In other words, a control strategy can include, but is not limited to, the following (ICH Q8 (R2):

- Control of input MAs (e.g. drug substance, excipients, in-process material and primary packaging materials) based on an understanding of their impact on processability or product quality.
- Product specification(s).
- Controls for unit operations that have an impact on downstream processing or product quality (e.g. the impact of drying on degradation and particle size distribution of the granulate or dissolution).
- In-process or real-time release testing in lieu of end-product testing (e.g. measurement and control of CQAs during processing).
- A monitoring programme (e.g. full product testing at regular intervals) for verifying multivariate prediction models.

The control strategy is originally developed to support clinical trial manufacture, but is refined during development for use in commercial manufacture. Analytical methodology or the acceptance criteria may change, for example, as more knowledge and understanding is gained. In the QbD paradigm, the control strategy is established by applying risk assessment that takes into account the criticality of the CQA and process capability. A minimal control strategy approach is the level of control traditionally used in the pharmaceutical industry, relying extensively on intermediate and end-product testing and tightly constrained MAs and process parameters. Applying an enhanced, QbD approach, quality controls are shifted upstream with reduced end-product testing and flexible MAs and process parameters within the established design space.

Figure 6.11 illustrates the differences between traditional controls and advanced controls for the manufacture of an oral solid dosage form.

Taking the enhanced approach to the extreme, automatic engineering controls can be used to monitor the CQAs of the output materials in real time. Typically, PAT, data management, and statistical tools are applied to the manufacturing process to measure and analyse parameters and attributes to verify continued operation within a state of control (refer to Chapter 9 for more details on PAT and real-time release testing). Input MAs are monitored, and process parameters are automatically adjusted to assure that CQAs consistently meet the established acceptance criteria. Feedback or feedforward can be applied technically in process control strategies and conceptually in quality management (ICH Q10) where

- Feedback is the modification or control of a process or system by its results or effects.
- Feedforward is the modification or control of a process using its anticipated results or effects.

Enabling real-time release testing is only possible when drug product quality can be ensured by risk-based control strategy for a well-understood product and process. In reality, there will probably be a combination of different control strategy approaches used for a particular drug product. There may be some cases where the product could not be well characterised and/or quality attributes are not readily measurable due to limitations of testing or detectability. In these cases, more emphasis should be placed on process controls (e.g. sterility). It is even possible for different control strategies to operate for the same product at different sites due to the availability of facilities, equipment, systems, and so on. In this case, the company should consider the impact of the control strategy implemented on the residual risk and the batch release process [25]. Most importantly, the output of the control strategy should ensure that the product CQAs remain within the design space.

Factors to consider when selecting a control strategy include the following:

- Minimising risk to the patient (linked to risk with product and process).
- Effectiveness of the control system.
- Cost of implementation.

It is recommended that controls are developed on the basis of previous experience and existing capabilities in collaboration with the commercial manufacturing site. To contain cost and ease implementation, the simpler the better, thus avoiding complications and ambiguity.

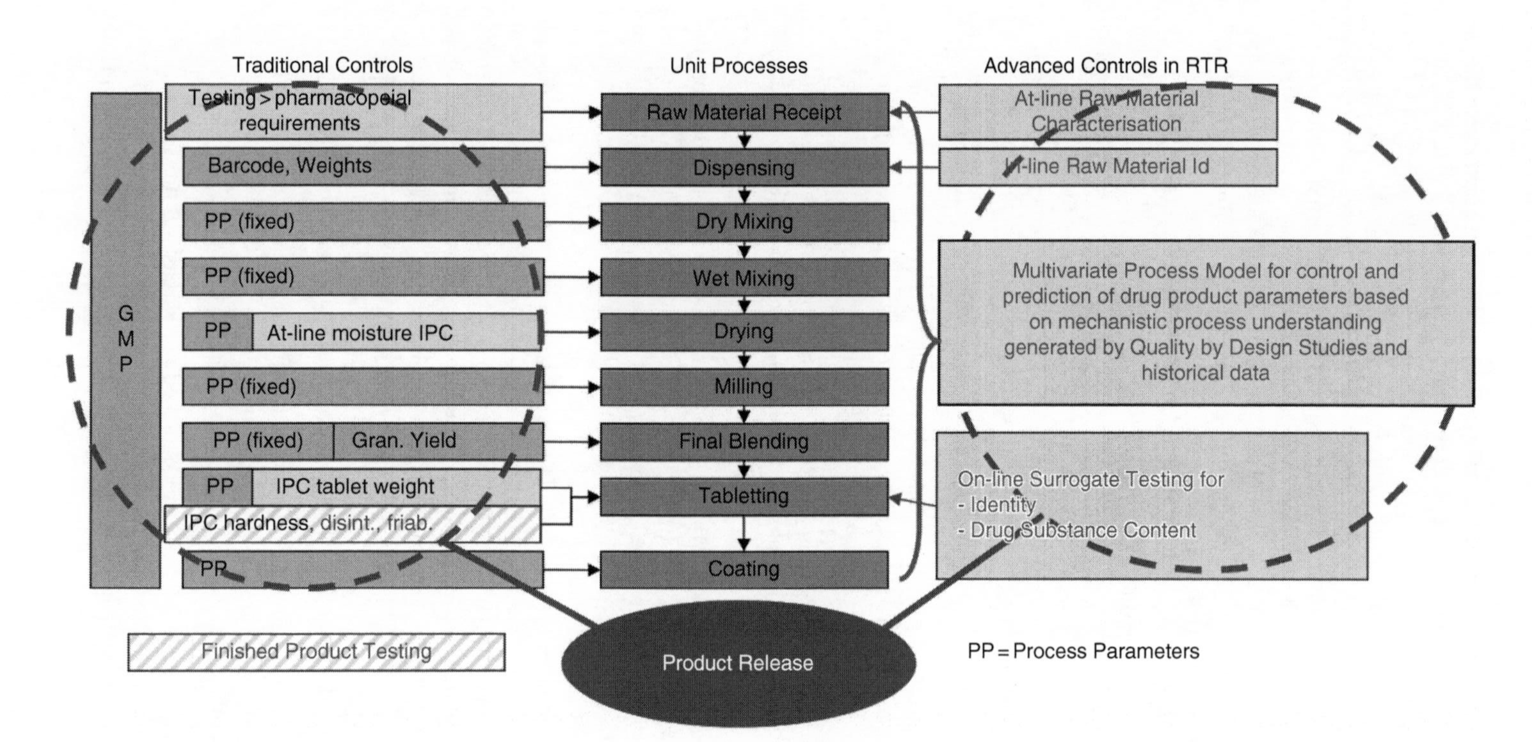

Figure 6.11 *Comparison of traditional controls with advanced controls for the real-time release of an oral solid dosage form.*

Table 6.14 Example of a control strategy for an oral solid dosage form.

CQA	Manufacturing stage	Unit operation	CPP or MA	Control strategy
Content uniformity	Drug Substance	Crystallisation	Particle size distribution (PSD) specification	Focussed beam reflectance measurement (FBRM) PSD determination
		Drying	Cooling rate	Automated process control
	Drug Product	Dispensing	API PSD	Drug substance certificate of analysis (CoA)
		Final blend	API blend uniformity	In-line near infra-red (NIR) analyser

Table 6.14 shows an example of a control strategy for the CQA content uniformity, in the manufacture of an oral solid dosage form, where the MAs and CPPs for both drug substance and drug product have been considered.

6.11 Continuous Improvement

Following the establishment of the process design space and control strategy, the next step for a successful product is process validation at commercial scale to demonstrate that the laboratory/pilot-scale systems used to establish design space will actually model the performance of the commercial-scale process; and that the commercial process will deliver a product of acceptable quality if operated within the design space. As the process design space should assure quality of the drug product, these limits should also provide the basis of the validation criteria. Refer to Chapter 11, and in particular, Section 11.3, for a detailed discussion of the validation requirements and lifecycle approach.

Stage 3 of the process validation lifecycle approach (Continued Process Verification) is the maintenance of the commercial production process in a state of control (the validated state) during routine commercial manufacture until the product is eventually discontinued. It is expected that an ongoing programme of continuous product and process monitoring, data collection and evaluation has been put in place to ensure that the process is performing within the defined acceptable variability declared in the defined space to the regulatory authorities. This should include process capability measurements based on the inherent variability for a pharmaceutical manufacturing process in a state of statistical control. Again, refer to Chapter 11 for the description and calculation formula for process capability. Process capability can be used to measure process improvement efforts that focus on removing sources of inherent variability from the process operation conditions and raw material quality. Ongoing monitoring of process data for Cpk and other measures of statistical process control will also identify when special variations occur that need to be identified and corrective and preventative actions implemented [20].

The trends and knowledge gained to understand variability in API, excipients, and the manufacturing process should be acted upon to improve product and supply for customers and manufacturing efficiencies. There may be opportunities for widening the operating space within the design space, for example, based on manufacturing experience. Another potential improvement opportunity could be to move from end-product testing to real-time release testing. Additional benefits of design space flexibility opportunities to manage minor changes post approval are well documented (ICH Q8 (R2)). Ongoing risk evaluation and data monitoring should trigger continuous improvement opportunities, so it may be possible to make further process improvements within the filed design space without the need for post-approval regulatory submissions. This could include making changes to the equipment type, for example, blenders, the equipment scale or the excipient suppliers and grades.

In conclusion, applying a QbD approach to product development should ultimately result in a more robust, quality product and process, and provide several advantages and opportunities to both industry and the regulatory authorities. It is recommended that QbD is implemented to increase manufacturing efficiencies and help reduce costs owing to fewer product failures and less waste. This is all achieved through building a stronger scientific knowledge base and understanding, and also, by incorporating risk management approaches throughout the product lifecycle. Although QbD can be applied to all product types with both small and large molecules, there are currently still challenges, particularly with complex biopharmaceutical products and processes, where complete characterisation is not practical [26]. The industry and the regulators must continue to work together to find ways forward for these difficult areas.

6.12 Acknowledgements

We would like to thank the following people for their contributions to this chapter with materials and practical examples taken from the De Montfort University, Leicester, distance learning MSc lecture series on Quality by Design:

Bruce Davis, Bruce Davis Consulting
Grisela Ferreira, MedImmune (including Figure 6.5)
Tim Lucas, Pfizer (including Figure 6.11)
David Holt and James Kraunsoe, AstraZeneca (including Figure 6.11)
Mathias Walther, Pfizer
Mikkel Nissum and Cristiana Campa, GlaxoSmithKline

6.13 References

[1] Gibson, M. (2001). Chapter 5, Early drug development: product design, pp. 157–173 in *Pharmaceutical Preformulation and Formulation: A Practical Guide from Candidate Drug Selection to Commercial Dosage Form*, 1st ed., Ed. M. Gibson. IHS Health Group, Englewood, CO, USA.

[2] Gibson, M. (2009). Meeting customer needs, pp. 174–180 in *Pharmaceutical Preformulation and Formulation: A Practical Guide from Candidate Drug Selection to Commercial Dosage Form*, 2nd ed., Ed. M. Gibson, Informa Healthcare, New York, USA.

[3] Juran, J.M. (1986). The quality trilogy: A universal approach to managing for quality. *Quality Progress*, Vol. **19**, No. 8, pp. 19–24.
[4] Juran, J.M. (1988). *Juran on Planning for Quality*, 1st ed., The Free Press, New York, USA.
[5] Juran, J.M. (1992). *Juran on Quality by Design: The New Steps for Planning Quality into Goods and Services*, 1st ed., The Free Press, New York, USA.
[6] DeFeo, J.A. and Juran, J.M. (2010). *Juran's Quality Handbook: The Complete Guide to Performance Excellence 6/e*. McGraw Hill Education, New York City, USA.
[7] FDA (2004). *Pharmaceutical cGMPs for the 21st Century – A risk-based approach, Final Report, Department of Health and Human Services*. US Food and Drug Administration. https://www.fda.gov/downloads/drugs/developmentapprovalprocess/manufacturing/questionsandanswersoncurrentgoodmanufacturingpracticescgmpfordrugs/ucm176374.pdf (accessed 30 August 2017).
[8] ICH Q8 (R2) (2009) *Pharmaceutical Development*, http://www.ich.org/fileadmin/Public_Web_Site/ICH_Products/Guidelines/Quality/Q8_R1/Step4/Q8_R2_Guideline.pdf (accessed 30 August 2017).
[9] ICH Q9 (2005) *Quality Risk Management*, http://www.ich.org/fileadmin/Public_Web_Site/ICH_Products/Guidelines/Quality/Q9/Step4/Q9_Guideline.pdf (accessed 30 August 2017).
[10] ICH Q10 (2008) *Pharmaceutical Quality System*, http://www.ich.org/fileadmin/Public_Web_Site/ICH_Products/Guidelines/Quality/Q10/Step4/Q10_Guideline.pdf (accessed 30 August 2017).
[11] Johnson, A. (2007). Insulin flop costs Pfizer $2.8 billion. *The Wall Street Journal*, October 19.
[12] Berliner, C. and Brimson, J. (Eds.) (1988). *Cost Management for Today's Advanced Manufacturing: The CAM-I Conceptual Design*. Harvard Business School press, Boston.
[13] Conformia CMC-IM Working Group (2008). Pharmaceutical Development Case Study "ACE Tablets", version 2.0, March 13, 2008, http://studylib.net/doc/11826308/%C2%A0-pharmaceutical%C2%A0development-case%C2%A0study-%C2%A0%E2%80%9Cace%C2%A0tablets%E2%80%9D%C2%A0 (accessed 30 August 2017).
[14] FDA (2013). *Quality by design for ANDAs: An example for immediate-release dosage forms*. https://www.fda.gov/downloads/drugs/developmentapprovalprocess/howdrugsaredevelopedandapproved/approvalapplications/abbreviatednewdrugapplicationandagenerics/ucm304305.pdf (accessed 30 August 2017).
[15] FDA (2011). Quality by design for ANDAs: An example for modified-release dosage forms. https://www.fda.gov/downloads/drugs/developmentapprovalprocess/howdrugsaredevelopedandapproved/approvalapplications/abbreviatednewdrugapplicationandagenerics/ucm286595.pdf (accessed 30 August 2017).
[16] A-Mab: A Case Study in Bioprocess Development (2009), CMC Biotech Working Group, version 2.1. http://c.ymcdn.com/sites/www.casss.org/resource/resmgr/imported/A-Mab_Case_Study_Version_2-1.pdf (accessed 30 August 2017).
[17] Gibson, M. (2014). Chapter 6, Chemical drug product development and technology transfer, pp. 161–218 in *Technology and Knowledge Transfer: Keys to Successful Implementation and Management*, Eds. Gibson, M. and Schmitt, S., PDA, Bethesda, MD, USA & DHI Publishing, LLC River Grove, IL, USA.
[18] Davis, B., Deshmukh, R., Lepore, J., Lundsberg-Nielsen, L., Nosal, R., Tyler, S. and Potter, C. (2012). Product Quality Lifecycle Implementation (PQLI) in Pharmaceutical Technology, April.
[19] CMC-Vaccines Working Group (2012). A-Vax: Applying Quality by Design Principles to Vaccines. Version 1, May, 2012 published on the PDA website. www.pda.org
[20] Yu, L.X., Amidon, G., Khan, M.A., *et al.* (2014). Understanding pharmaceutical quality by design review article, *The AAPS Journal*, Vol. **16**, No. 4, July.
[21] FDA (2011). Guidance for Industry, Process Validation: General Principles and Practices. US Food and Drug Administration, January.https://www.fda.gov/downloads/drugs/guidances/ucm070336.pdf (accessed 30 August 2017).
[22] PDA (2013). *Technical Report No*. 60. Process Validation: A Lifecycle Approach. Parenteral Drug Association, Bethesda, MD, USA. https://store.pda.org/tableofcontents/tr6013_toc.pdf (accessed 30 August 2017).

[23] Mitchell, M. (2013). Determining critical-process parameters and quality attributes Part I: Criticality as a continuum. *BioPharm International*, Volume **26**, Issue 12.

[24] Gibson, M., Ashman, C.J. and Nelson, P. (2013). *OpEtimizEation Methods in Encyclopedia of Pharmaceutical Science and Technology*, 4th edition, Ed. James Swarbrick, CRC Press, North Carolina, USA.

[25] ICH Quality Implementation Working Group Points to Consider (R2), ICH Endorsed Guide for ICH Q8/Q9/Q10 Implementation (2011). http://www.ich.org/fileadmin/Public_Web_Site/ICH_Products/Guidelines/Quality/Q8_9_10_QAs/PtC/Quality_IWG_PtCR2_6dec2011.pdf (accessed 30 August 2017).

[26] Rathore, A.S. and Winkle, H. (2009). Quality by Design for biopharmaceuticals. *Nature Biotechnology*, Vol. **27**, pp. 26–34.

7

Design of Experiments

Martin Owen[1] *and Ian Cox*[2]

[1] *Insight by Design Consultancy Ltd, United Kingdom*
[2] *JMP Division, SAS Institute, United Kingdom*

7.1 Introduction

Using actual and simulated data, this chapter takes a pragmatic view of the cost, risk and benefits of collecting process data by actively manipulating process settings according to a pre-specified plan or design. We start by contrasting theory and practice, which introduces the challenge of variation. We illustrate how, using statistically designed experiments (often called DoE for short), we can build up useful information about how a process works. Specifically, we will look at how laboratory data, collected according to certain classical designs, allow us to understand how process variation influences our ability to detect real effects and so control a process effectively. We then step back and look at how statistical models are useful in handling data, and how DoE fits into the more general Quality by Design (QbD) picture.

Next, we use a simulation to compare the utility of a traditional sequential approach to experimentation (using a classical screening design followed by a response surface design) with a more recent design approach (the definitive screening design). Finally, we touch briefly on the use of computer-generated, or custom, designs. In following this path, we will demonstrate that both definitive and custom designs usefully complement and extend the repertoire of DoE practitioners beyond the more well-known classical designs, allowing them to more quickly, easily and uniformly realize all the benefits that DoE is justifiably famous for.

Pharmaceutical Quality by Design: A Practical Approach, First Edition.
Edited by Walkiria S. Schlindwein and Mark Gibson.

7.2 Experimental Design in Action

"In theory, theory and practice are the same. In practice, they are not."

This quote has been attributed to multiple people, including Albert Einstein. Irrespective of who actually made this statement, what is undoubtedly true is that it is certainly applicable in the case of DoE. There are numerous texts on the theory of DoE, but far fewer on how to maximize the chances of a successful practical implementation. The aim of this chapter is to help you look at your products and processes through the eyes of an experienced DoE practitioner. Through our own experiences and the experience of others, acquired by conducting consultancy sessions, running workshops and performing after-action reviews, we have found that there are many pitfalls for the unwary and hence many opportunities to improve your chances of successful practical implementation.

Although we have seen the profound impact that DoE can deliver again and again in a staggeringly wide variety of applications, we also come across scientists who tell us that 'DoE has let us down'. Given that the challenges we are dealing with are often complex and multifaceted, it is not altogether surprising that occasionally we meet scientists who feel DoE has not delivered quite what they had hoped for.

On closer inspection, it is almost always the case that it is not DoE per se that was the problem; it is the way it was applied. So:

- What is it we look for when we approach a new piece of science to investigate?
- What would we talk about if we were discussing your piece of research and why should you care?

The pattern we see is this: DoE is not guaranteed to give the right answer. But it gives richer insights than unstructured approaches based on scientific intuition alone. It shines a spotlight on the underlying process. Sometimes not everything it tells us is what we want to hear. But the outcomes of a process cannot be controlled successfully without understanding the patterns of uncertainty which underlie them.

In this chapter, we will cover the challenges of DoE deployment and measures to prevent some pitfalls of practical implementation. Although the focus will be on the application of DoE, first we have to understand the impact of variation on our ability to control our process and the subsequent risk to quality. Unexplained variation adds complexity. To understand complexity, we need to focus on things that reduce that complexity. And now for something controversial: Although DoE itself is a tool for objective analysis, this does not always mean that scientists themselves are always objective. We need ways of working that help scientists be more objective, and to do that effectively, the combination of scientific curiosity and statistical thinking is our greatest ally.

7.3 The Curse of Variation

There is a very popular television programme called *The Great British BakeOff*. The premise is that each week, contestants face a number of challenges in order to win the accolade of Star Baker. In one of the tasks, the Technical Challenge, contestants are given a succinctly worded recipe to make, for example, a dozen specialty biscuits. An hour later they present their achievements to the panel of judges, who then rank their efforts in terms of quality and consistency. Without fail, at the highest end of the scale there will be a plate of pristine, identically sized biscuits, immaculately presented and cooked to perfection. At the other end of the scale is a sorry mess of ill-shaped biscuits, some raw and some burnt round the edges. All the contestants are experienced bakers. So:

- How come this happens again and again, week after week?

The answer is that despite there being only one recipe, there are many sources of variation present. Some contestants will have experience of making these particular type of biscuits before. On the other hand, some contestants are using types of ingredients they have never previously encountered or are unaware of a particular trick in the procedure that is not specifically mentioned in the recipe. Some of the bakers may be naturally more precise in weighing out and measuring volumes. In addition, instead of working in their own kitchens, they are all cooking with ovens they are unfamiliar with, in a marquee which is more prone to changes in temperature and humidity.

The position each contestant finds themselves in, is similar to that of a scientist investigating a new product or process. It is a great example of where a fishbone or Ishikawa diagram would be useful in elucidating potential sources of variation and determining the factors to control, to experiment with or let vary as noise (e.g. see Chapter 6, Figure 6.4). However, such intellectual exercises only take us so far. At some point, we will actually need to implement an experimental procedure to investigate which factors are indeed most important and ultimately determine which factors help us achieve our goal.

The focus of this chapter is on DoE, but before we get to that we really have to understand a little more about experimental variation. For the past 15 years or so we have been running workshops with participants from universities, from undergraduates through to lecturers and professors. Naturally we cover the theory of DoE, we give case studies and we run simulations of chemical reactions. Although participants acknowledge the existence of variation, it is not really until we task them with a practical exercise that they appreciate just how much this impacts whether they can detect important factors. This is critical, because although DoE is a great tool, it is even more effective once we understand the impact of unwanted variation. So let us start with the basics.

7.3.1 Signal-to-Noise Ratio

Our first workshop exercise could not be simpler. We wanted to measure the time it takes for an effervescent tablet to disintegrate when dropped into a glass of water. We had two different formulations of the tablet; Original and XS. The goal was to find out whether there is a statistically significant difference between the disintegration time for the two formulations. We split the room into eight teams: four teams investigated the Original formulation, the other group studied the XS formulation. Each team performed two experiments (runs). Each run was measured by two different people within the same team. We then collated the results and asked the team to draw conclusions on the basis of their data (Table 7.1).

- How easy is it to see whether there is a difference between the two formulations from the table on the left in Table 7.1?
- Is there anything that looks odd about the data?

Table 7.1 *Typical Microsoft Excel data:table (left) and reformatted for analysis (right).*

<table>
<tr><th>Run Number</th><th>Team</th><th>Tablet</th><th>Time (s)</th></tr>
<tr><td>1</td><td rowspan="2">A</td><td rowspan="8">Original</td><td>51.203</td></tr>
<tr><td>2</td><td>49.78</td></tr>
<tr><td>3</td><td rowspan="2">B</td><td>51 seconds</td></tr>
<tr><td>4</td><td>51 seconds</td></tr>
<tr><td>5</td><td rowspan="2">C</td><td>39.94</td></tr>
<tr><td>6</td><td>39.72</td></tr>
<tr><td>7</td><td rowspan="2">D</td><td>ca. 40</td></tr>
<tr><td>8</td><td>ca. 40</td></tr>
<tr><td>9</td><td rowspan="2">E</td><td rowspan="8">XS</td><td>57</td></tr>
<tr><td>10</td><td>102</td></tr>
<tr><td>11</td><td rowspan="2">F</td><td>55</td></tr>
<tr><td>12</td><td>48</td></tr>
<tr><td>13</td><td rowspan="2">G</td><td>64</td></tr>
<tr><td>14</td><td>61</td></tr>
<tr><td>15</td><td rowspan="2">H</td><td>57.7</td></tr>
<tr><td>16</td><td>56.5</td></tr>
</table>

Run Number	Team	Tablet	Time (s)
1	A	Original	51.203
2	A	Original	49.78
3	B	Original	51
4	B	Original	51
5	C	Original	39.94
6	C	Original	39.72
7	D	Original	40
8	D	Original	40
9	E	XS	57
10	E	XS	102
11	F	XS	55
12	F	XS	48
13	G	XS	64
14	G	XS	61
15	H	XS	57.7
16	H	XS	56.5

One thing that appears worthy of note is that the data is collected to different numbers of significant places, suggesting there may be something different about the measurement systems or data entry between the teams.

Rather than reading a data table, it is often far easier to gain insight by visualizing the data. One of the ubiquitous problems with capturing data in Microsoft Excel is that scientists love to merge cells and put numerical data into the same cell as text data (Table 7.1, left-hand table). To analyse and visualize the data, it needs to be in a different format (Table 7.1), right-hand table).

In the right-hand table, each row contains its own descriptive data (in this case, the name of the Team, the Run Number and Tablet (formulation) that the measured Time relates to). Text and numerical data are in different cells. A really important point we stress to the workshop participants is that if we get data formatting right first time, we can avoid the countless hours of grindingly dull and tedious effort needed to rework badly formatted Excel data.

One useful way to view the data is to use a variability gauge chart. To construct this chart using the data from Table 7.1, we need to cast the selected columns into roles. We have shown this in Figure 7.1 using a data analysis package called JMP [1]. The exact format and terminology may be slightly different if other analysis packages are used, but the principle should be the same.

By visualizing the data in various ways, we can gain different insights. So, what does the Variability Gauge view Figure 7.2 tell us?

First, if you compare the Group Mean between the two formulations, it looks like on average there is a clear difference between the Original and XS formulations. There also seems to be something different about team 5. The circled datapoint from Run 10 looks particularly suspect.

It is worth pointing out that, at the workshop, this was not picked up by the participants from the table itself, and it was only after this graph was displayed that the 'analyst' for

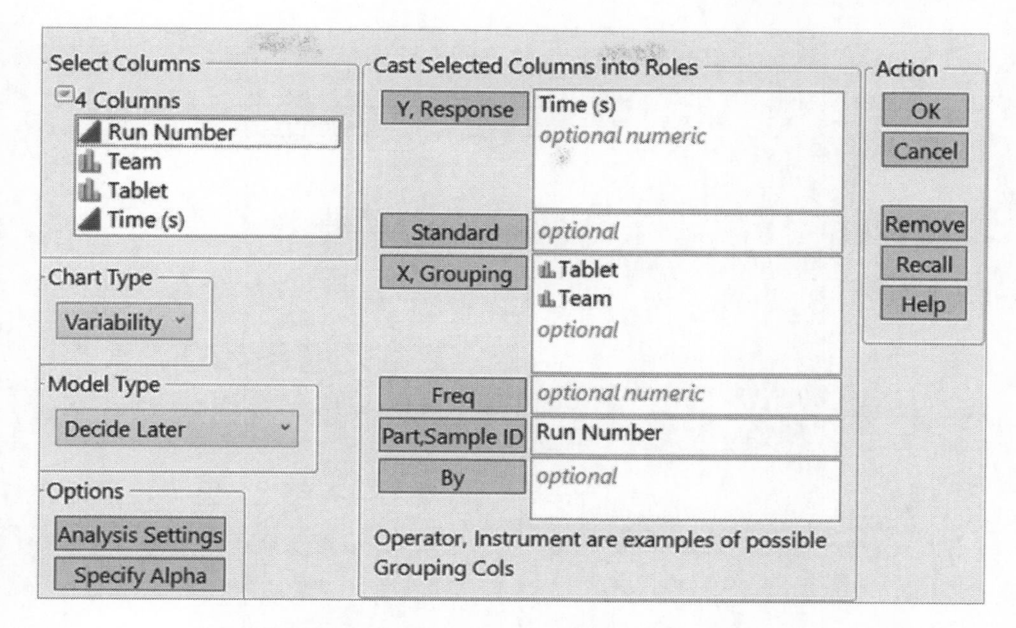

Figure 7.1 *Constructing a variability gauge chart.*

Run 10 realized she had announced the result as 'one zero two', intended to mean one minute and two seconds (that is, 62 s). This mistake is an example of an outlier, a rogue result, caused in this case by incorrect data capture.

In this case, we can provide a rational explanation for the outlier and take preventative steps to stop this kind of error happening again. We can legitimately correct the outlier data and produce the plot in Figure 7.3. Note that participants may also want to remove outlier datapoints for no other reason than they do not appear to fit the hypothesis. This is *not* a legitimate way of dealing with outliers!

The next thing we may want to do is to fit a model to see whether the effect of Tablet (that is, formulation type) is statistically significant.

7.4 Fitting a Model

The model graphically illustrated in Figure 7.4 is produced by calculating the mean of the 'Original Formulation' disintegration time values, which is 45.33 s. This is then compared with the mean of the 'XS Formulation' values, 57.65 s. The two means are connected by a line, and the error bars represent the variation of results for each Tablet type.

A more detailed analysis report and some interpretation is given in Figure 7.5.

There is a lot of information here, wrapped up in statistical jargon, which can appear overwhelming at first sight. Let us translate this into what it means for the scientist.

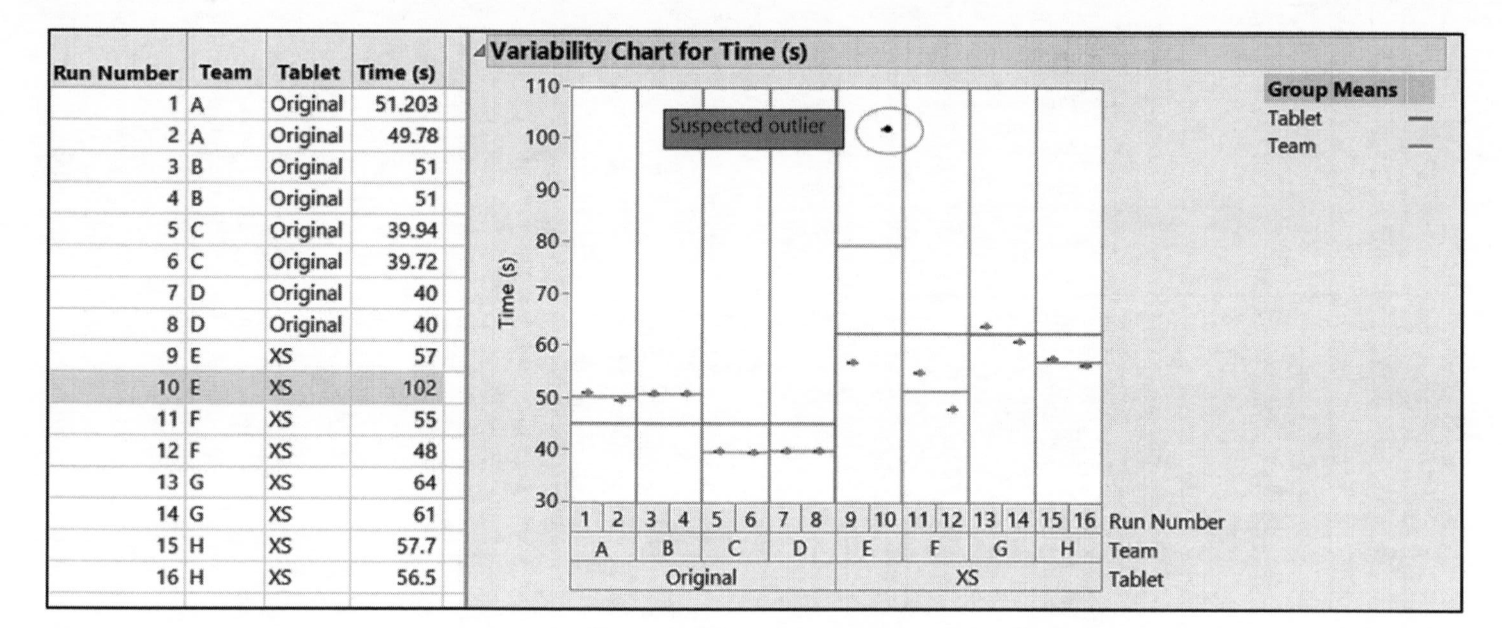

Run Number	Team	Tablet	Time (s)
1	A	Original	51.203
2	A	Original	49.78
3	B	Original	51
4	B	Original	51
5	C	Original	39.94
6	C	Original	39.72
7	D	Original	40
8	D	Original	40
9	E	XS	57
10	E	XS	102
11	F	XS	55
12	F	XS	48
13	G	XS	64
14	G	XS	61
15	H	XS	57.7
16	H	XS	56.5

Figure 7.2 *Impact of formulation of disintegration time with outlier.*

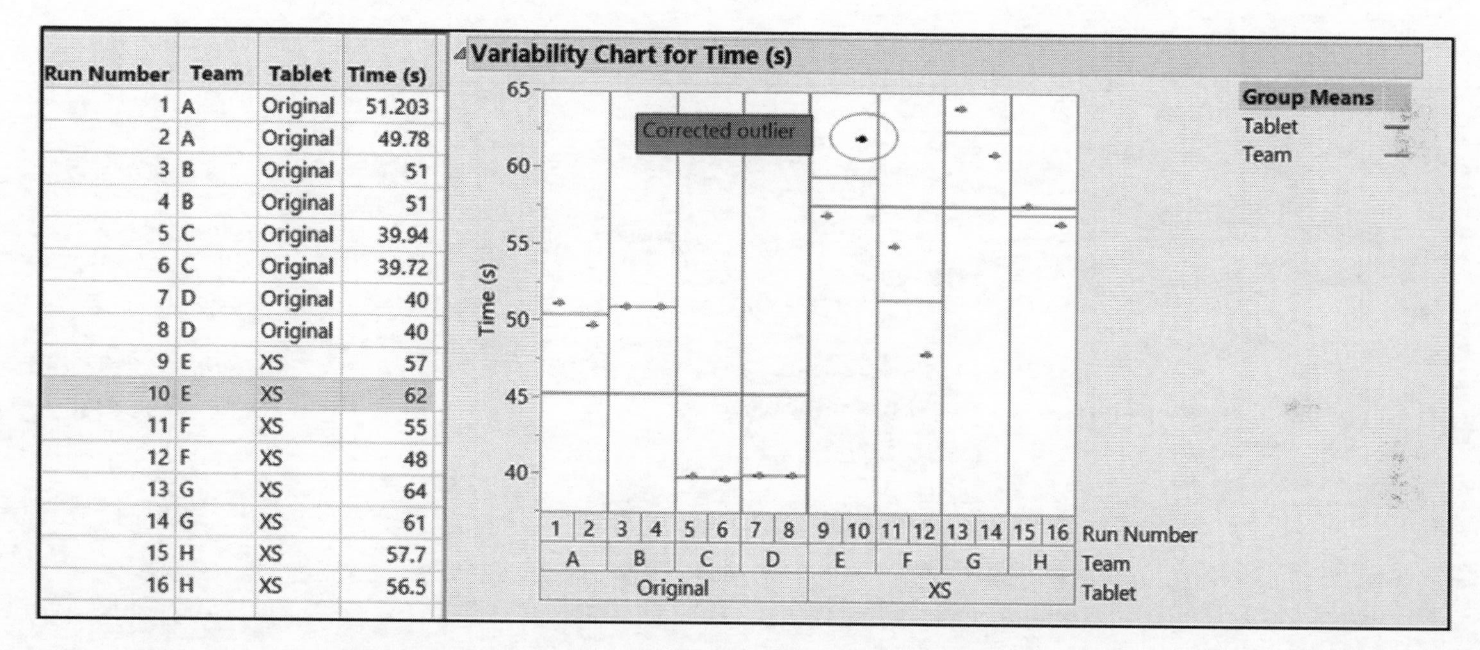

Run Number	Team	Tablet	Time (s)
1	A	Original	51.203
2	A	Original	49.78
3	B	Original	51
4	B	Original	51
5	C	Original	39.94
6	C	Original	39.72
7	D	Original	40
8	D	Original	40
9	E	XS	57
10	E	XS	62
11	F	XS	55
12	F	XS	48
13	G	XS	64
14	G	XS	61
15	H	XS	57.7
16	H	XS	56.5

Figure 7.3 *Impact of formulation on disintegration with the corrected data point.*

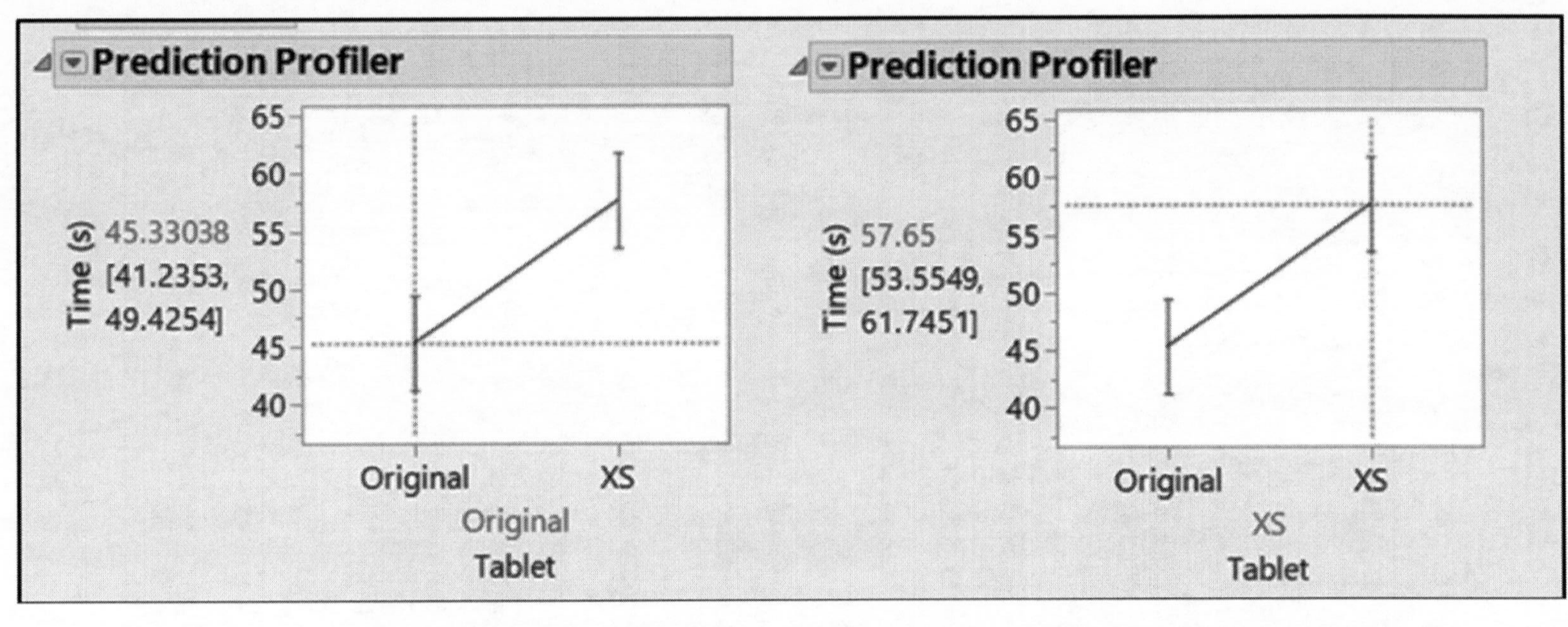

Figure 7.4 *Prediction profilers showing the impact of formulation on disintegration time.*

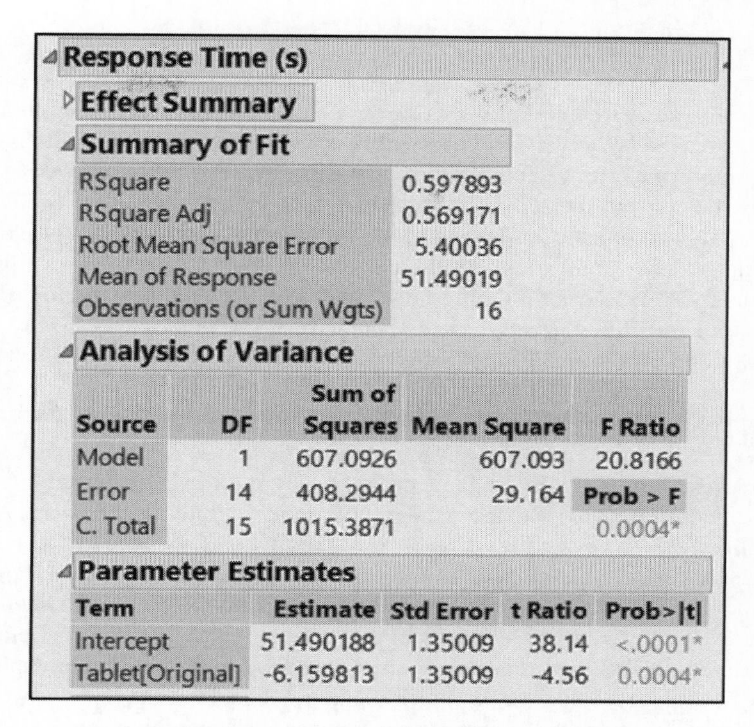

Response Time (s)

Effect Summary

Summary of Fit

RSquare	0.597893
RSquare Adj	0.569171
Root Mean Square Error	5.40036
Mean of Response	51.49019
Observations (or Sum Wgts)	16

Analysis of Variance

Source	DF	Sum of Squares	Mean Square	F Ratio
Model	1	607.0926	607.093	20.8166
Error	14	408.2944	29.164	Prob > F
C. Total	15	1015.3871		0.0004*

Parameter Estimates

Term	Estimate	Std Error	t Ratio	Prob>|t|
Intercept	51.490188	1.35009	38.14	<.0001*
Tablet[Original]	-6.159813	1.35009	-4.56	0.0004*

Figure 7.5 *Statistical analysis of the disintegration time data.*

7.4.1 Summary of Fit

This gives an idea of how good our model is and how useful it is likely to be in predicting future results. *RSquare* estimates the proportion of variation in the response that can be attributed to the model rather than to random error. An *RSquare* closer to 1 indicates a better fit to the data than does an *RSquare* closer to 0. An *RSquare* near 0 indicates that the model is not a much better predictor of the response than is the response mean. Note that RSquare increases as you add terms. *RSquare Adj* accounts for the number of terms in your model and hence facilitates comparisons among models with different numbers of parameters.

The *root mean square error* represents the sample standard deviation of the differences between the observed value and the predicted value – the smaller the better.

7.5 Parameter Estimates

The intercept is the mean of all the readings (51.49 s). The parameter estimate is the impact that Tablet has:

- Original Formulation decreases the disintegration time by –6.16 s (to 45.33 s)
- XS Formulation increases the disintegration time by +6.16 s (to 57.65 s)

7.6 Analysis of Variance

The *Analysis of Variance* report provides the calculations for comparing the fitted model to a simple mean model. *Prob>[t]* is a probability or *p-value*: If this probability is low, then we have a significant effect. In our example, the value for *Prob>[t]* of *<0.0004** indicates that we have demonstrated that the difference in disintegration time caused by changing the formulation from Original to XS is statistically significant and is unlikely to have occurred just by chance. The * indicates that the calculated value is less than a value of 0.05 (conventionally taken as the threshold below which we discount the possibility that chance was the cause of the difference we have determined).

We arrived at this conclusion using a study with 16 data points.

- What if we had carried out the same investigation using two studies of four data points?
- Would it give the same outcome?

We can give a flavour of what might happen if we look at two different subsets of the 16 datapoints. Teams C and G have been randomly assigned to Table A and Teams A and F to Table B Table 7.2.

We can now compare the analyses of the two smaller datasets. Figure 7.6 shows that in Study A, the signal-to-noise ratio much greater than for Study B. In this case, the teams that carried out Study A would be able to claim they have a statistical significant effect as the *Prob>|t| value* is 0.0044, which means there is a greater than 99% probability this is a real effect. The *RSquare Adj* value is 0.98, which suggests a model with high predictive power. It looks like a great result. Instead of using 16 experiments to achieve this result, we only needed 4 experiments. We have cut the resource by a quarter – well done, teams C and G, great job!

In contrast, the teams which carried out Study B do not see a statistically significant effect as the p-value is large, 0.8042. Because we know the outcome of the larger study, we can see they have missed what appears to be a critical factor. Also, the *RSquare Adj* is –0.44, indicating low predictive power. As the teams are oblivious to the results from Study A, they come to a very different conclusion. Ironically if the teams were trying to prove that there is no impact on disintegration time when switching formulation, this is the result they want. From their perspective, they think 'great job, we have a robust process – it doesn't matter whether we use Original or XS, we still get the same result'. Let us implement an expensive manufacturing campaign based on this result. What could possibly go wrong!

Table 7.2 *Data subsets (Study A and Study B).*

Table A				Table B			
Run Number	Team	Tablet	Time (s)	Run Number	Team	Tablet	Time (s)
5	C	Original	39.94	1	A	Original	51.203
6	C	Original	39.72	2	A	Original	49.78
13	G	XS	64	11	F	XS	55
14	G	XS	61	12	F	XS	48

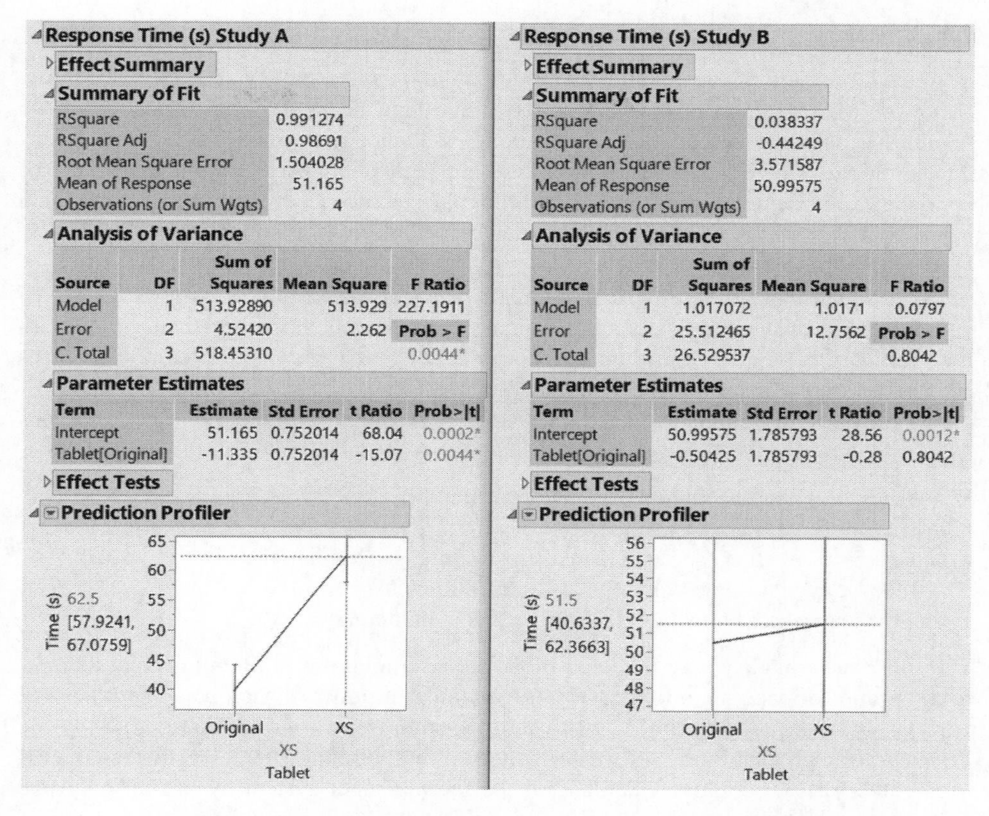

Figure 7.6 *Comparison of the statistical analysis of Study A and Study B.*

This is indeed a disturbing finding. These are real results we observed in the workshop. Within the same room, on the same day, using the same simple experimental and measurement methods, we have both confirmed and rejected our hypothesis!

- How many times in your scientific career do you think you might have incorrectly rejected or accepted a hypothesis in this way?

The key message here is that our ability to detect effects is both impacted by the signal-to-noise ratio and the number of experiments used. We will see this crop up again as we discuss DoE.

So, in summary, to implement a successful DoE, in practice we need to determine:

- What attribute responses are we going to measure (and what are we not going to measure)?
- What factors are we going to study (and what are we not going to study)?
- What ranges of these factors are we going to study? (Note: For continuous factors, the wider the range, the more likely we are to detect a signal).
- How much residual noise is there compared to the effect of the factors (the signal)?
- How can we increase the signal-to-noise ratio?
- How much resource are we going to use?
- If resources are constrained, what is the risk of not knowing?

7.6.1 Reflection

Before we go further, we should look back at our tablet study and ask the question, 'Knowing what we know now, what might we have done differently'. This is a question we relentlessly ask in workshops, while consulting, and when performing any type of experimental investigation or analysis. This is a key habit to develop when building experience of applying experimental design. An ability to humbly recognize where mistakes have been made or where things have been overlooked is crucial to continuous improvement and accelerating the learning journey.

Take another look at Figure 7.7: What might be a possible source of variation?

Let us perform an analysis of variance, considering the tablet type and the team. We can now see that the team variance component is 27.72%, that is, team-to-team differences account for over a quarter of the total variation in the data.

So how does this help us reduce the variation? One approach might be to get the teams together and review their procedures for disintegrating the tablets and measuring the results:

- Are they dropping the tablets from the same height?
- Is the volume of water the same?
- Is the source of water the same (same temperature, same mineral content, and pH?)
- Are they using the same type of measuring equipment?
- Can the teams agree on when the disintegration endpoint occurs?

In this case, an easy way to check the latter is to record a video of a tablet disintegrating, and for each team to measure the disintegration time. That way any observed variation can entirely be attributed to the measurement variation, rather than by the disintegration process itself. Or we could accept that to reduce the team-to-team measurement error, we could standardize on the measurement system by having only one

Analysis of Variance

Source	DF	SS	Mean Square	F Ratio	Prob > F
Tablet	1	607.0926	607.093	9.9786	0.0196*
Team	6	365.0378	60.8396	11.2518	0.0016*
Run Number	8	43.25666	5.40708	.	.
Within	0	0	0		
Total	15	1015.387	67.6925		

Variance Components

Component	Var Component	% of Total	20 40 60 80	Sqrt(Var Comp)
Tablet	68.28163	67.3		8.263
Team	27.71627	27.3		5.265
Run Number	5.40708	5.3		2.325
Within	0.00000	0.0		0.000
Total	101.40498	100.0		10.070

Figure 7.7 Identifying sources of variation.

Table 7.3 An alternative data collection plan.

Run Number	Team	Tablet	Time (s)
1	A	Original	
2	A	XS	
3	B	XS	
4	B	Original	
5	C	Original	
6	C	XS	
7	D	XS	
8	D	Original	
9	E	Original	
10	E	XS	
11	F	XS	
12	F	Original	
13	G	Original	
14	G	XS	
15	H	XS	
16	H	Original	

measurer and one stopwatch. These are all methods of controlling the things which are not under investigation.

An alternative approach is to accept we do have team-to-team variation and treat this as noise. Rather than attempt to control this, we get each team to analyze the two different types of tablets (see Table 7.3). Note that the order in which the type of tablet is studied is also systematically varied. When we executed this design at a follow-up workshop, the team-to-team differences dropped from 27% to less than 3%.

7.7 'To Boldly Go' – An Introduction to Managing Resource Constraints using DoE

So far we looked at studying one factor using 16 runs to do so.

- What if we have more factors?
- How can we explore where we can operate our processes with constrained resources?

Imagine you are in a rowing boat, and have been given the task to survey a lake by taking depth soundings using a rope marked every metre and with a heavy weight on the end. You row to a spot, throw the weight over the side and then count the number of marks as you pull the weight back up to find out how deep the lake is at this point. But this takes time and energy, and so, given the task at hand, two questions soon arise:

- How many depth soundings can you take?
- Where should you take them?

You have some prior information (the lake has a depth of zero all round its edge), but answering these questions needs a bit more thought because it will depend on what you want to use the final survey for. For example, if you are interested in fishing, you might want to locate deep parts of the lake which you can cast to, so knowing the depth at the centre, which you cannot reach from the shore, might not be of much use.

In this example, we have an 'investigation space' or 'opportunity space' that we need to probe to get useful information: The places we decide to row to constitute a 'design', and the opportunity space is two-dimensional because we need a longitude and latitude to specify each point in that design. When we measure the lake depth at a design point, we get a 'response'. For those that know the topic already, this is, of course, a gentle introduction to DoE: Classical DoE can provide an efficient way to probe an opportunity space to gather useful new knowledge, and there have been some recent important advances in design. Here the word 'recent' is in relation to the history of DoE, which is generally considered to have emerged in 1926 with the publication of 'The Arrangement of Field Experiments' by Sir Ronald Fisher [2].

There can be no doubt of the practical value of DoE: Whenever any product is in development, testing or manufacturing, DoE finds application to assure that patients, customers, consumers or users get what they want, and do not get what they do not want (and at the lowest cost to you). Industry journals are full of DoE examples (*The Journal of Quality Technology* from American Society for Quality Control, *International Journal of Experimental Design and Process Optimisation* from InterScience Publishers), and initiatives such as Six Sigma and more recently QbD promote usage. All leading companies outside of the pharmaceutical sector whose markets demand sustained innovation use DoE at all stages of the product lifecycle.

7.8 The Motivation for DoE

But there are some problems. When we run experimental design workshops in the academic community, many participants have not come across DoE before. They have had the principle of 'only study one factor at a time' drilled into them throughout their scientific education. Many of them feel they have done great science already without having to resort to such techniques. If DoE really is such a good thing, why was it not it routinely taught in mainstream chemistry undergraduate degrees? Motivating scientists to come to what they may anticipate will be a dull lecture on the theory of statistics is a really hard challenge.

When we teach these modelling tools, we use simulations to help scientists and engineers better understand the regularities in their systems in a 'safe' environment. Simulation builds experience of dealing with complexity and uncertainty. The GlaxoSmithKline (GSK) Reaction Simulator (Figure 7.8) has the advantage of helping workshop participants – both from industry and academia – explore the consequences of their decisions without burning laboratory research time and failing large-scale batches.

The chemistry chosen for the GSK reaction simulator is relatively simple [3], with just four process parameters and three critical quality characteristics – yield, impurities, and unreacted starting material. The chemistry is straightforward, and the *qualitative* effect of changing each of the process parameters is very much what an experienced chemist might expect. The simulated effects are simply main effects or two factor interactions. The noise

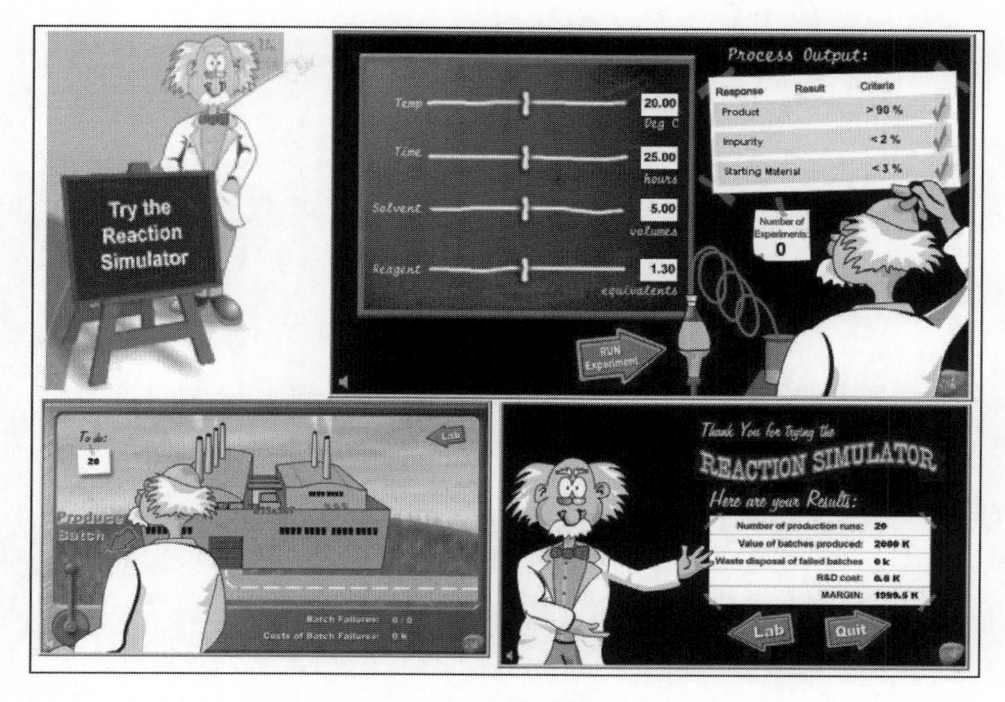

Figure 7.8 *The GlaxoSmithKline reaction simulator.*

variability is similar to that typically observed. And the goal is relatively straightforward – to define a process where the yield is maximized while minimizing the amount of unreacted starting material and formation of the impurity.

And yet, using this particular simulation over the past 14 years in over 20 universities and over 15 GSK development and manufacturing sites, we have observed how difficult it is for scientists and engineers to successfully tackle this problem using their intuition and one-factor-at-a-time methods alone.

7.8.1 How Does the Workshop Exercise Work?

Participants work individually or in small teams. Using their knowledge of the chemistry, they perform experiments until they arrive independently at different sets of conditions which pass all the criteria. Some of the participants arrive at a 'solution' in a handful of experiments; some take as many as 50to 100. So, the resources to solve the problem using a non-structured approach is very variable.

We then ask the workshop participants: 'What is the main factor that controls the process?' If there are eight teams in the class then, typically, at least one team nominates each one of the less important factors as 'most important'. So clearly, as the teams have come to such different conclusions, the process understanding within the room is suspect.

Having produced a set of conditions that 'work', participants are then asked to transfer their processes into the plant. Occasionally, one of the teams may stumble upon a solution close to the optimum and upon transfer to production, the process runs smoothly and all 20batches

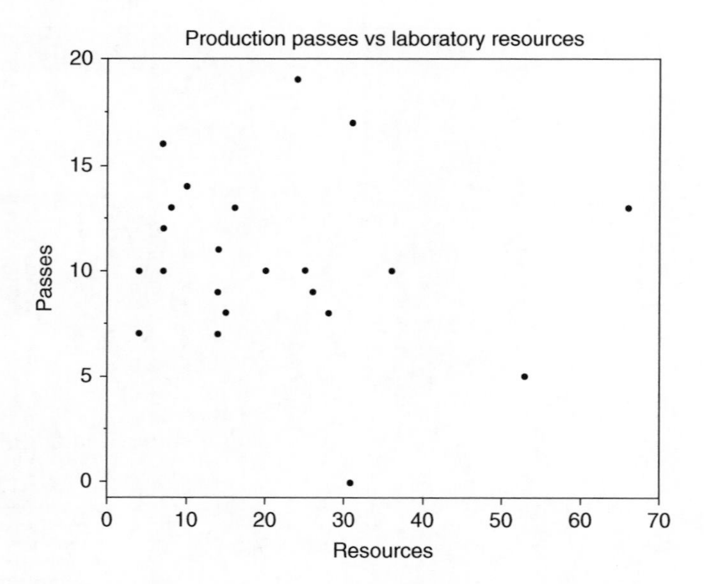

Figure 7.9 *'Understanding and Optimizing Chemical Processes' 2015 Workshop results.*

pass. However, this is a rare event. Usually, disaster strikes, and there is a wide variation in the success rates of teams, spanning the entire range from 0 to 20 failing batches (Figure 7.9).

On average, across all teams, approximately half the production batches do not meet the process specifications. Perversely, it is often the teams who have used the most resources who fail most batches in production!

- So why do good scientists using their preconceptions, logic and rationale, generally do such a poor job in meeting this challenge?

The answer lies in the fact that the complexity of dealing with a noisy system of four factors and three responses makes it difficult for even the most experienced of scientists to stumble across a combination of settings that will routinely pass all the criteria. Workshop participants tell us the GSK reaction simulator smashes, once and for all, the delusion that it is possible to solve such problems using a mixture of intuition and a traditional one-factor-at-a-time approach. So:

- Is there a better way of solving this problem? Fortunately, the answer is yes!

7.8.2 DoE Saves the Day!

At this point in the workshop, we step back and describe the theory of fractional designs. We do this to demonstrate that if we perform a 10 run half-fractional factorial, we can obtain a useful model. Participants simulate their own results, and because of the noise in the simulation, the participants generate slightly different models. Despite these small discrepancies, they all now broadly agree on which factors and interactions are most important. When they use their models to obtain settings, they are now able to routinely pass all the desired specifications.

7.9 Classical Designs

There are many excellent textbooks about the construction and analysis of classical designs, including the probably familiar full factorial, fractional factorial, Plackett–Burman, Taguchi and response surface designs [4–6]. If these are not familiar, we encourage you to the look at the references provided, or the user manual of your favourite DoE software.

7.9.1 How Do Resource Constraints Impact the Design Choice?

Here we are going to illustrate the different costs, risks and benefits associated with three types of classical designs: fractional factorial, full factorial and response surface designs. For simplicity, in Figure 7.10 we will schematically represent these for three factors only.

The resource requirements are indicated in Table 7.4. Fractional factorial designs use a fraction (half, quarter, eighth, etc.) of the resources of full factorials. The combination of amount of resource and the number of factors investigated determines how well the design can resolve main effects and interactions between the main effects (Resolution III, IV and V designs).

7.9.2 Resource Implications in Practice

Although theory, case studies and simulations are useful teaching tools, there is no substitute for actually implementing a practical session where participants experience implementing a design to get a more realistic appreciation of variation. Over the years, we have incorporated 'laboratory of life' exercises into our workshops where participants investigate very familiar procedures. We have run experiments, collected and analyzed data and generated models for making toast, boiling eggs and (most popularly) pouring beer.

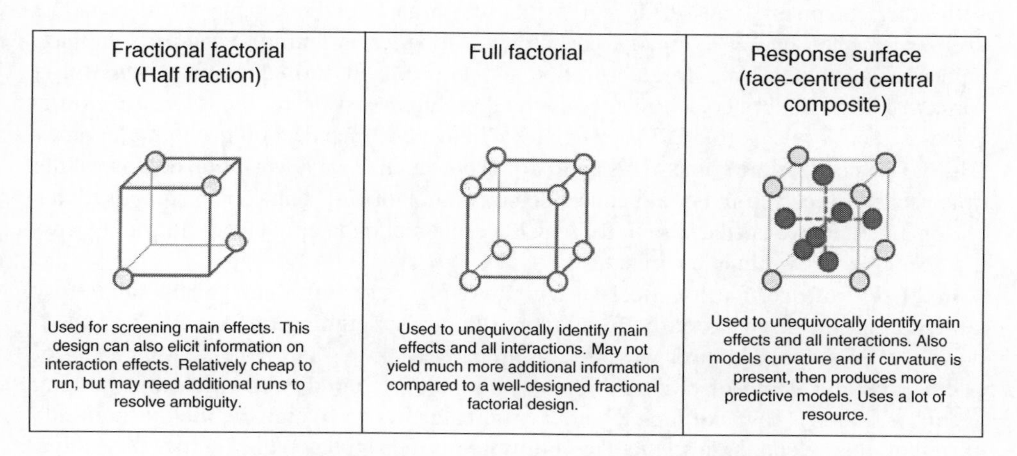

Figure 7.10 *Costs, risks and benefits of classical designs.*

Table 7.4 *The relationship between factors and experiments.*

Experiments \ Factors	2	3	4	5	6
4	Full	Half Res IV			
8	Central Composite	Full	Half Res IV	Quarter Res III	Eighth Res III
14		Central Composite			
16			Full	Half Res IV	Quarter Res III
24			Central Composite		
32				Full	Half Res V
42				Central Composite	
64					Full
76					Central Composite

7.10 Practical Workshop Design

We wanted to show how we could investigate tableting using DoE. In pharmaceutical drug development, tablets are produced by blending an active pharmaceutical ingredient (API) with other excipients (non-active ingredients which are 'bulking agents or fillers'). The excipients, when chosen correctly, can promote long-term stability, enhance solubility, reduce viscosity, facilitate drug absorption and so on). New formulations are tested in a variety of ways, and one such method is the dissolution test. Here, the response investigated is the % release of API measured by high performance liquid chromatography (HPLC) at specific time points. This approach is not suitable for a workshop because of the time needed for each run. For a practical demonstration of DoE that can be completed in a reasonable time, we need a system for which we can manipulate several factors and measure responses all within a few minutes.

In 2015, Professor Patrick Steel, of Durham University, introduced a new laboratory tableting optimization exercise. To avoid participants weighing out an API, he substituted a yellow dye. The task was then to prepare a 100 mg tablet containing 2 mg of yellow dye and 98mg of other excipients (non-active ingredients which are 'bulking agents or fillers'). To make such tablets, participants have to weigh out the dye and each excipient, then blend them so that the yellow dye is distributed evenly across the excipient mixture. They then take 100mg aliquots of this powder, put them into a tableting press and produce the final tablets. For each DoE run, the percentages of some of the excipients were varied, and processing parameters such as the load and speed settings of the tablet press were manipulated according to the chosen design.

Prior to the workshop, Patrick had a team of expert chemists spend some time familiarizing themselves with the process and measuring devices. After careful consideration, this team chose to study four continuous factors: speed, load, %starch and %lactose. They determined the release of the dye by measuring the %absorbance (a measurement that can be taken in a few seconds) at a given time point. The greater the absorbance, the more dye has been dissolved in the water. They also measured the tablet hardness by squeezing it between two jaws until the tablet broke, at which point the hardness was read against a sliding scale. After performing a few scoping experiments to see how they could standardize the measurement and production procedures, they performed a 30 run central composite design. The goal was to make tablets with a hardness value of greater than 50 and a %absorbance greater than 0.5 at 4 min.

7.10.1 Choice of Factors and Measurements

It is important to note there is no definitive model. DoE is partly about empirical modelling, partly about articulating basic scientific principles and patterns which operate behind procedures such as tableting. A model which tried to describe all aspects of tableting would be no simpler than reality itself and would add nothing to our understanding. Sometimes the things models leave out are more important than those they include. If applied correctly, DoE is a profoundly important tool which can explain past results and, in some cases, predict the future.

7.10.2 Data Collection and Choice of Design

The experiments were performed and the data collected.

Analyzing the Four Different Designs

Table 7.5 was deconstructed into four different sets of data based on the Block value in the last column:

- Half fraction: Block A
- Alternate half fraction: Block B
- Full fraction: Blocks A + B
- Central composite design: Blocks A, B + C

Each Design (half fraction, alternate half fraction, full fraction, central composite) was then individually analyzed. Initially, we will look at just the analysis for the half fraction. Later we will compare the prediction profiles for each of the four designs.

7.10.3 Some Simple Data Visualization

A common objection to DoE from advocates of one-factor-at-a-time of thinking is that a lot of the experiments we perform in DoEs fail, so we have wasted resources. They could certainly point out that looking at the Table 7.5 we only pass the criteria in 6 out of 30 runs. The next thing they might do is shrug their shoulders and 'pick the winner', i.e. just choose a setting where both the criteria are met (e.g. run 3) and go with that. Sadly, one-factor-at-a-time advocates who do this are missing the point, because a formal statistical analysis will reveal much richer information.

Table 7.5 *The tableting results data table.*

	Run number	Speed	Load	% Starch	% Lactose	Hardness	Absorbance @ 4min	Target Pass/ Fail	Block
1	1	40	350	15	60	42	0.946	Fail	A
2	2	50	425	9	40	92	0.896	Fail	A
3	3	60	350	15	20	118	0.802	**Pass**	A
4	4	50	425	9	40	92	0.876	Fail	A
5	5	60	500	15	60	59	0.869	Fail	A
6	6	40	500	3	60	69	0.808	Fail	A
7	7	60	350	3	60	41	0.847	Fail	A
8	8	40	350	3	20	127	0.695	**Pass**	A
9	9	60	500	3	20	165	0.246	Fail	A
10	10	40	500	15	20	137	0.395	Fail	A
11	11	60	350	15	60	54	0.882	Fail	B
12	12	50	425	9	40	95	0.893	Fail	B
13	13	40	350	15	20	101	0.896	**Pass**	B
14	14	50	425	9	40	94	0.87	Fail	B
15	15	40	500	15	60	67	0.893	Fail	B
16	16	60	500	3	60	74	0.816	Fail	B
17	17	40	350	3	60	48	0.83	Fail	B
18	18	60	350	3	20	131	0.693	**Pass**	B
19	19	40	500	3	20	162	0.273	Fail	B
20	20	60	500	15	20	133	0.36	Fail	B
21	21	50	425	9	40	97	0.923	Fail	C
22	22	50	425	9	40	100	0.903	**Pass**	C
23	23	60	425	9	40	95	0.924	Fail	C
24	24	40	425	9	40	81	0.93	Fail	C
25	25	50	500	9	40	106	0.856	**Pass**	C
26	26	50	350	9	40	85	0.9	Fail	C
27	27	50	425	15	40	90	0.865	Fail	C
28	28	50	425	3	40	97	0.705	Fail	C
29	29	50	425	9	60	52	0.858	Fail	C
30	30	50	425	9	20	143	0.449	Fail	C

The 30 run design used by our group of expert scientists was built up by starting with a half fraction (eight experiments with two centre points), augmenting that to give a full factorial (an additional eight experiments with two centre points) and finally adding eight star-points and two more centre points. In practice we could run all 30 experiments on one day or in three separate blocks on different days. The nice thing about this approach is that we can analyze each block of data and see what information we get from the half fraction and then what additional information is added at each stage.

Note: Experimental design purists could argue that in the discussion that follows we could have taken blocking into account and/or used a mixture design [7]. We have elected not to discuss these subtleties to keep the focus on the practical aspects and the consequences of design choices.

Simple data visualization (Figure 7.11) can give greater insight than just looking at the data table alone. By plotting hardness on the Y axis and absorbance on the X axis, we can see six data points fall into the acceptable zone (top-left quadrant) where both specification criteria are met. Of these, three are right on the edge of failure (runs 13, 22 and 25); leaving three preferred batches (runs 3, 8 and 18). Now, if we create an interactive distribution plot

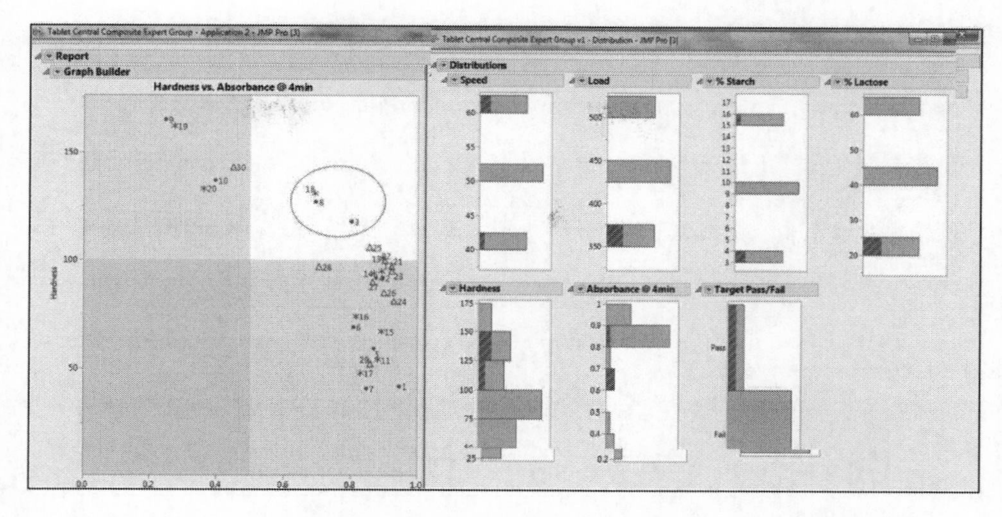

Figure 7.11 *Data visualization. (See insert for color representation of the figure.)*

and select these three batches, we see that they all have low load and low %lactose settings. Note we have not actually built a model at this stage, we have just visualized the data. What we do not know is (a) whether these results are due to chance alone and (b) how well we are likely to be able to predict the future. So let us look at what additional information we get when we build an empirical model.

7.10.4 Analysis of the Half Fraction

Once again the details of the statistical analysis may appear overwhelming at first sight, but the principles are similar to those discussed previously. We now have two models, one for hardness and one for the absorbance (Figure 7.12).

The sorted parameter estimates order the parameters according to statistical significance. For hardness, there are three main effects (% lactose, load and % starch) and a joint or interaction effect [load * % lactose] that are statistically significant. For absorbance, only two main effects, % lactose and load, are statistically significant.

7.10.5 How to Interpret Prediction Profiles

The prediction profile (see Figure 7.12 and Figure 7.13) is a matrix in which each cell shows the relationship between a factor (speed, load, % starch and % lactose) and a response (hardness and absorbance). The steeper the slope, the more important the effect is, because by altering the setting of the factor, the response can be increased or decreased. If the slope is horizontal, the response will be the same independent of the factor setting. The shaded boxes indicate that the effects that are statistically significant. When we are optimizing the process, we want to identify the important factors that will be critical to controlling the process. When we want to show that the process is in control, we want to show that the factors have little impact within the ranges defined in the control strategy.

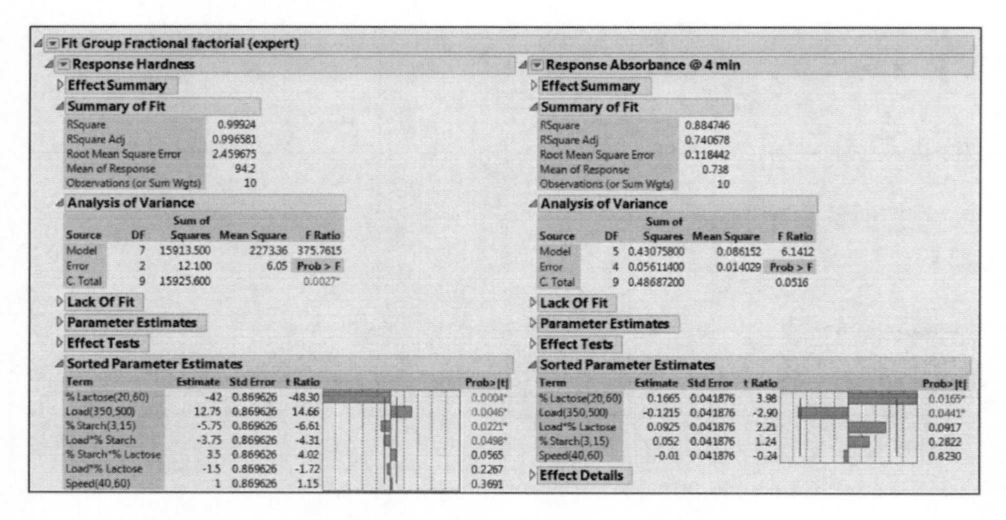

Figure 7.12 *Analysis of the half fraction.*

The dotted lines either side of the slope are the confidence limits which give a measure of variation. The narrower the confidence intervals, the better the model. If the confidence intervals are wide, then this indicates there is a lot of uncertainty associated with a prediction.

Let us use the profiler output to walk through the outcomes of the different designs one by one.

7.10.6 Half Fraction and Alternate Half Fraction

Both of these data sets are half fractions containing 10 runs which, when combined together, produce the full factorial design. How the design is setup in the first place determines which runs fall into which fractions. This is often a source of concern for teams, who may want to know which fraction to run. In reality, in the absence of outliers, the fractions often give very similar results, as is shown here.

Where the signal-to-noise ratio is high, as in the hardness response, the models for the two fractions are nearly identical. For the absorbance response, the signal-to-noise ratio is lower, so % starch was not statistically significant in either fraction. The centre-point predictions for hardness (94.2 and 95.9) and absorbance (0.738 and 0.741) are almost identical for each fraction.

As always, it is useful to check that the model makes scientific sense. For example, as we increase the load, we would expect the tablet should get harder. The harder the tablet, the slower the dissolution. This is indeed what we observe.

7.10.7 Interaction Effects

Figure 7.14 highlights the interaction effects, which are characterized by diverging lines. Where no interactions are present, then the lines are parallel. For the absorbance response, the % lactose factor is set high at 60, and the impact of the load is negligible (as it is an

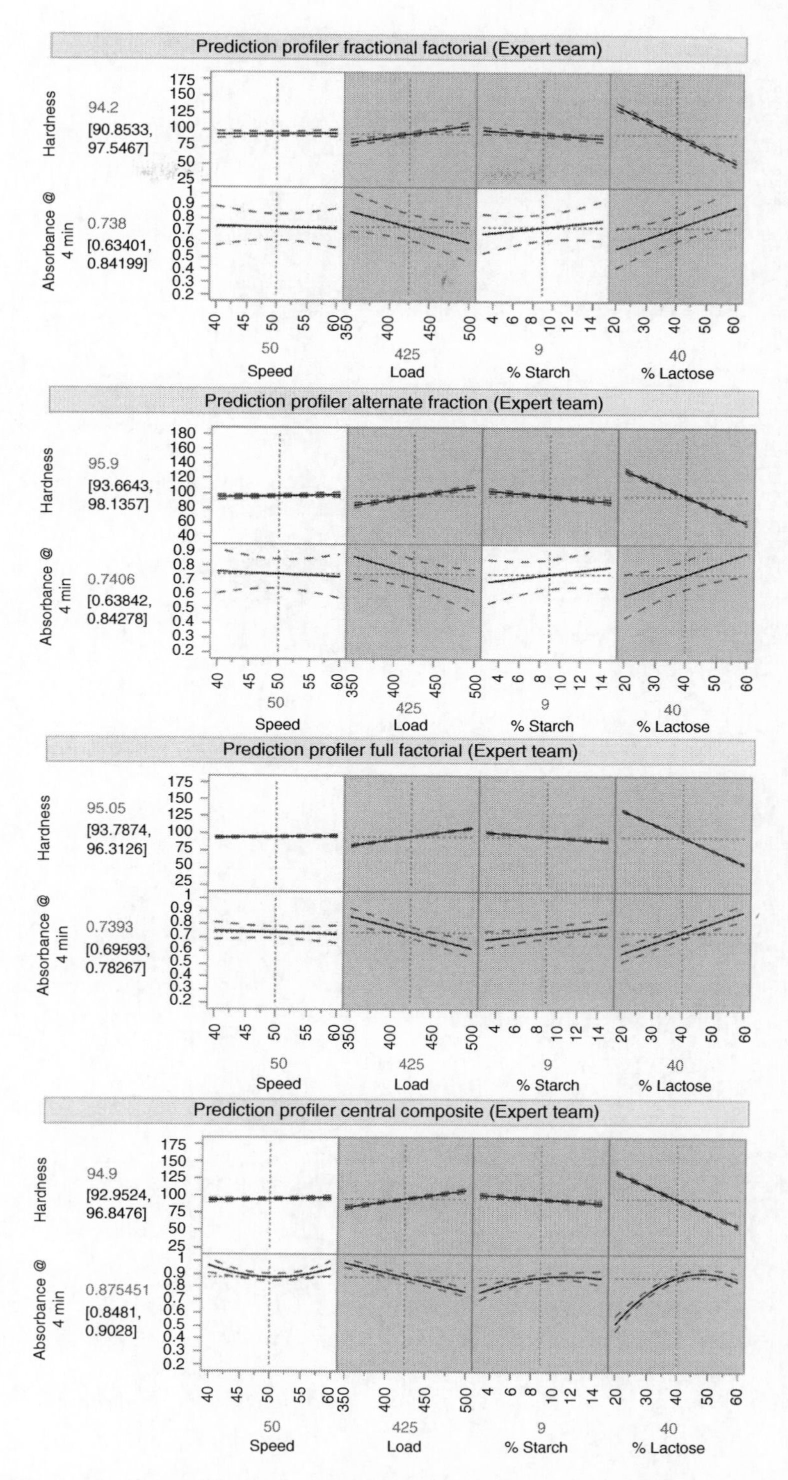

Figure 7.13 *Prediction profiles for the four different designs (shaded boxes indicate statistically significant effects).*

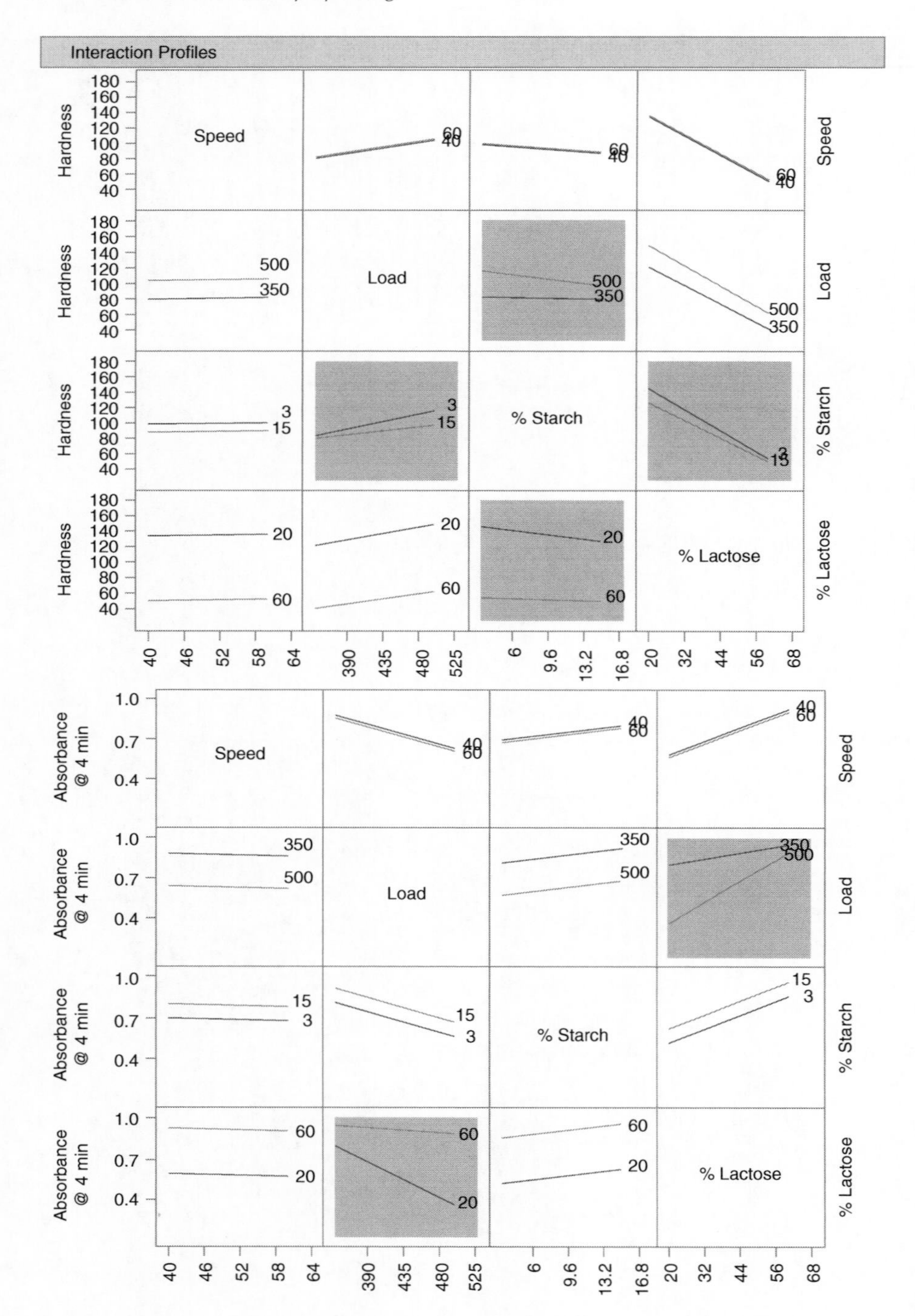

Figure 7.14 Interaction effects.

almost horizontal line). If the setting for % lactose is at 20, then the line is inclined, meaning that the setting for the load then becomes important. This is referred to as an interaction effect. It is often through exploiting interaction effects that problems with conflicting responses can be solved.

With full factorials, all interactions can be unambiguously identified. With fractionated designs, there is ambiguity, and the degree of ambiguity is known as the resolution of the design:

- Resolution V: Main effects are aliased with uncommon four factor interactions. Two factor interactions are aliased with three factor interactions.
- Resolution IV: Two factor interactions are aliased with other interaction effects. Main effects are aliased with three factor interactions.
- Resolution III: With more heavily fractionated designs, main effects may be aliased with main effects.

In this case, the [load * %lactose] interaction is aliased with the [speed * %starch] interaction. As the load and %lactose main effects are significant, then it is more probable that the first interaction is real. In other cases, it can be harder to resolve the ambiguity without performing more experiments.

7.10.8 Full Factorial

When the two half fractions are combined, the confidence intervals are narrower. Now % starch has a statistically significant effect on absorbance.

In the fractional factorial absorbance response, the [load * % lactose] interaction is aliased with the [speed * % starch] interaction. We had proposed that if load and %lactose main effects are significant, then the first interaction is more likely. By performing more runs in the full factorial, the interaction effects are now fully resolved (i.e. not confounded). We can see that our interaction designation was correct.

7.10.9 Central Composite Design

Finally, when all blocks of data are combined, we have a central composite design, and we can now model the curvature which appears to be present in the absorbance response. Note that with more experiments the confidence intervals become even narrower.

Note that although the hardness prediction of 94.9 is similar to the value predicted by the half fraction and full factorial designs, the absorbance centre-point prediction is now about 0.87, compared with predictions of about 0.74, obtained from the fractional and full designs.

7.10.10 How Robust Is This DoE to Unexplained Variation?

Now, what if a less experienced team perform the same study? As in The Great British BakeOff challenge referred to previously, in the workshop setting the participants are stepping into the unknown (Figure 7.15 They are working on a tableting process they have never performed before, using unfamiliar processing and measuring equipment and new materials. They are in newly formed teams working under acute time pressure with very sparse instructions. All factors which are likely to increase background variation. Now what happens when we apply DoE?

Figure 7.15 *Explaining the experimental procedure to the 'novice' workshop delegate.*

George Box, [8] one of the great industrial statisticians of the 20th century, is justly famous for the quote 'All models are wrong, but some are useful'. Let us assume for the purposes of this discussion that the expert team have produced a useful model which will enable us to explain the past and predict the future. They have found they can control the tableting process by adjusting the settings of load, % starch and % lactose. They understand how to alter each of the input factors to make one response better without making any of the other responses worse (to the extent it fails to meet the specification). Even if models are not perfect, they can convey tremendously complex information. This team should be rewarded because they are offering something that is both scarce and highly valued; high-quality process understanding.

The results for Novice Teams A and B expose several important issues (Figure 7.16).

Let us deal with Team A first. Their model for hardness has a poor signal-to-noise ratio relative to the expert group, and as a consequence only % lactose is found to be statistically significant. For the absorbance response, no significant factors have been identified. This suggests that Team A can only control hardness with one factor, % lactose. The effects of the remaining factors are swamped by other things that control quality, and they have missed things that may be important such as the blend time, weighing errors, measurement equipment calibration, data transcription errors, and the stability of the API.

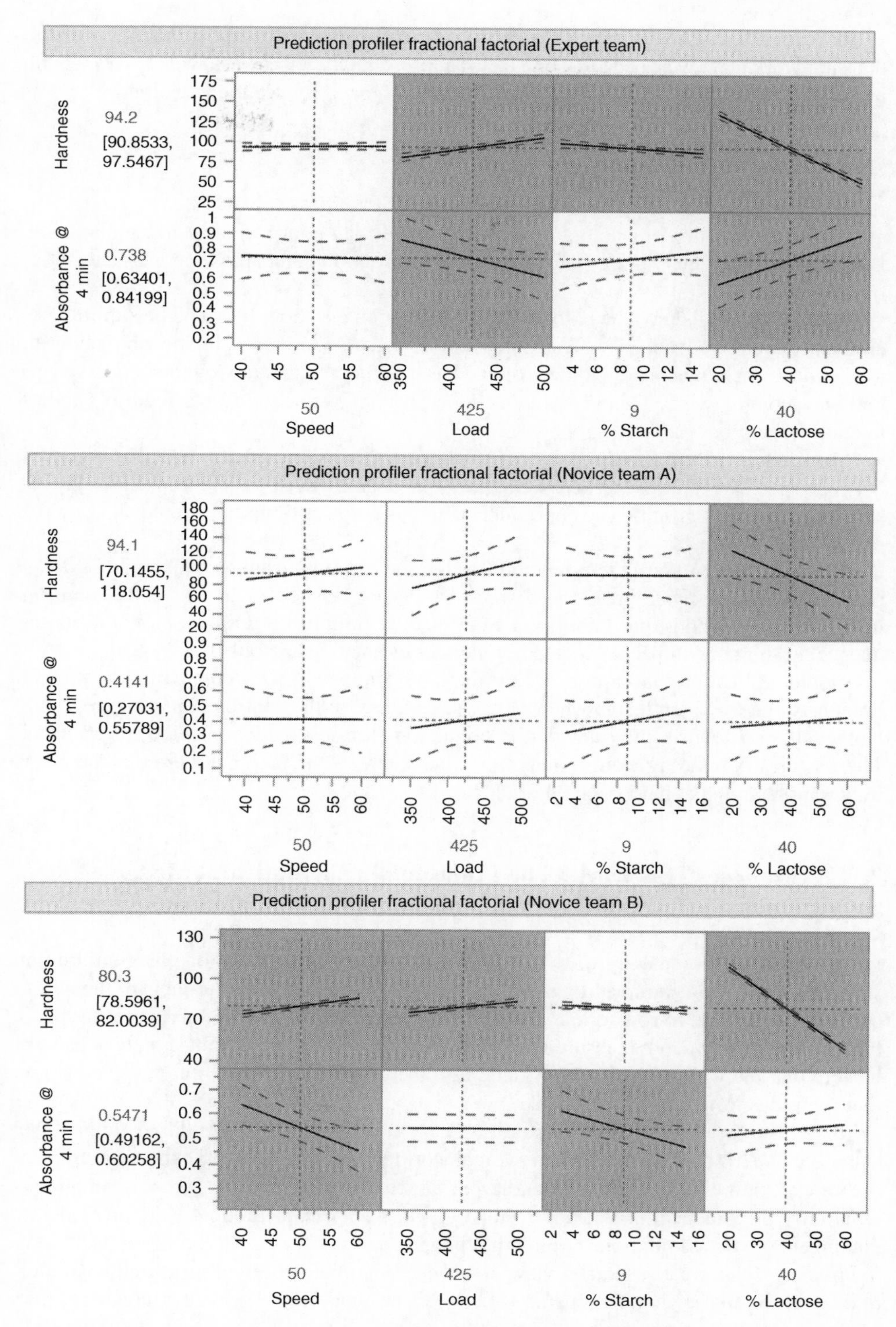

Figure 7.16 *Comparison of 'expert' and 'novice' findings for the half fraction design.*

So Team A is without a credible control strategy. They can complain about how DoE does not work and revert back to a one factor approach. Or they can ask what is wrong with my world? If I improve my experimental procedures, can I reduce the variation by

- Calibrating the balances and measuring equipment?
- Ensuring good data stewardship?
- Extending the blending time?

Good technical implementation is more powerful than poor technical implementation. DoE is brutally efficient at exposing the effect of variation compared to a one-factor-at-a-time approach.

A good hypothesis must be based on a good research question. It should be simple, specific and stated in advance [9]. Team A have not been able to show a relationship between load and hardness where we believe (on the basis of the bigger study by the expert group) that one actually exists. This is problematic because it adversely impacts Team A's ability to control the process.

Conversely, Team B has done the opposite. They have shown that a relationship exists between speed and hardness where we believe one does not exist (on the basis of the bigger study by the expert group). The consequence here is not as serious as far as 'quality on the patient' is concerned.

However, it unnecessarily imposes additional resourcing implications. This could be doing more work to investigate press speed when we do not need to, or increased costs raising deviation reports about speed not being controlled when in fact it does not have any consequence. In short, the impact affects the economics of manufacture rather than quality.

Another indication that more work needs to be done is given by comparing the predictions between the three teams when all the settings are at the centre points (Figure 7.16). Novice Team B consistently gets lower predictions for absorbance than the expert team. There is clearly a strong team-to-team bias, indicative of the fact that there may be other important factors that have been ignored.

7.11 How Does This Work? The Underpinning of Statistical Models for Variation

Let us return to the analogy of mapping the lake. The practicalities of doing so are not straightforward. For example, if we chose a spot at which to measure the depth of the water, how can we be sure to row to that spot exactly, particularly if we want to later return and take more than one depth measurement there? Even if we get to the right spot, how can we be sure that the weight has actually made it to the bottom, and is not stuck on something that protrudes from the bed of the lake?

To make progress in situations like this, we have to abstract the essential features of the situation, and in so doing move from the real world, with all of its complexity, to a simpler, theoretical view that is easier to handle but which inevitably makes some assumptions. Obviously the practical use of any results we get from our theoretical view or model is contingent on the assumptions roughly holding.

Figure 7.17 shows a schematic view of a model involving different measured quantities of interest. We make a logical separation between the quantities or *variables* into things that

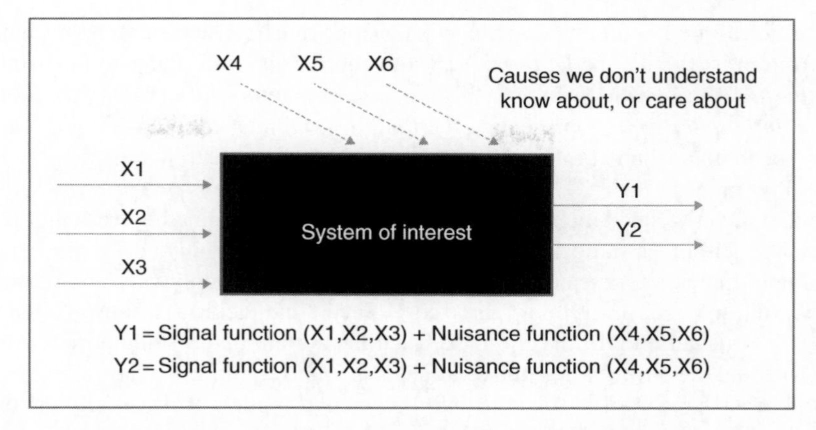

Figure 7.17 *Theoretical model of a specific real-world situation.*

we can change at will (inputs, or Xs for short) and variables that we think will change as a result (outputs, or Ys for short). For illustration, in Figure 7.17 we have threeXs and twoYs.

However, by the way it is constructed, our model is always incomplete, in the sense that there will be other variables that we do not know (or perhaps care) about that also, in reality, affect our Ys. Such variables are usually called *nuisance variables* or *noise variables* (again, for illustration we suppose there are three of these, shown in Figure 7.17 as dotted arrows).

Note that, even if the system under study is *deterministic* in the sense that the values of X1 to X6 uniquely fix the values of Y1 and Y2, because the variables X4 to X6 are unobserved and unmeasured, the values of Y1 and Y2 will appear to exhibit *variation* that cannot be accounted for by the changes we make to the inputs. The defining characteristic of a *statistical model* is that the noise variables have an important rather than a minor influence on the outputs, and such models have a particular utility for the study of technological and biological systems in which we rarely have the chance to fully understand all of the details.

Figure 7.17 also shows how the different types of variables are related via a 'signal function' and a 'nuisance function': Given that X4 to X6 are unobserved, we can simplify this to

$$\mathrm{Y1} = \mathrm{f}\left(\mathrm{X1}, \mathrm{X2}, \mathrm{X3}\right) + \varepsilon \tag{7.1}$$

$$\mathrm{Y2} = \mathrm{g}\left(\mathrm{X1}, \mathrm{X2}, \mathrm{X3}\right) + \varepsilon \tag{7.2}$$

where the notation 'f()'means 'depends on', and ε is the variation due to the noise variables. In general, the functions f() and g() might be expected to be complicated, and a further assumption that is made in DoE is that they are *polynomials* of fairly low order [8]. For example, we might assume the model

$$\mathrm{Y1} = \mathrm{a} * \mathrm{X1} + \mathrm{b} * \mathrm{X2} + \mathrm{c} * \mathrm{X3} + \mathrm{d} * \mathrm{X1} * \mathrm{X2} + \varepsilon \tag{7.3}$$

$$\mathrm{Y1} = \mathrm{e} * \mathrm{X1} + \mathrm{f} * \mathrm{X2} + \mathrm{g} * \mathrm{X3} + \mathrm{h} * \mathrm{X1} * \mathrm{X1} + \varepsilon \tag{7.4}$$

where 'a' to 'h' are fixed numbers or *coefficients*. Each expression on the right-hand side (apart from ε) is called a model *term*: X1 (with coefficient 'a') is called a l*inear term* (or main effect), X1*X1 (with coefficient 'h') is a *quadratic term* and X1*X2 (with coefficient 'd') is an *interaction term*, representing the joint or combined effect of X1 and X2 on Y1. In many technological and biological applications, the joint effect of variables can be particularly important.

Note that the associated interaction terms can only be determined by varying the variables involved within the same design. This can seem counter-intuitive if you have had a lot of training in fundamental science, in which the tendency is to only vary one variable at a time and (try to) keep everything else fixed. Such 'one-factor-at-a-time' (often called OFAT) experimentation [10] makes sense when the system under study is relatively well understood, and the effect of any noise variables is relatively minor.

On the other hand, DoE is used to build *empirical* models that usefully capture the relationship between Xs and Ys. In so doing, they may alert you to the fact that you have overlooked some important Xs, signpost the way to additional experiments that will further your understanding and ultimately give some insights into the deterministic mechanisms at work.

Given that many chemical, pharmacological and biological phenomena are non-linear, the restriction to linear functions only may seem limiting, but how well such a model can capture the true relationship between Xs and Ys is related to the range over which the Xs are varied: We can approximate the shape of any reasonable function using a Taylor series expansion with an appropriate number of terms [11], and we might hope the number of terms will be small for relatively small ranges.

No matter what approach we use for mapping the lake, the chemistry reaction simulation, or the tableting process (that is, how we determine a useful design), ultimately we will be faced with a set of data built up from the measured values of the Xs and Ys under study e.g. Table 7.5). This data can be arranged in a table where the measured values of a variable appear in a column, and there is a row for each run in the design.

Note that, *when we have defined the polynomial model we are interested in* (such as the one in equation (2)), we may be in a position to estimate the values of the associated coefficients from the data we have measured and collected (in this case, the values of 'a' to 'h'). Once we have these coefficients, we can predict values for Ys given values of Xs, at least within the uncertainty implied by the inevitable noise term ε in the model. The process of estimating the coefficient of each model term from a set of data is well understood by statisticians [12] – it relies essentially on picking coefficient values that make the chosen model fit the measured data as closely as possible [13].

The difference between what the model predicts for a Y variable and the value that was actually recorded is known as a *residual*, and this is a manifestation of the noise term ε. There are statistical procedures for distinguishing the *significant* (and hopefully important) terms (the 'signal') in relation to the noise ε, but these procedures also rely on the assumption that the residuals contain nothing other than random variation. Checking that the residuals do indeed have this property is an important part of assessing whether the model being considered is likely to be useful.

One way to estimate ε is to repeatedly take measurements at the same design point (repeatedly drop the plumb line at the same point in the lake or make several tablets). Although appealing because it removes the impact of the Xs on the Ys, such *replicates* only give us information about the variation at the chosen design point. Related to the assumption mentioned immediately above is the idea that *size* of the random variation in the residuals does not depend on the point chosen, and if we only take replicates at a single point we have no way to check this.

Note also that there is a clear dependence between the design we choose and the models that we can estimate from data collected using that design. To take an obvious example, if we pick a design in which, say, X1 only occurs at two values or *levels* as we look across the runs of the design, we cannot hope to estimate the coefficient of the quadratic term X1*X1, since (whatever the value of the output) we can draw an infinite number of quadratic curves between the two levels and we have no way to tell them apart. Saying it differently, if we think that the dependence of Y on X1 is not simply linear, we need to be sure that we vary X1 at three or more levels in our design.

For those who already know the topic, this section has done nothing more than remind you of the rudiments of fitting statistical models to data using linear regression [14]. Note that, in this framework (and as equations (1) and (2) imply), we fit each of the Ys separately. Generally, most systems of interest will have more than one Y, and (following Derringer and Suich [15]) common DoE practice is to use a combined or composite response (the harmonic mean of the Ys) along with the desirability function associated with each Y. Alternative methods of handling multiple Ys, such as partial least squares [16], are also available.

7.12 DoE and Cycles of Learning

As has been well publicized, the FDA believe we absolutely need 'a maximally efficient, agile, flexible pharmaceutical manufacturing sector that reliably produces high-quality drug products without extensive regulatory oversight' [17]. This view is shared by other regulatory bodies. Fostering and promoting QbD, with its shift from a conformance-based to an evidence-based approach, is intended as an important way to engender the desired state.

In any situation and in any setting, acquiring new process understanding and making improvements continually is hard – but, whatever point of view you care to take, whichever quality guru you follow or whatever improvement paradigm you work within, it is fair to say that you cannot realistically expect to make improvements without understanding, and also that, somewhere along the line, data from real-world measurements have a crucial role to play.

Leveraging data to acquire new process understanding has many facets, including but not limited to, conceptual, technical, cultural and organizational. As we will see, these different aspects are also somewhat interdependent. To deal with the organizational aspect first, there is often a division of labour between analysts on the one hand and scientists, engineers and technicians on the other. Unfortunately, this division is at odds with the obvious point that all data is contextual, so that the chances of discovering relevant and important features in the data are much reduced if the person interacting with it does not 'know' the data intimately.

This brings us to a manifestation of a cultural influence (or rather, the lack of it), namely, the relative failure of the statistical community to adequately communicate the relevance and power of what is usually called 'statistical thinking' [18] for real-world applications.

Generally, scientists, although numerate, can have less appreciation of the importance of variation since the systems they have studied have a restricted domain within which it is not crucial (so the influence of the noise term ε is relatively minor). Saying it differently, they are sometimes at risk of having precise answers, but to questions that may not be of practical importance in an operational setting. This raises a further issue, namely, that unlike in science per se, the acquisition of new process understanding is *not* necessarily intrinsically good – it is only good to the extent that it can be used to make valuable operational or procedural improvements.

The judgement of what is valuable or not is exactly that – a judgement in relation to a shared set of values and commitments that express how we want to measure and run our business [19].

Figure 7.18 (adapted from Box, Hunter and Hunter [4]) gives a schematic view of cycles of learning. The key point is that learning is, or should be, iterative, accumulating useful knowledge that narrows the gap between what we think is happening and what is really happening. In turn, this raises issues of knowledge management [20], particularly important in the QbD setting when time to market is usually measured in years or decades and organizations are medium- to large-sized and multi-faceted. We do not consider these aspects further except to highlight the fact that useful statistical models provide a very concise way to encapsulate and codify understanding. However, as well as the model itself, it is important to record the process by which it was built, and Figure 7.18 can help here.

Building on the point that learning is iterative, then (assuming that there are good reasons to know more), the use of DoE is also sequential. Historically, the use of the classical designs mentioned earlier has encouraged or even required the use of two distinct steps, usually known as 'screening' followed by 'optimization'.

Classical screening designs are constructed to weed out Xs that actually are not influential [21], and designs aimed at optimization then give more details about the relationships

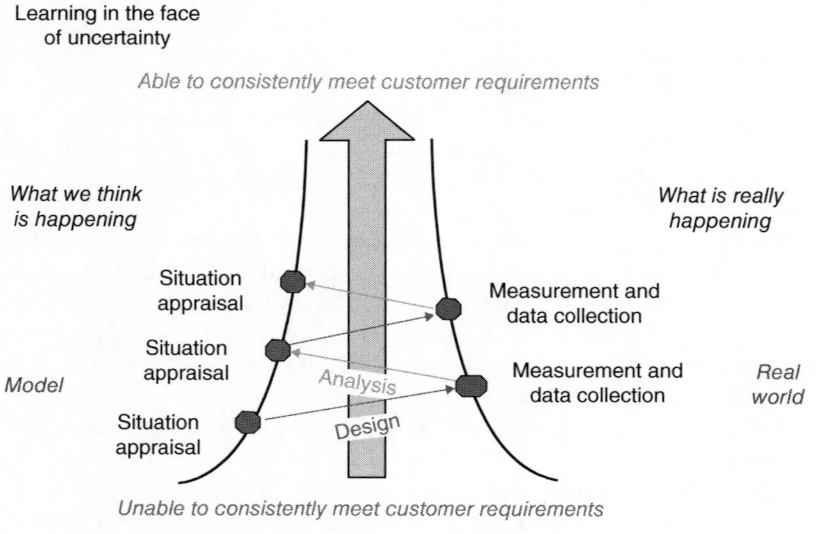

Figure 7.18 *Cycles of learning iterate between the real world and our model of it.*

involving the Xs that remain. As we shall see, the new definitive screening designs offer the possibility to collapse screening and optimization into a single step. Although this has definite advantages in some situations, it does not dilute the general point that learning should be iterative.

7.13 Sequential Classical Designs and Definitive Screening Designs

So far we have looked at classical design options for just four factors. What if we find ourselves in a position where we have even more factors? As the number of factors increases, the number of experiments starts to escalate, typified by Figure 7.19.

Let us suppose our tablet example has six factors, and we only have enough resources to run thirty experiments. We think we have curvature present.

- What might our strategy be now?

With six factors, a full factorial design would require 2^6 = 64 runs. But that would not tell us anything about curvature. If we want a central composite design, then we have 2^6 = 64 factorial points, 2*6 =12 axial points and 4 centre points; a total of 80 runs. That is over twice our budget. The standard approach is to perform a fractional factorial approach to screen for the most important factors. Assuming (as most practitioners would) that not all the factors are active, and faced with the fact that 64 runs is almost certainly too expensive, many practitioners would use a fractional factorial design to isolate the active factors.

Then they would perform a central composite design on only the most important factors. If our first screening design is an eighth fraction factorial Resolution III design, we require eight runs to identify the main effects. Suppose we find that only three main effects are important. We then do a central composite design, which is another 16–20

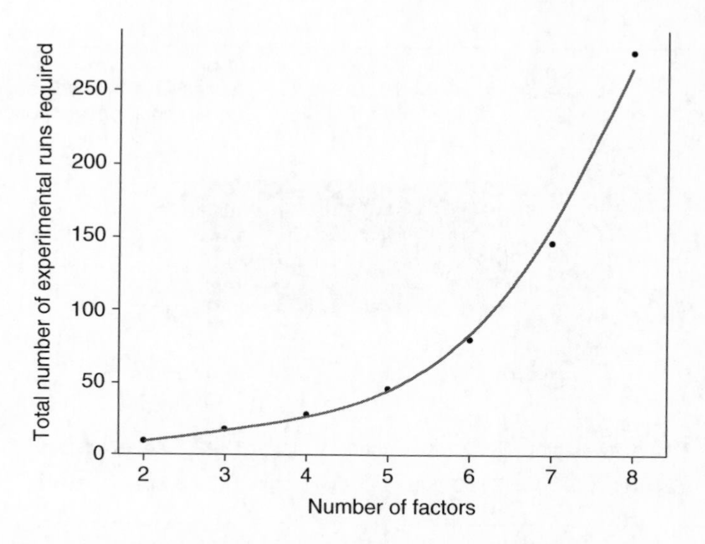

Figure 7.19 *Resource requirements for central composite design.*

experiments on the three active factors; making a total of 26–30 experiments (the range is dependent on the number of centre points chosen).

Definitive screening designs (DSDs) are a new type of design that is just appearing in software. In this regard, [1] leads the way. DSD provide the opportunity to combine screening and optimization into a single step. With six factors, this DSD would only require 13 runs! Sounds too good to be true? It would be really interesting to compare a DSD with the central composite using actual data. However, we have seen that any differences we observe in actual results may be due to other factors we are not aware of. In any case, we are not in a position to run another design for the tableting investigation.

When we handled the measured data, we deduced the underlying model, a process generally called *induction* [22]. However, to see how well different approaches are likely to perform, we can reverse this process, using a given theoretical model to generate or simulate data that we then analyze. We can then compare the results from the two simulated approaches with the theoretical model to compare how well we did.

7.14 Building a Simulation

Let us begin by building a simulation of this scenario (Figure 7.20):

We can arbitrarily create two models, which are described as

$$Y1 = 2*X1 + 3*X2 + 1*X3 + 2*X1*X1 + 1*X1*X2 + \varepsilon(0,0) \quad (7.5)$$

$$Y1 = 3*X1 + 2*X2 + 1*X4 + 1*X2*X2 + 2*X2*X4 + \varepsilon(0,0) \quad (7.6)$$

For those readers less comfortable with reading equations, here is a quick translation of the equation and what to look out for.

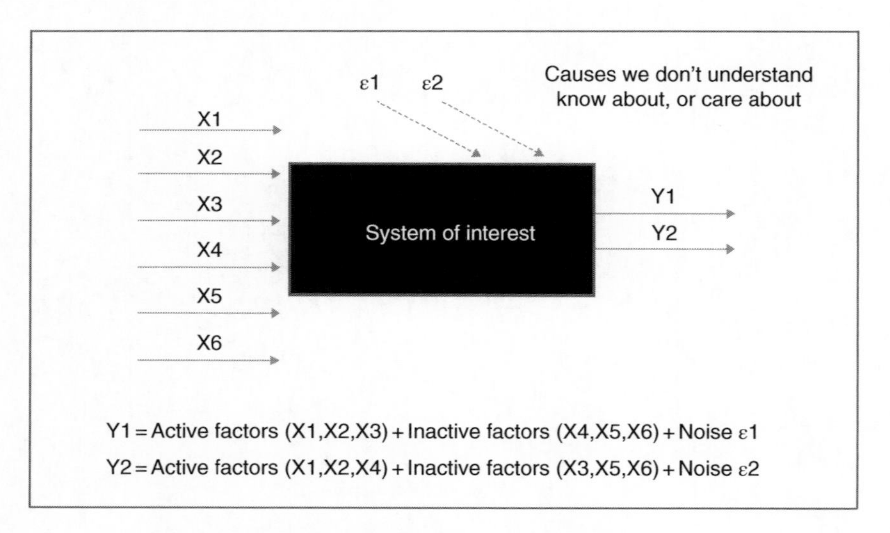

Figure 7.20 *Simulation scenario schematic.*

We have six factors and two responses:

- X1 and X2 impacts both Y1 and Y2.
- X3 impacts on Y1 only, and X4 impacts on Y2 only.
- X5 and X6 impact on neither Y1 nor Y2 and so are not represented in the model, because they are inactive for both responses.

Some of the factors appear in main effects, in interaction terms and in curvature terms:

- X1 represents a linear term.
- X1*X1represents a quadratic term (measures curvature).
- X1*X2 represents an *interaction term*, representing the joint or combined effect of X1 and X2 on Y1.
- The X1*X2 interaction term impacts on Y1, while X2*X4 impacts on Y2.
- The quadratic term X1*X1 impacts on Y1, and X2*X2 impacts on Y2.

We can see from our equations some terms are more important than others:

- The numbers in front of the linear, quadratic or interaction terms are fixed numbers or coefficients. The larger the number, the bigger the effect
- For example, X2 has three times the effect of X3 on Y1.

To simplify things, suppose also that $\varepsilon = 0$ for both responses. Although apparently a little extreme, this approximation just makes the results easier to interpret. We always have the option to introduce sensible levels of noise once we have seen what happens with no noise.

7.14.1 Sequential design, Part 1: Screening Design (10 Runs)

Many choices are available for our first design. We have six factors, and at this stage we do not know which the active and non-active factors are. The tactic here is to use a highly fractionated design to identify the main effects before proceeding to a central composite design.

We will perform an eight run Resolution III design with two centre points (Table 7.6). The response values for each run are provided by equations 7.5 and 7.6.

The results in Figure 7.21 and Figure 7.22 show that

- X1, X2 have a statistically significant impact on Y1.
- X1, X2, X4 and the interaction X2*X4 have a statistically significant impact on Y2.
- Neither X5 nor X6 impact Y1 and Y2.

We now know we can drop X5 and X6 in the next study. But what about X3? We have set the criterion for statistical significance conventionally at 0.05. Note that X3 and the X1*X2 both narrowly miss this target with values of 0.0545. So we will have to make a judgement call on whether to take X3 into the next study, based on whether we think X3 could still have a practical importance in controlling Y1.

7.14.2 Sequential Design, Part II: Optimization Design (26 Runs)

Armed with this knowledge, we can proceed to the second optimization step. Although other choices are possible, we use a central composite design (sometimes called a response

Table 7.6 The first experiment in the sequential approach – a 10 run fractional factorial Resolution III design.

	Pattern	X1	X2	X3	X4	X5	X6	Y1	Y2
1	−+−+−+	-1	1	-1	1	-1	1	1	3
2	000000	0	0	0	0	0	0	0	0
3	−++−+−	-1	1	1	-1	1	-1	3	-3
4	+−+−−+	1	-1	1	-1	-1	1	1	3
5	+−−++−	1	-1	-1	1	1	-1	-1	1
6	++++++	1	1	1	1	1	1	9	9
7	++−−−−	1	1	-1	-1	-1	-1	7	3
8	−−++−−	-1	-1	1	1	-1	-1	-1	-5
9	000000	0	0	0	0	0	0	0	0
10	−−−−++	-1	-1	-1	-1	1	1	-3	-3

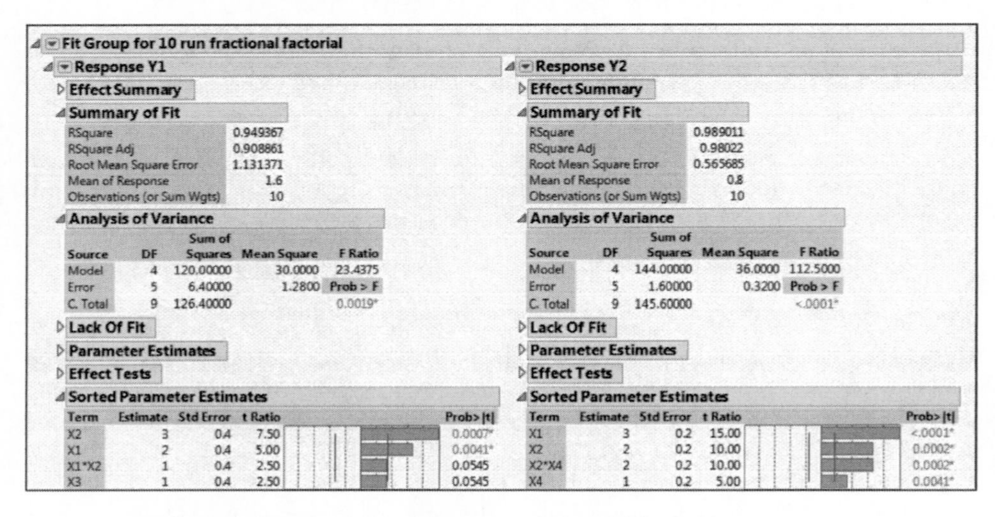

Figure 7.21 Analysis of the 10 run fractional factorial experiment.

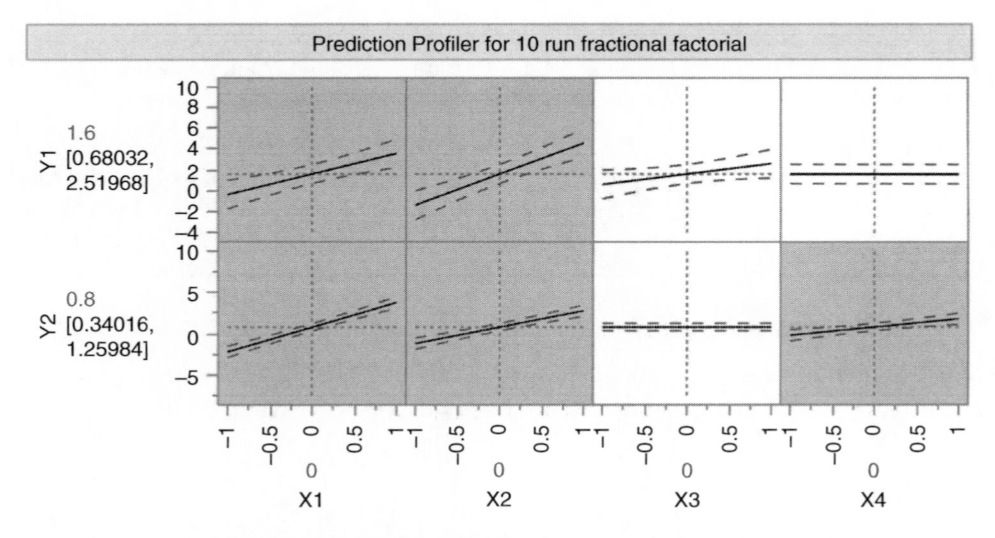

Figure 7.22 Prediction profiler for the 10 run fractional factorial.

surface or RSM design) to estimate main effects, interactions and quadratic effects. The intent is to make predictions over the region spanned the factor ranges.

If we drop some factors, we save resources in the next part of the sequential approach. For example, if we study only the three statistically significant effects identified (X1, X2 and X4), then we only need 20 experiments for the central composite design. X3 is not statistically significant, but there are indications it may be practically important. If we drop X3, we will not learn anything more about it. If we include X3, the 26 run central composite design will mean we go over our experimental budget as we already used 10 runs on the fractional factorial. The questions we need to ask ourselves are

- What do we need to know?
- What is the cost of not knowing?

Whether we push ahead and go over budget by including X3 depends on the cost of not knowing. In this situation, assuming the investigation will lead to increased quality for patients, we will take the decision to include the extra factor.

Table 7.7 shows the 4 factors, 26 run central composite design. Again, the response values for each run are provided by equations 7.5 and 7.6.

Table 7.7 *Second experiment in the sequential approach – a 26 run central composite design.*

	Pattern	X1	X2	X3	X4	Y1	Y2
1	00a0	0	0	-1	0	-1	0
2	---+	-1	-1	-1	1	-3	-5
3	+---	1	-1	-1	-1	-1	3
4	+-++	1	-1	1	1	1	1
5	----	-1	-1	-1	-1	-3	-3
6	+--+	1	-1	-1	1	-1	1
7	-+++	-1	1	1	1	3	3
8	--+-	-1	-1	1	-1	-1	-3
9	--++	-1	-1	1	1	-1	-5
10	++++	1	1	1	1	9	9
11	0000	0	0	0	0	0	0
12	0a00	0	-1	0	0	-3	-1
13	-+-+	-1	1	-1	1	1	3
14	a000	-1	0	0	0	0	-3
15	0A00	0	1	0	0	3	3
16	000a	0	0	0	-1	0	-1
17	-+--	-1	1	-1	-1	1	-3
18	+++-	1	1	1	-1	9	3
19	++--	1	1	-1	-1	7	3
20	A000	1	0	0	0	4	3
21	0000	0	0	0	0	0	0
22	++-+	1	1	-1	1	7	9
23	000A	0	0	0	1	0	1
24	00A0	0	0	1	0	1	0
25	-++-	-1	1	1	-1	3	-3
26	+-+-	1	-1	1	-1	1	3

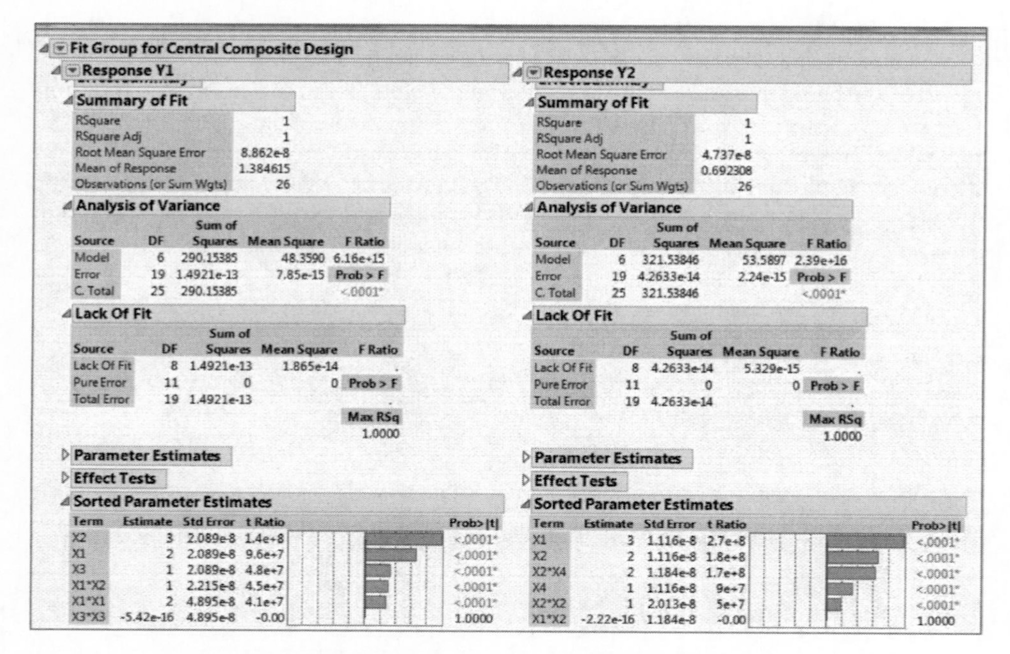

Figure 7.23 *Analysis of the central composite design.*

Figure 7.23 shows the analysis of these results. Looking at the section of the figure under the 'Sorted Parameter Estimates' heading we see that we have been able to correctly identify the model that generated the data, along with (nearly) exact estimates of the coefficients involved. So, as might be expected, the sequential approach has done well, but it needed 36 runs in total to gain or recover the required knowledge.

7.14.3 Definitive Screening Design

Now we repeat the exercise using the DSD shown in Table 7.8. Note that, like all DSD, this design involves factor settings at three levels, and a centre point (row 2) is included. Again, the response values are calculated from equation (3), skipping any factors not involved.

The analysis of the DSD is shown in Figure 7.24. Stepwise regression [23] was used to identify the important terms, and then a simple regression model built using just these terms. Looking at the section of the figure under the 'Sorted Parameter Estimates' header, we again see that we have been able to correctly identify the model that generated the data, along with exact estimates of the coefficients involved.

Finally, we can compare the prediction profilers for the fractional factorial/central composite and the DSD (Figure 7.25). In the absence of residual background noise, they both give almost identical models. The big difference is that we only needed 13 runs to gain or recover this knowledge using the DSD, compared with 36 for the sequential approach. The practical impact of such savings can be considerable!

7.14.4 Robustness Design

We have shown that screening and optimization designs generate process understanding. Our manufacturing colleagues and indeed the regulators also want us to show the process is in control. As usual, there are several approaches to doing this.

Table 7.8 *The experiment in the definitive screening approach – a 13 run definitive screening design (DSD).*

	X1	X2	X3	X4	X5	X6	Y1	Y2
1	-1	1	0	1	-1	-1	2	3
2	0	0	0	0	0	0	0	0
3	-1	1	-1	-1	1	0	1	-3
4	1	0	-1	1	1	-1	3	4
5	1	1	-1	0	-1	1	7	6
6	-1	0	1	-1	-1	1	1	-4
7	-1	-1	1	0	1	-1	-1	-4
8	1	-1	1	1	-1	0	1	1
9	0	-1	-1	-1	-1	-1	-4	0
10	0	1	1	1	1	1	4	6
11	1	-1	0	-1	1	1	0	3
12	1	1	1	-1	0	-1	9	3
13	-1	-1	-1	1	0	1	-3	-5

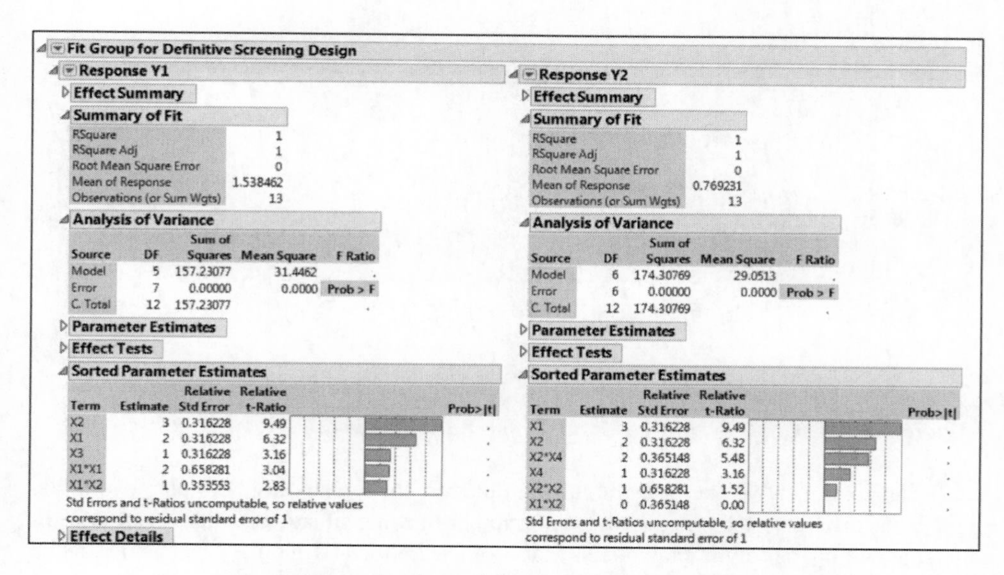

Figure 7.24 *Analysis of the 13 run DSD experiment.*

As mentioned previously, DSD can also be used as part of a sequential approach. Given that we have a budget of 30 experimental runs and we have used just 13, what might we do next? We could do another fractional factorial Resolution III design, which would use just 8 runs and still leave 9 runs for later. Through our understanding of the DSD, we can use that design to help define operating ranges in which all possible runs are now expected to pass. It is likely that the factor ranges will be narrower than those used in a screening design and that they will be shifted relative to the first study.

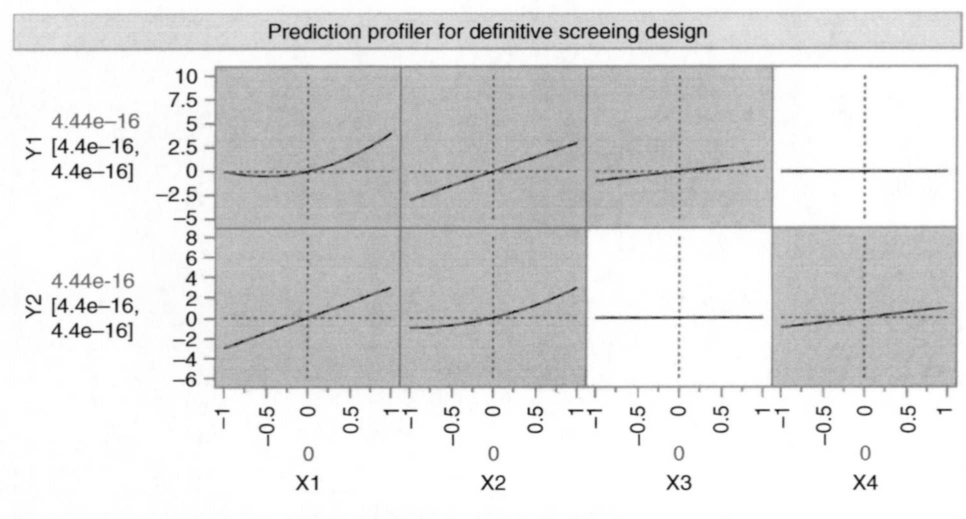

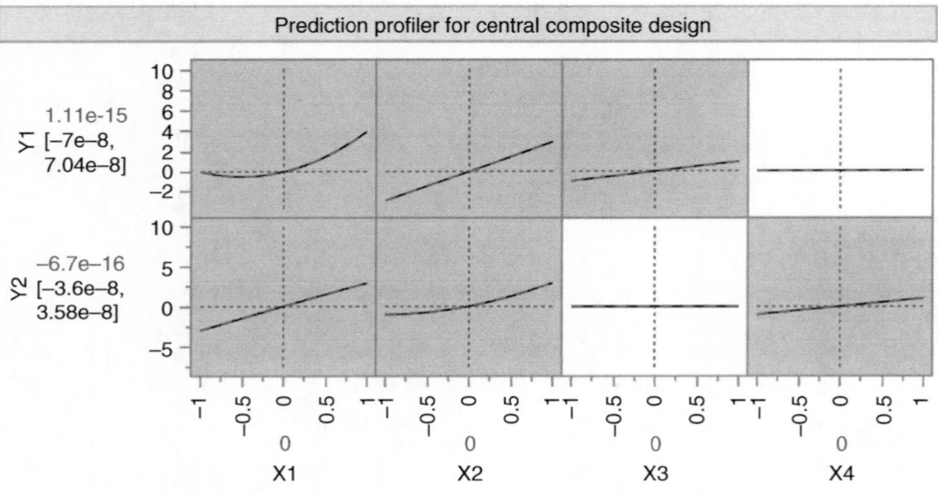

Figure 7.25 *Comparison of the central composite and definitive screening design (DSD).*

Of course, deductive exercises like the simulation above can only be illustrative rather than exhaustive – with six factors (a small number in some situations), there is an incredibly large number of polynomial models that could be defined (without even considering the relative sizes of the coefficients of the active terms). Another aspect is how the size of the noise term ε compares with the contribution from the factors, since we have seen previously that this is an essential aspect of empirical modelling. However, we can repeat the exercise here with different reasonable levels of noise in the model and see the practical consequences impact the amount of resource used, the results, the decision making and the ability to actually control the process.

Another aspect is that the example relies on what is normally called sparsity of effects (in this case, only three of the six factors were active for each model). Both DSDs and fractional factorials rely on this phenomenon to work well (though when it does not hold, they fail in different ways). Here, though, we come up against the problem of induction again – we can only take a view on whether or not sparsity is likely to be true in the specific situation we are confronted with, and plan accordingly. This is a good example of the truism that characterizes the use of DoE generally, namely, that you can design the best experiment only after you have its results.

7.14.5 Additional Challenges

For example, you may have noticed that not all lakes are square. DoE originally was developed with the requirement that designs be made 'by hand'. This heritage is still obvious in most software that DoE practitioners use today: These so-called 'classical designs' have been compiled into libraries that help set up how an experiment will be run. Although well tested and proven, these designs assume that the opportunity space is always square. An additional problem is that classical designs can be inflexible in relation to the new knowledge you are after, simply because they cannot fit every case exactly.

These problems lead to compromises like probing a smaller region of opportunity space than you want to, and having too many or too few design points, sometimes in the wrong places. Aside from the impact of having too many points (increased cost, cycle time and so on), the lake example makes clear that the design should be tied to your objectives and the usefulness of the new knowledge you hope to gain.

Custom designs are just like classical designs in the sense that they require you to consider a model of interest. However, they are more overtly 'model driven' in that they require you to explicitly state what that model is before you design the experiment. Once the model is defined, the workflow is essentially the same as when using a classical design. However, the ability to tailor the design to your specific situation, rather than risk doing more work and incurring more time and expense than necessary because you need to force fit your situation to a pre-existing design clearly gives you more flexibility for efficient learning.

To take just one example, you may have sound technological or chemical reasons for believing that you have specific interaction or quadratic terms, you can construct a design that will allow you to efficiently estimate the coefficients of the terms of interest, almost always with a smaller number of runs than the 'nearest' classical design.

In addition, by imposing a set of linear constraints between the Xs, you can easily probe an irregularly shaped opportunity space without having to 'shrink' a symmetrical classical design to fit. The latter procedure introduces more risk of failing to detect active terms against the background of noise for a given investment of resources. Custom designs are computer-generated using the so-called optimality criteria [24], and so require appropriate software (for example, [1]). The cost of acquiring and learning to use such software is usually trivial compared to the likely benefits.

7.15 Conclusion

More than 50 years of real-world usage have demonstrated the power of DoE to explore opportunity spaces to efficiently gain valuable new knowledge. From a historical perspective, one could argue that the pharmaceutical and biopharmaceutical sectors have been relatively slow adopters of DoE, for a variety of reasons. However, QbD and initiatives like it, with their insistence on an evidence-based approach rooted in science and technology rather than on a well-intentioned conformance-based approach, are rapidly driving more usage.

Every individual choice on factors, ranges, responses to be measured, designs used that are made implies a judgement has been made on the value of information even if nobody has been honest enough to own up to it or even admit it to themselves.

If the scientists, engineers and senior managers place only a small value on process understanding, then experimental design may not add much value compared to arriving at an answer by altering one factor at a time. DoE tells you how things work and what is likely to happen if you change them. What organizations and leaders do with that information is another matter.

Although there are theoretical aspects, it should be clear that DoE is rooted in practicalities that can only properly be appreciated by those who know the application area well. We hope we have given you a sense of some of the many interesting decisions and judgement calls needed to set up good experiments.

Since the issue of design is the cornerstone of DoE, it is important that scientists, technologists and engineers are aware of the potential benefits that relatively recent advances in this field can offer. The custom and DSDs introduced in this chapter clearly fall in this category. Because custom designs are model driven, it can be argued that they provide a better balance of responsibility in the way experiments are set up.

They offer a more tailored approach, but within a more unified conceptual framework that can be easier to get to grips with for those less familiar with statistical modelling. In addition, although necessarily a little artificial, the example in Section 7.14.3 shows the potential of this new design type to make a difference.

7.16 Acknowledgements

Patrick Steel, Matt Linsley, Dennis Lendrem, Iain Grant, Walkiria S. Schlindwein and participants of our workshops who generated the data.

7.17 References

[1] JMP (2016) *Statistical Discovery Software from* SAS, http://www.jmp.com(accessed March 1, 2016).
[2] Fisher, R.A. (1992) The arrangement of field experiments in *Breakthroughs in Statistics* (eds. Koltz, S. and Johnson, N.), Springer, New York, pp. 82–91.
[3] Owen, M.R., Luscombe, C.L., Lai, L., Godbert, S., and Crookes, D., and Emiabata-Smith, D. (2001) Efficiency by design: Optimisation in process research. *Organic Process Research & Development*, **5**, 308–323.
[4] Box, G.E.P., Hunter, J.S., and Hunter, W.G. (2005) *Statistics for Experimenters*, 2nd edition, Wiley, New Jersey.
[5] Deming, S.N. and Morgan, S.L.(1993) *Experimental Design: A Chemometric Approach*, 2nd edition, Elsevier, Amsterdam.
[6] Montgomery, D.C.(2008) *Design and Analysis of Experiments*, John Wiley &Sons, New Jersey.
[7] Cornell, J.A.(2011) *Experiments with Mixtures: Designs, Models, and the Analysis of Mixture Data*, John Wiley & Sons, New Jersey.

[8] Quotations of George Box, https://en.wikipedia.org/wiki/All_models_are_wrong
[9] Hulley, S.B., Cummings, S.R., Browner, W.S., Grady, D., Hearst, N., and Newman, T.B. (2001). Getting ready to estimate sample size: Hypothesis and underlying principles in *Designing Clinical Research–An Epidemiologic Approach*, 2nd edition, Lippincott Williams and Wilkins, Philadelphia, pp. 51–63.
[10] Frey, D.D., Engelhardt, F., and Greitzer, E.M. (2003) A role for "one-factor-at-a-time" experimentation in parameter design, *Research in Engineering Design*, **14**(2), 65–74.
[11] Wolfram Web Resources, http://mathworld.wolfram.com/TaylorSeries.html (accessed August 30, 2017)
[12] Wolfram Web Resources, http://mathworld.wolfram.com/LeastSquaresFitting.html(accessed August 30, 2017).
[13] *NIST/SEMATECH e-Handbook of Statistical Methods*, http://www.itl.nist.gov/div898/handbook/pmd/section1/pmd141.htm(accessed August 30, 2017).
[14] *NIST/SEMATECH e-Handbook of Statistical Methods*, http://www.itl.nist.gov/div898/handbook/pmd/section4/pmd431.htm (accessed August 30, 2017).
[15] Derringer, G. and Suich, R. (1980) Simultaneous optimization of several response variables, *Journal of Quality Technology*, **12**(4), 214–219.
[16] Cox, I. and Gaudard, M. (2013) *Discovering Partial Least Squares with JMP*, SAS Press, Cary, NC.
[17] Woodcock, J. (slides: Evolution in FDA's Approach to Pharmaceutical Quality), http://www.gphaonline.org/media/cms/Janet_Woodcock.pdf (accessed August 30, 2017).
[18] Wild, C. and Pfannkuch, M. (1998) What is statistical thinking? *ICOTS*, **4**, 335–341.
[19] Rivett, P. (1994) *The Craft of Decision Modelling*, Wiley, New Jersey.
[20] Davenport, T.H. and Prusak, L. (1997) *Working Knowledge: How Organizations Manage What They Know*, Harvard Business School Press, Boston
[21] Moen, R.D., Nolan, T.W., and Provost, L.P. (1991) *Improving Quality Through Planned Experimentation*, McGraw Hill, New York
[22] Sloman, S.A. and Laguado, D.A. (2005) The problem of induction in *The Cambridge Handbook of Thinking and Reasoning* (eds. K.J. Holyoak and R.H. Morrison), Cambridge University Press, Cambridge, pp. 95–116.
[23] JMP Statistical Discovery, Stepwise Regression Models, http://www.jmp.com/support/help/Stepwise_Regression_Models.shtml (accessed August 30, 2017).
[24] Goos, P. (2002) *The Optimal Design of Blocked and Split-Plot Experiments*, Springer Lecture Notes in Statistics.

8

Multivariate Data Analysis (MVDA)

Claire Beckett,[1] Lennart Eriksson,[2] Erik Johansson,[2] and Conny Wikström[2]

[1] OSIsoft, London, United Kingdom
[2] Sartorius Stedim Data Analytics AB, Sweden

8.1 Introduction

This chapter introduces the concept of MVDA based on projection methods. To this end, we shall describe three multivariate techniques and illustrate how they may be used within Quality by Design (QbD) and process analytical technology (PAT). The data analytical methods that will be described are principal component analysis (PCA), partial least squares (PLS) and orthogonal partial least squares (OPLS® multivariate software). PCA, PLS and orthogonal PLS (OPLS® multivariate software) apply well within the areas of PAT and QbD, and our discussion will target their merits and possibilities within the pharmaceutical industry.

PAT is founded on chemometric cornerstones like MVDA, multivariate statistical process control (MSPC) and batch statistical process control (BSPC). A major benefit of PAT is the use of multivariate process data to provide more information and a better understanding of manufacturing processes than conventional approaches (such as univariate SPC). Another benefit is the reduction of idle time at the end of a process step when product is often held awaiting laboratory results. This second benefit is a concept called "parametric release."

QbD is founded on similar tools as PAT, but in addition emphasizes the need for design of experiments (DoE) to achieve trustworthy results. A major benefit of QbD is the development of a sound scientific basis for a process design space that accommodates a range of defined variability in commercial process materials and operations and still produces the right product quality outcomes. A closely related idea to the design space concept is the

Pharmaceutical Quality by Design: A Practical Approach, First Edition.
Edited by Walkiria S. Schlindwein and Mark Gibson.

notion of proven acceptable ranges (PARs) for the factors. In fact, the PAR notion fits well into the topics discussed in this chapter. Different ways to state and communicate PAR of process factors are discussed.

The remainder of this chapter is structured as follows: In Section 8.2, an account of PCA is given, which is followed by an application of PCA for raw material characterization in Section 8.3. PLS is then introduced in Section 8.4, and a continuous process optimization example is used as an illustration of PLS in Section 8.5. This process optimization example is based on a DoE, and as a consequence the optimization results will also end up in a discussion of identifying a design space for this process, where also PARs for the DoE factors are considered. Section 8.6 then takes up a presentation of orthogonal PLS (OPLS® multivariate software), which is a recent development of PLS where interpretation simplicity is at the heart of the model formalism. Section 8.7 discusses a batch process example comprising both regular process factors and spectral measurements measured together. The batch example is used to give a contrast of classical PLS and orthogonal PLS (OPLS® multivariate software), and as will be shown, orthogonal PLS offers some distinct diagnostic and interpretational advantages. The final section, Section 8.8, provides a discussion and gives a position overview of how multivariate data analytical methods can facilitate and amplify applications like PAT, continued process verification (CPV) and QbD.

8.2 Principal Component Analysis (PCA)

PCA forms the basis for MVDA by projection methods [1–3]. As shown by Figure 8.1, the starting point for PCA is a matrix of data with N rows (observations) and K columns (variables), here denoted by X. The observations can be analytical samples, chemical compounds or reactions, process time points of a continuous process, batches from a batch process, biological individuals, trials of a DoE protocol and so on. In order to characterize the properties of the observations, one measures variables. These variables may be of spectral origin (NIR, NMR, IR, UV, X-ray, ...), chromatographic origin (HPLC, GC, TLC, ...) or they may be measurements from sensors in a process (temperatures, flows, pressures, curves, etc.).

PCA goes back to Cauchy, but was first formulated in statistics by Pearson, who described the analysis as finding lines and planes of closest fit to systems of points in space [1]. The most important use of PCA is indeed to represent a multivariate data table as a

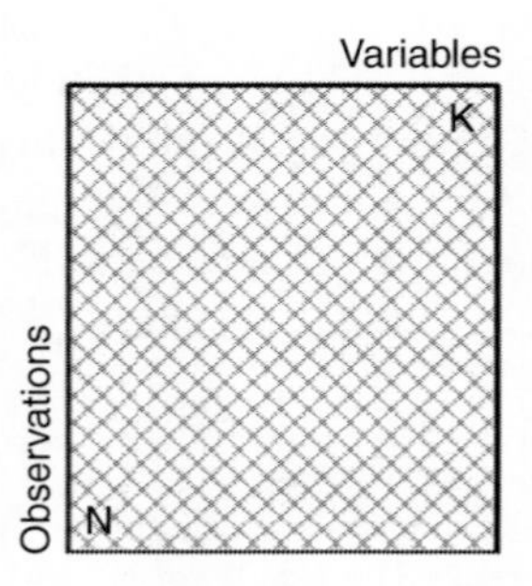

Figure 8.1 Notation used in principal components analysis (PCA).

low-dimensional plane, usually consisting of two to five dimensions, such that an overview of the data is obtained. This overview may reveal groups of observations, trends, and outliers. This overview also uncovers the relationships between observations and variables, and among the variables themselves.

Statistically, PCA finds lines, planes and hyperplanes in the K-dimensional space that approximate the data as well as possible in the least squares sense. It is easy to see that a line or a plane that is the least squares approximation of a set of data points makes the variance of the coordinates on the line or plane as large as possible (Figure 8.2).

By using PCA, a data table **X** is modeled as

$$X = 1 * \bar{x}' + T * P' + E \tag{8.1}$$

In the expression above, the first term, $1 * \bar{x}'$, represents the variable averages and originates from the pre-processing step. The second term, the matrix product **T*P´**, models the structure, and the third term, the residual matrix **E**, contains the noise. The principal component scores of the first, second, third, …, components ($\mathbf{t}_1$, $\mathbf{t}_2$, $\mathbf{t}_3$, …) are columns of the score matrix **T**. These scores are the coordinates of the observations in the model (hyper-)plane. Alternatively, these scores may be seen as new variables which summarize the old ones. In their derivation, the scores are sorted in descending importance (t_1 explains more variation than t_2, t_2 explains more variation than t_3 and so on). Typically, two to five principal components are sufficient to approximate a data table well.

The meaning of the scores is given by the loadings. The loadings of the first, second, third, …, components ($\mathbf{p}_1$, $\mathbf{p}_2$, $\mathbf{p}_3$,.) build up the loading matrix **P**. The loadings define the orientation of the PC plane with respect to the original X-variables. Algebraically, the loadings inform how the variables are linearly combined to form the scores. The loadings unravel the *magnitude* (large or small correlation) and the *manner* (positive or negative correlation) in which the measured variables contribute to the scores.

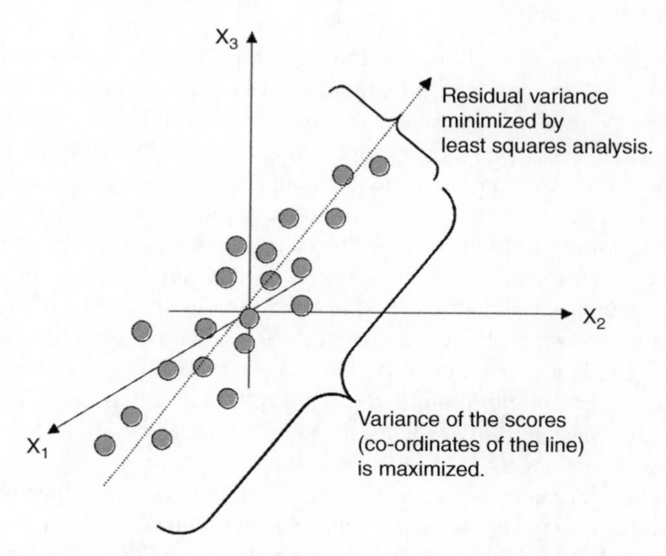

Figure 8.2 *PCA derives a model that fits the data as well as possible in the least squares sense. Alternatively, PCA may be understood as maximizing the variance of the projection coordinates.*

In other words, using PCA, the data in the matrix X are transferred into a new coordinate system. This coordinate system is defined by the principal components. The direction in variable space occupied by the most varying data points will define the location of the first PC, and the second PC will coincide with the largest variation orthogonal (at right angle) to the first component. New PCs can be extracted in this way until only minor variation is left unexplained by the PC model. Each consecutive PC (a) consists of a score vector (t_a) and a loading vector (p_a'). The score vector contains one numerical value (a "score") for each observation that tells how far from the origin the observation lies in the direction of the loading vector. A PCA score plot shows the location of the observations relative to each other, and a loading plot displays how important the variables are for the different directions in a score plot. Observations close to each other in a score plot have similar properties, and variables close to each other in a loading plot are also correlated.

Scores plotted in a scatter plot are useful for detecting strong outliers, clustering among observations and time trends. Strong outliers may have a degrading impact on model quality and should normally be discarded. Grouped data may make it difficult to understand properties inside a cluster. Hence, a splitting of data into smaller groups according to the nature of the clustering is often done. Separate PCA models may then be fitted to better resolve the structure inside each cluster. In addition to a score plot, a plot of model residuals (DModX) is useful for uncovering weakly deviating observations that do not conform well with the dominant correlation structure.

8.3 PCA Case Study: Raw Material Characterization using Particle Size Distribution Curves

8.3.1 Dataset Description

The example chosen by us to illustrate the potential of PCA relates to raw material characterization based on particle size distribution measurements. Variation in raw material properties can be one reason for problems during manufacturing. Such undesired variability may arise because different suppliers supply raw material that is not the same, or that, within the batches of raw material from one single supplier, there is significant batch-to-batch variability. In this case, it is recommended to collect a large number of measurements on the raw material using,for example, spectroscopic methods, particle size distribution curves, and wet-chemistry measurements, indirectly and jointly capturing the dominant properties of the raw material batches. This is a process usually known in the literature as *multivariate characterization*. PCA is an integral part of multivariate characterization and is used to quantify and understand the variation that occurs among the investigated samples or batches. After establishing the common variability over an extended period of time, the long-term objective is using this knowledge about raw material variability for classification of future batches of raw material.

The particle size distribution dataset comes from a pharmaceutical company and relate to all incoming batches during approximately two years of production. There are in total 250+ variables corresponding to particle size intervals ranging from 2.5 to 150 μm (Figure 8.3). The readout represents the number of particles in each bin (size interval). In the dataset there are 45 batches constituting the training set and 13 batches constituting the test set.

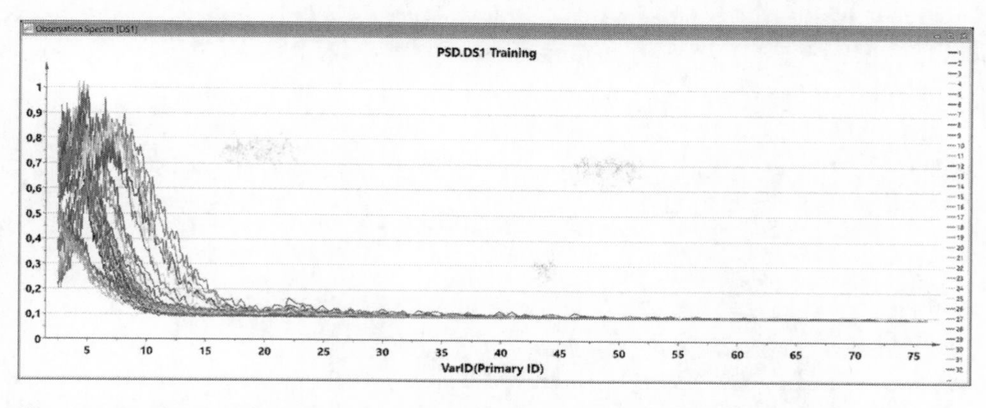

Figure 8.3 The raw data curves of the 45 batches of the training set. (See insert for color representation of the figure.)

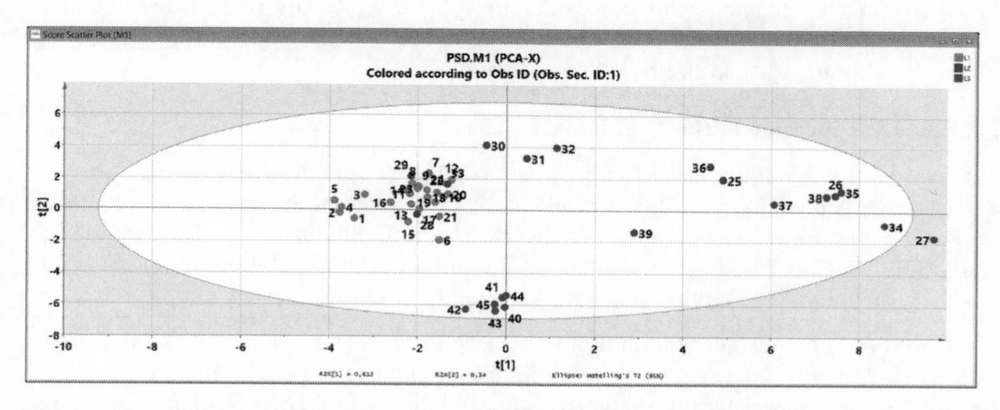

Figure 8.4 Scatter plot of the two principal components. Each point represents one batch of raw material. The plot is color-coded according to supplier. (See insert for color representation of the figure.)

8.3.2 Fitting a PCA Model to the 45 Training Set Batches

The PCA model of the pareto-scaled particle size distribution data of the training set batches yielded a two-component model explaining 95.2% and predicting via cross-validation 94.7% of the variability in the data. Figure 8.4 shows the PCA score plot from the analysis of the particle size distribution curves. In the score plot, there are as many points as there are particle size distribution curves. Thus, we realize that curve form data and spectra can be understood as points in the score space. Similar spectra are situated close together in the score plot, whereas spectra of clearly different shape are located far away from each other. In the example, there are three vendors of excipients. The plot is color-coded according to the information on the vendors. The score plot reveals that suppliers L1 and L3 provide the most homogeneous and consistent starting material with little quality variation over time.

In order to interpret the score plot, it is instructive to look at the loadings. With curve form data and spectra, loading line plots are usually used as they benefit from the physical ordering structure (e.g. wavelengths) that exists among the variables. Scatter plots of

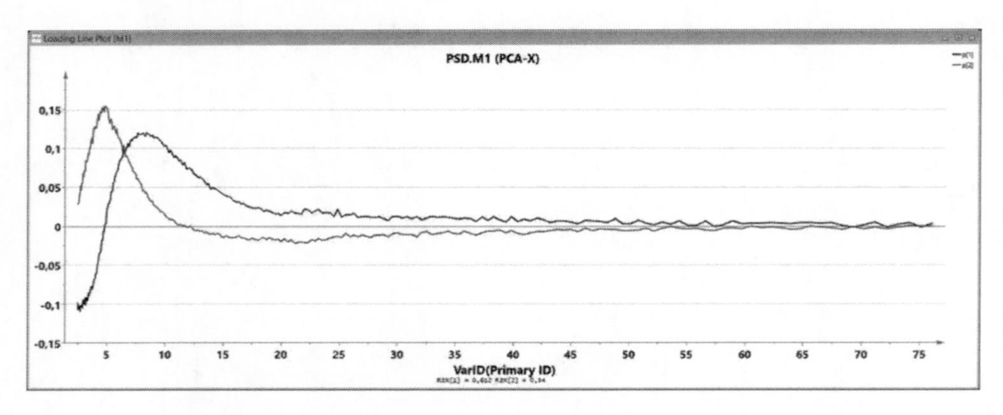

Figure 8.5 Loading line plots of the two principal components.

loadings are less common in this context. Figure 8.5 displays the loading profiles of the two PCA components. Evidently, the two loading spectra suggest two types of peak-like structures for the particle size distribution curves.

8.3.3 Classification of the 13 Test Set Batches

After some time, additional batches of raw material became available. We can use the existing PCA model as a quality control chart to investigate whether the new batches are similar or dissimilar to the training set batches. This is accomplished by classifying the 13 additional test set batches of raw material into the PCA model. Interesting questions to address are the following:

- How do the test set batches conform with the PCA model?
- Are there any mismatches, that is, batches that do not fit the model well?
- Are the three suppliers still shipping raw material batches of the same quality as before?
- Do raw material quality characteristics change with time for any of the three suppliers?
- If so, how is the raw material composition changing with time?

In order to classify the batches of the test set into the existing PCA model, we start by examining the predicted score plot (Figure 8.6, top left). For clarity, the score plot of the training set is reproduced (Figure 8.6, top right). A general comment is that all the prediction set batches are inside the score space spanned by the training set batches. Hence, we can state that the prediction set batches fit the model well and that the projections are reliable. Also, when looking at the corresponding plots of DModX (Figure 8.6, bottom left and right), we can conclude that the training and prediction sets conform reasonably well to one another.

However, on a more detailed level, the classification results mean different conclusions for the different suppliers. The classification results indicate very stable quality for L1 with time, somewhat changing quality for L2 (but still within the domain of the training set variability for that particular supplier), but a substantial shift in raw material quality for L3. Batches 111–113 are systematically shifted from batches 40–45 and appear more in line with L2 qualities. This systematic shift in raw material composition must be more closely examined. The closer scrutiny of why there is a systematic shift in raw material qualities

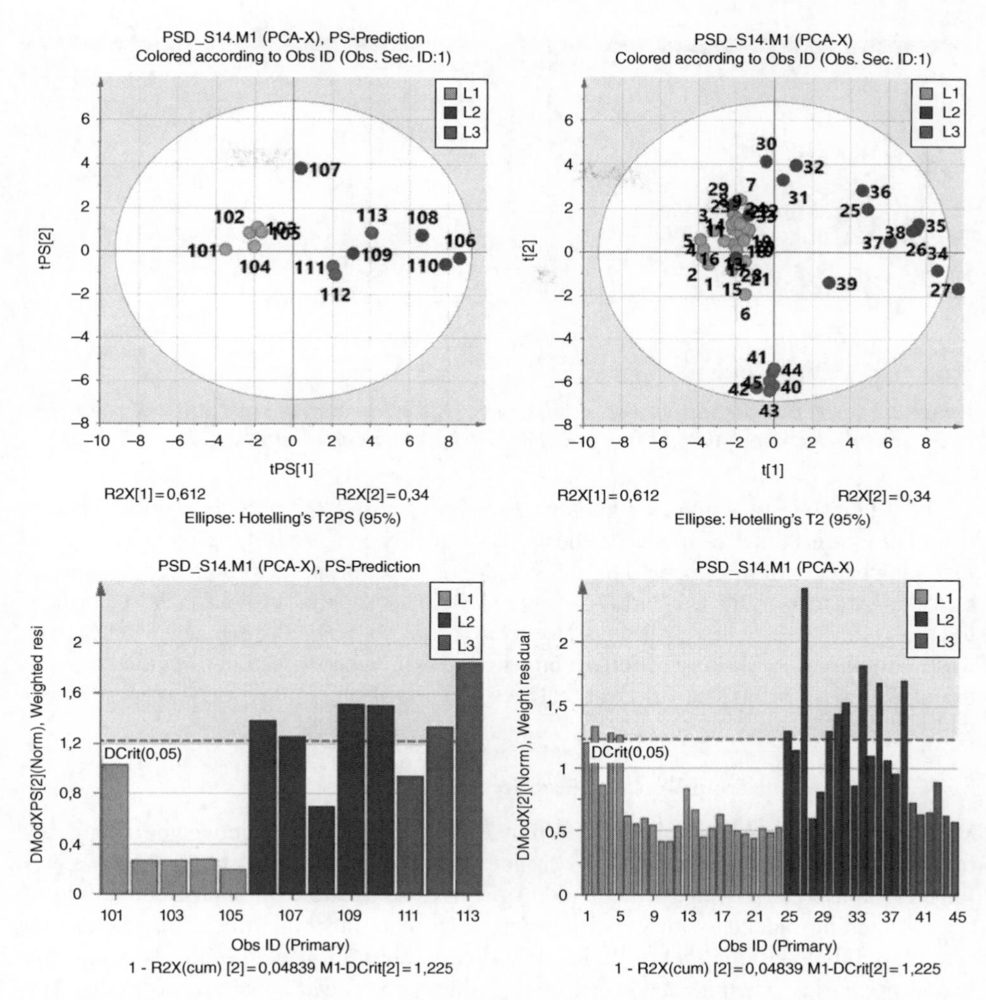

Figure 8.6 *Scores and DModX plots for the particle size distribution data set. Top left: Predicted scores for the test set batches. Top right: Scores for the training set batches. Bottom left: Predicted DModX for the test set batches. Bottom right: DModX for the training set batches.*

for the L3 supplier can be investigated using contribution plots and/or line plots of the underlying data. Figure 8.7 shows line plots of the six early and three late L3 batches; these line plots of the raw data show the completely different shapes of the power spectral density (PSD) curves.

The main conclusion from this example is that variation in raw material properties can be accounted for by collecting a large number of measurements describing the raw material. PCA on raw material batches is an efficient way to summarize such variation. Multivariate characterization of raw materials using appropriate techniques add to the already used specification data of the raw material properties.

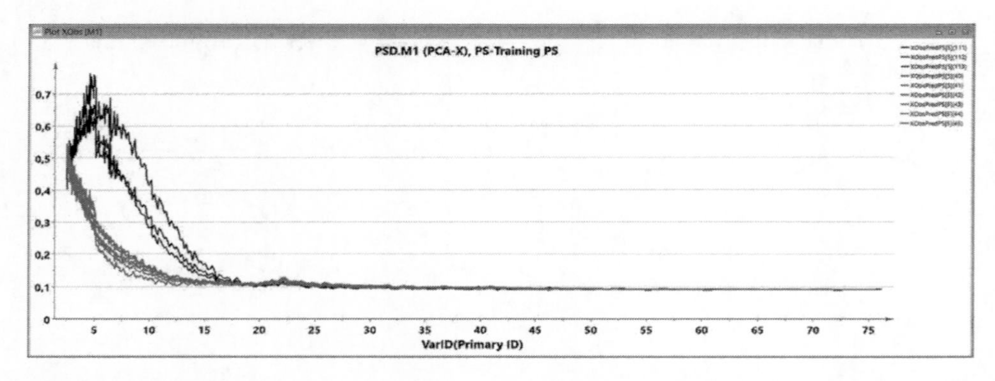

Figure 8.7 Line plot of the power spectral density (PSD) curves of the six early (red color) and three late (black color) batches of supplier L3. (See insert for color representation of the figure.)

The established PCA model explains the major trends in the particle size distribution data. The established PCA model can be used as a quality control model to uncover whether incoming batches of raw material have similar or different properties compared with previously used batches. This is clearly to ensure that the L1 supplier provides raw material of the most consistent quality over time. However, the model flagged some new batches from the L3 supplier as being very different compared with previous shipment from this supplier, the main reason being that the fresh batches comprised much higher fractions of larger particles.

8.3.4 Added Value from DoE to Select Spanning Batches

With such insights at hand, the next step investigates correlations between existing raw material variabilities and problems in production. A supplementary step paving the way for more cause-and-effect-oriented studies involves using DoE for selecting a subset of space-spanning batches with greatly varying raw material properties. The scatter plot (Figure 8.8) below suggests how to identify such a subset of batches with systematically varied properties. Such a smaller set of spanning batches can be examined in detail in order to ensure production robustness and in order to develop mitigation strategies in case batches of raw material with very intractable properties are being delivered to the manufacturing facility.

8.4 Partial Least Squares Projections to Latent Structures (PLS)

In the case of PCA, there is only one matrix (X) to consider. When making, for example, a calibration model or a batch evolution model, one also has to consider a single response variable (y) or a matrix of responses (Y). The idea is that these response data should be modeled and predicted from the X-data. The X- and Y-matrices are decomposed in a similar way as in PCA. One fundamental difference is that while PCA captures maximum variance directions in the X-space, PLS finds directions in the X- and Y-spaces corresponding to the maximum covariance [4].

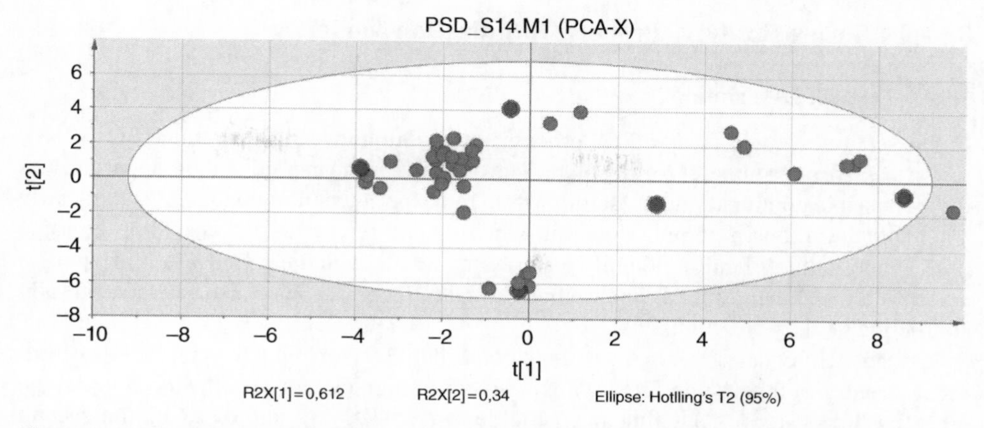

Figure 8.8 *Spanning batches can be selected using DoE. Such spanning batches can be subjected to a thorough investigation in order to ensure production robustness in all part of the PCA score space. (See insert for color representation of the figure.)*

One interpretation of PLS is that it forms "new X-variables," t_a, as linear combinations of the old ones, and thereafter uses these new t's as predictors of Y. Only as many new t's (components) are formed as are needed, as are predictively significant by cross-validation. For each component (a), the parameters, t_a, u_a, w^*_a, p_a, and c_a, are determined by a PLS algorithm. For the interpretation of the PLS model, the scores, t and u, contain information about the observations, and their similarities/dissimilarities with respect to the given problem and model. The weights w* and c give information about how the variables combine to form the quantitative relation between X and Y. Hence, these loading weights are essential for understanding which X-variables are important (numerically large w-values), which X-variables provide the same information (similar profiles of w_a-values), and for interpretation of the scores, t, and so on.

PLS modeling of the relationship between two blocks of variables can be described in different ways. Perhaps the most straightforward way is to state that it fits two "PCA-like" models at the same time, one for **X** and one for **Y**, and simultaneously aligns these models. The objectives are (a) to model **X** and **Y,** and (b) to predict **Y** from **X**, according to

$$\boldsymbol{X} = 1\overline{\boldsymbol{x}}' + \boldsymbol{TP}' + \boldsymbol{E} \tag{8.2}$$

$$\boldsymbol{Y} = 1\overline{\boldsymbol{y}}' + \boldsymbol{UC}' + \boldsymbol{F}\left(= 1\overline{\boldsymbol{y}}' + \boldsymbol{TC}' + \boldsymbol{G}, \textit{ due to inner relation}\right) \tag{8.3}$$

In these expressions, the first terms, $\mathbf{1}\,\overline{\boldsymbol{x}}'$ and $\mathbf{1}\,\overline{\boldsymbol{y}}'$, represent the variable averages and originate from the pre-processing step. The information related to the observations is stored in the scores matrices **T** and **U**; the information related to the variables is stored in the X-loading matrix **P′** and the Y-loading matrix **C′**. The variation in the data that was left out of the modeling form the **E** and **F** residual matrices. In summary, PLS forms "new X-variables," **t**a, as linear combinations of the old ones, and thereafter uses these new **t**'s as predictors of **Y**. Only as many **t**'s (components) are formed as are predictively significant.

8.5 PLS Case Study: A Process Optimization Model

8.5.1 Dataset Description

In this section, we will study a PLS process optimization investigation called SOVRING, which arises from a mineral sorting plant. The objective is (1) to predict the response values for all possible combinations of factors within the experimental region, and (2) to identify a suitable and robust setpoint. However, when several responses are treated at the same time, it is usually difficult to identify a single setpoint at which the goals for all responses are fulfilled, and therefore the final setpoint often reflects a compromise between partially conflicting goals.

The SOVRING dataset comprises three controllable factors which were varied according to a central composite design in 17 experiments. These factors were the feed rate of iron ore (Ton_In), speed of the first magnetic separator (HS_1), and speed of the second magnetic separator (HS_2). Among the 17 experimental runs, eight experiments arose from the factorial part of the design, six from locating experiments on the three factor axes, and three from doing replicated trials at the design center.

One way to study the shape of the resulting experimental design consists of plotting the raw data. Figure 8.9 (left) shows a scatter plot of two of the designed factors, HS_1 and HS_2. The regularity of the points is apparent. Hence, the influence of the underlying design cannot be mistaken. Similar scatter plots can be generated for HS_1 versus Ton_In, and HS_2 versus Ton_In, but such pictures are not given here. It is interesting that the MVDA – PLS modeling carried out in the next section – recognizes and utilizes this regularity expressed by the X-matrix (Figure 8.9, right).

Each factor combination in the SOVRING design was preserved for 20 minutes to better capture normal process variation. The impact on the manufacturing performance of each experimental point in the central composite design was encoded in the registration of six response variables (= six Y-variables). The concentrated material is divided into two

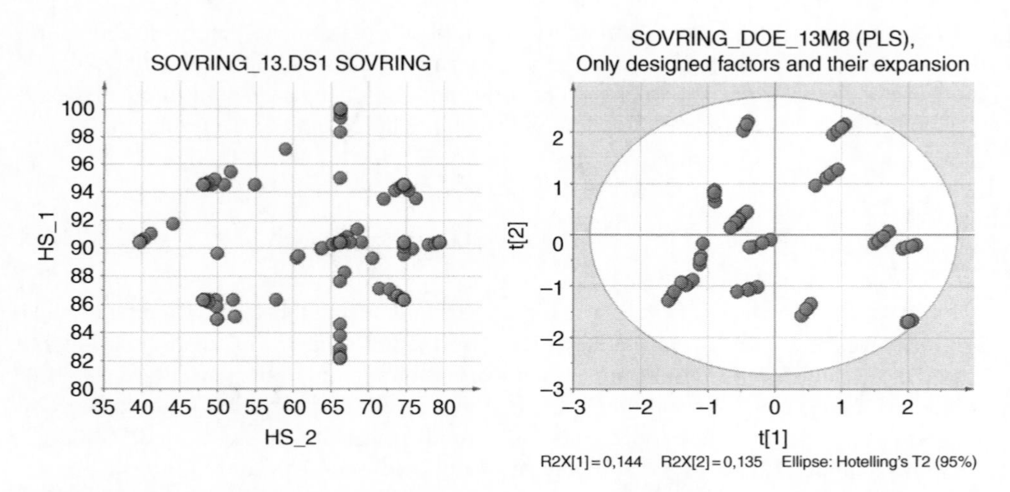

Figure 8.9 *(Left) Scatter plot of HS_1 against HS_2, (right) PLS t_1/t_2 score plot (note the resemblance to Figure 8.8).*

product streams, one called PAR and the other called FAR. The amounts of these two products form two important responses (they should be maximized). The other important responses are the percentages of P in FAR (%P_FAR), and of Fe in FAR (%Fe_FAR), which should be minimized and maximized, respectively.

Five observations with complete Y-data were registered for each factor setting. Thus, the SOVRING subset with complete Y-data contained 85 observations (the product of 17 design runs and 5 time points). This is the dataset considered in the next section.

8.5.2 PLS Modeling of 85-Samples SOVRING Subset

A PLS model was established in the three designed factors (Ton_In, HS_1, HS_2) plus their six expanded terms. This model contained five components and utilized 61% of X ($R^2X = 0.61$) for modeling 76% ($R^2Y = 0.76$) and predicting 70% ($Q^2Y = 0.70$) of the response variation. Figure 8.10 (left) displays the evolution of R^2Y and Q^2Y as a function of increasing model complexity. Figure 8.10 (right) displays how well each individual response variable is modeled and predicted after five components. We can see that the two important responses PAR and FAR are excellently described and predicted by the model. These two responses should be maximized. The other two important responses, %P_FAR and %Fe_FAR, which should be minimized and maximized, respectively, are fairly well modeled.

Figures 8.11 (left) and (right) give scatter plots of the PLS weights. The two amount of product responses (FAR and PAR) are greatly influenced by the load (Ton_In), which is logical. However, what is really interesting is that the two qualities of product responses (%Fe_FAR and %P_FAR) are strongly influenced by the operating performance of the second magnetic separator (HS_2). Hence, by adjusting HS_2, it is possible to influence

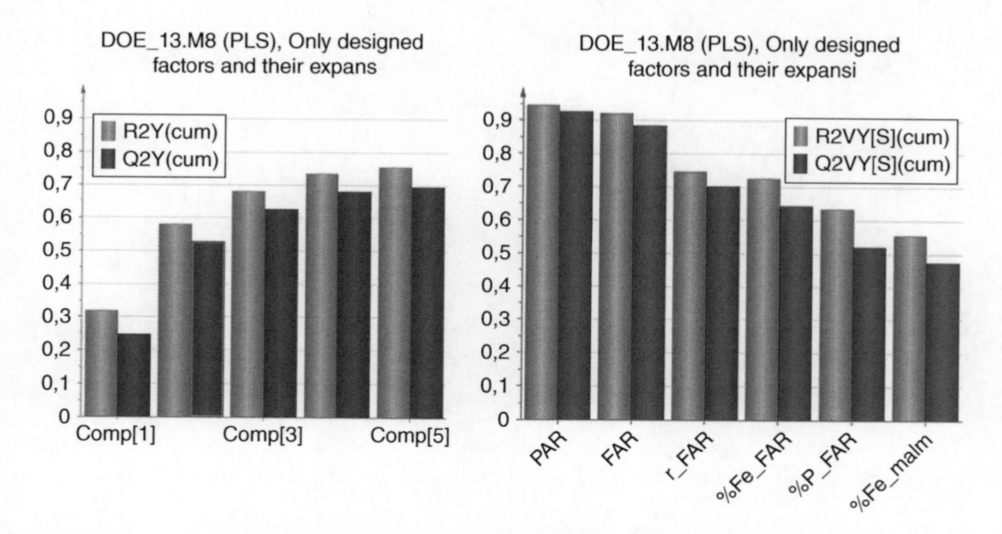

Figure 8.10 *(Left) Summary of fit plot of the PLS model of the SOVRING subset with complete Y-data. Five components were significant according to cross-validation. (Right) Individual R^2Y- and Q^2Y-values of the six responses. The most important responses are PAR, FAR, %Fe_FAR and %P_FAR.*

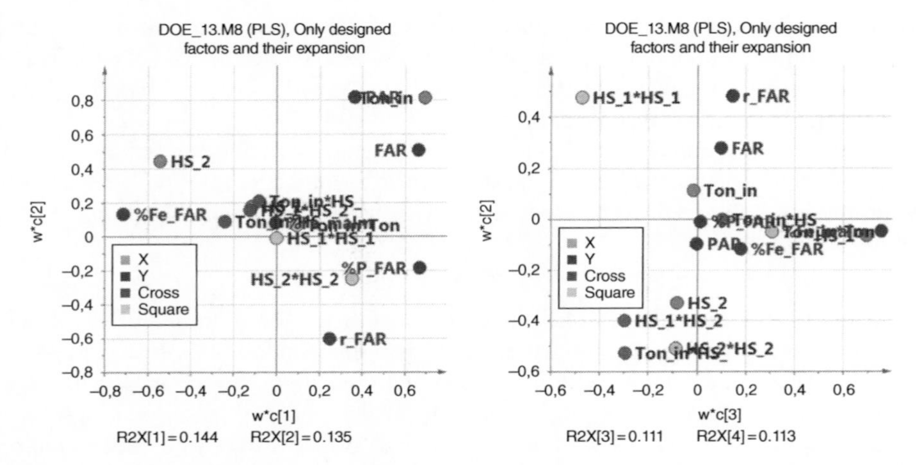

Figure 8.11 (Left) PLS w^*c_1/w^*c_2 loading weight plot. Ton_In is the most influential factor for PAR and FAR. The model indicates that by increasing Ton_In, the amounts of PAR and FAR increase. There is a significant quadratic influence of HS_2 on the quality responses %Fe_FAR and %P_FAR. This nonlinear dependence is better interpreted in a response contour plot, or a response surface plot. (Right) PLS w^*c_3/w^*c_4 weight plot. (See insert for color representation of the figure.)

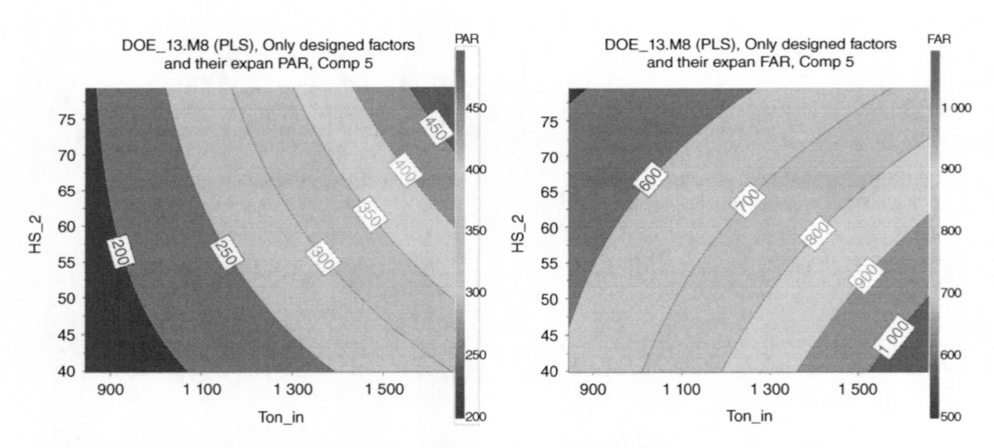

Figure 8.12 (Left) Response contour plot of PAR, where the influences of Ton_In and HS_2 are seen. (Right) Response contour plot of FAR. (See insert for color representation of the figure.)

the quality of product without sacrificing the amount of product. It is also noticeable that there is some quadratic influence of HS_2^2 with respect to %Fe_FAR and %P_FAR.

8.5.3 Looking into Cause-and-Effect Relationships

In order to better assess the importance of the higher-order model terms of the PLS model (cross-terms and square-terms), we decided to inspect response contour plots. Figure 8.12 and Figure 8.13 show response contour plots of the important responses

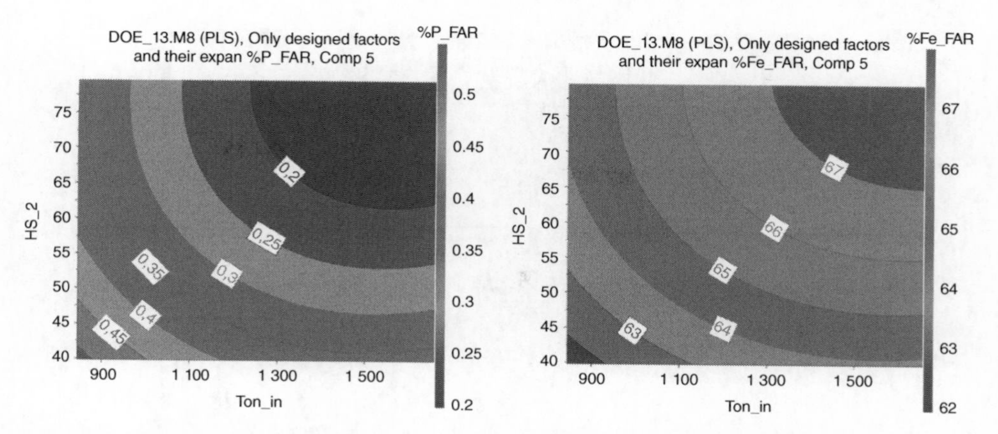

Figure 8.13 (Left) Response contour plot of %P_FAR. (Right) Response contour plot of %Fe_FAR. (See insert for color representation of the figure.)

PAR, FAR, %P_FAR, and %Fe_FAR, respectively. These plots were produced by using Ton_In and HS_2 as axes, and by setting HS_1 at its high level (setting HS_1 high was found beneficial using plots of regression coefficients).

The evaluation of the four response contour plots represents a classical example of how the goals of many responses rarely harmonize completely. For three responses, PAR, %P_FAR, and %Fe_FAR, the model interpretation suggests that the upper-right-hand corner (high HS_1, high HS_2, and high Ton_In) represents the best operating conditions. At the same time, however, it appears optimal for the response FAR to run the process according to conditions represented by the lower-right-hand corner. Staying somewhere along the right-hand edge thus seems as a reasonable compromise.

8.5.4 Making a SweetSpot Plot to Summarize the PLS Results

The SweetSpot plot is a practical way to overview optimization results when several response contour plots have to be considered simultaneously. Figure 8.14 shows such a plot for the SOVRING dataset, which was created from the underlying PLS model using the MODDE® DoE software. The SweetSpot plot can be thought of as a superimposition of several response contour plots, that is, an overlay which is color-coded according to how well the specifications on the responses are met. If the plot contains a green area, the investigator knows there is a SweetSpot. Figure 8.14 shows that a SweetSpot region does indeed exist for the SOVRING process.

A limitation of the SweetSpot plot is that it does not display the risk of failure to comply with the response specifications. However, by using additional QbD-directed technology inside the MODDE® software, it is possible to evaluate such a risk of failure. Figure 8.15 shows a design space plot –from the Design Space Explorer tool – showing the risk of failure for the SOVRING process. The green area (light grey color) corresponds to a low-risk area with less than 1% probability to fail. The interpretation of the design space results of the PLS model is discussed in the next session.

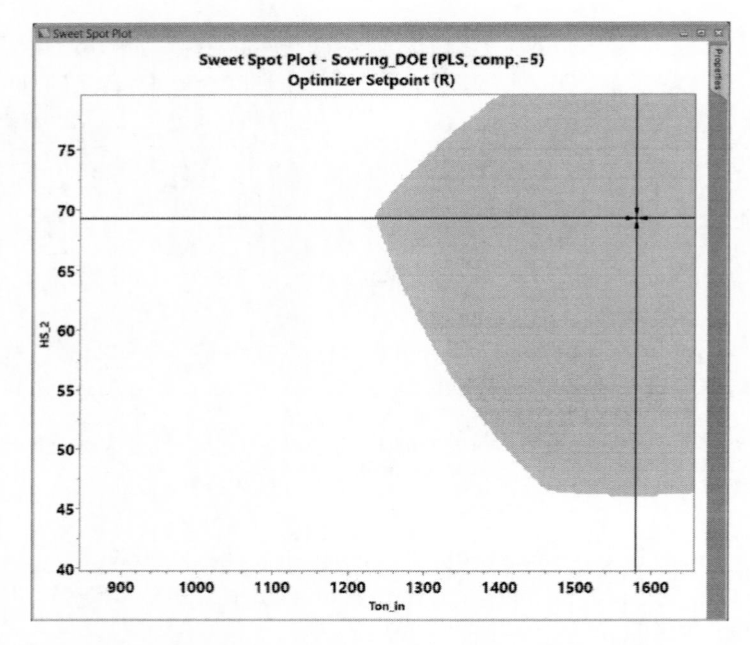

Figure 8.14 SweetSpot plot of the SOVRING example, suggesting the SweetSpot should be located in the upper and right-hand part.

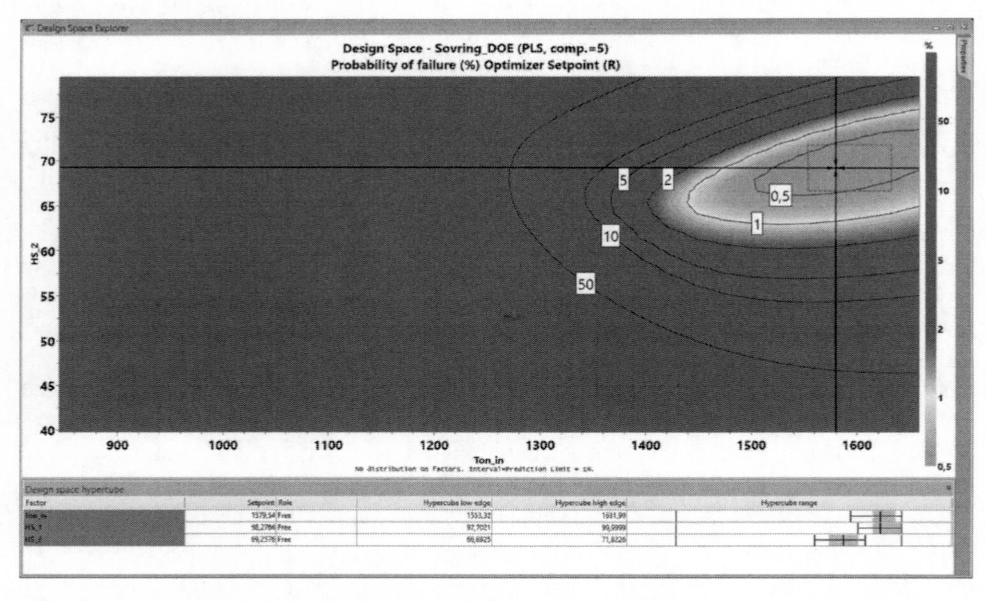

Figure 8.15 *A design space plot of the SOVRING example. The green area corresponds to design space with a low risk, 1%, of failure. (See insert for color representation of the figure.)*

8.5.5 Using the PLS-DoE Model as a Basis to Define a Design Space and PARs for the SOVRING Process

The Design Space Explorer is used for a graphical appraisal of the design space and to understand how it extends in multidimensional factor space. We can use its output as a means to bridge the two concepts of "design space" and PAR for the factors.

Figure 8.16 presents a schematic representation of a design space and how it relates to other concepts often encountered in the context of optimization multiresponse PLS models. First, we have the notion of "knowledge space." In our interpretation, the knowledge space is the region in which you have done experiments using an experimental design. Second, we have the design space, which is located somewhere inside the knowledge space. And, finally, inside the design space we sometimes also talk about the "normal operating region," which represents a volume in which the process is usually operated (control space).

Moreover, the resulting design space is usually a highly irregular volume. This is because it is defined by one or more PLS or multiple linear regression (MLR) regression models, and their properties, and a number of constraints arising from for example, response specifications and the chosen statistical acceptance criteria. Regardless of the size and shape of the design space, it represents an area or volume in which all responses are according to specifications and also when we consider uncertainty. In order to get an in-depth assessment of a design space, it is recommended to work with the tool called Design Space Explorer.

The Design Space Explorer is accessible after running the Optimizer. If a robust setpoint has been identified, that coordinate will momentarily be taken into account when launching the tool. The main part is a 2D plot (see Figure 8.15) showing the size and shape of the design space as viewed in the subspace defined by the two factors chosen for the X- and Y-axes. Underneath this there is a Design space hypercube overview functionality, which is useful when trying to comprehend the size and shape of the investigated design space in the dimensions beyond the 2D representation of the corresponding plot.

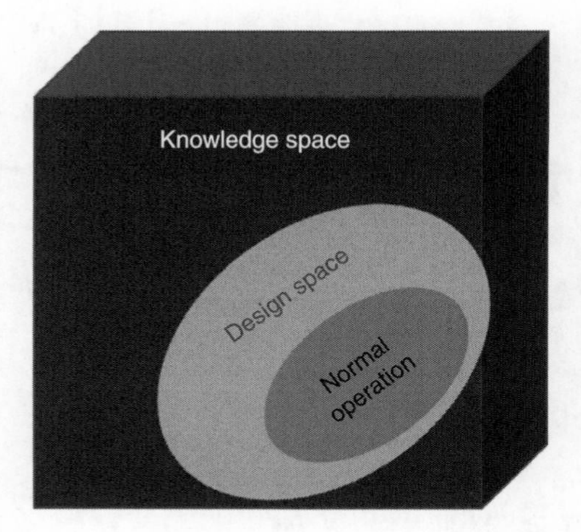

Figure 8.16 *A schematic representation of how the concepts knowledge space, design space, and normal operation (control space) region relate to one another.*

Apart from a visualization of the design space in two dimensions, the Design Space Explorer (Figure 8.15) provides two additional graphical tools: (1) a cross-hairs symbol that indicates the position of the robust setpoint and (2) a dotted frame that shows the placement of the so-called design space hypercube. The inscribed design space hypercube corresponds to the volume in which all factor combinations can be used without compromising the response specifications. Its extension in the 2D plot is given by the dotted frame, and its elongation across all dimensions is shown by the green color (light grey color) in the Design space hypercube overview.

More specifically, the green color in the Hypercube range field of Figure 8.15 designates the *mutual* ranges within which all factors can be changed at the same time and without further restrictions. Slightly wider *individual* ranges are patterned by the black T-lines, which represent the allowable range of a process factor, while keeping all other factors constant at their setpoint value. The mutual factor ranges provided by the Hypercube range field correspond to the largest regular volume that can be inserted into the irregular design space. It should be noted that the hypercube range can be changed interactively by clicking and dragging the low or high end for the range of a particular factor. The regular hypercube volume that then arises is always smaller that the initial regular volume, however (Figure 8.17).

A closely related idea to the design space concept is the notion of PAR for the factors. In the first approach, the information from the Hypercube range field of the Design Space Explorer is used. The black T-lines extending out from the robust setpoint coordinate indicate the individual ranges for the factors, that is, the largest allowable range of a process parameter, while keeping all other parameters constant at their setpoint value. Also, the second approach is based on the information available in the Hypercube range field, but in this case the green bars are considered. The green bars mark the mutual ranges within which all factors can be changed at the same time and without further restrictions.

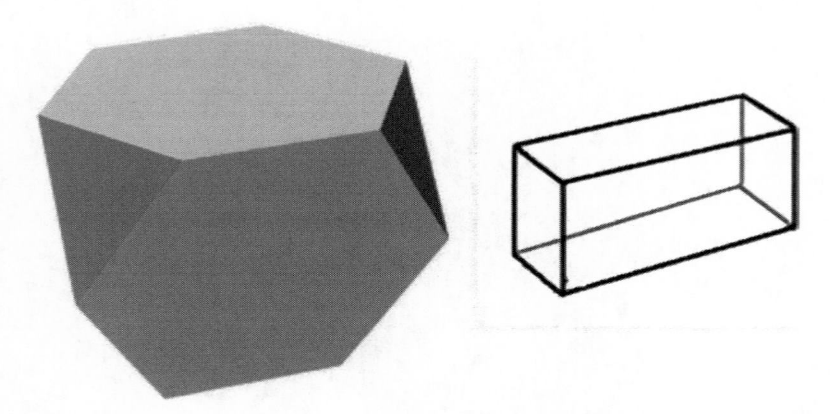

Figure 8.17 *The design space hypercube represents the largest regular geometrical structure that can be inserted into the irregular design space.*

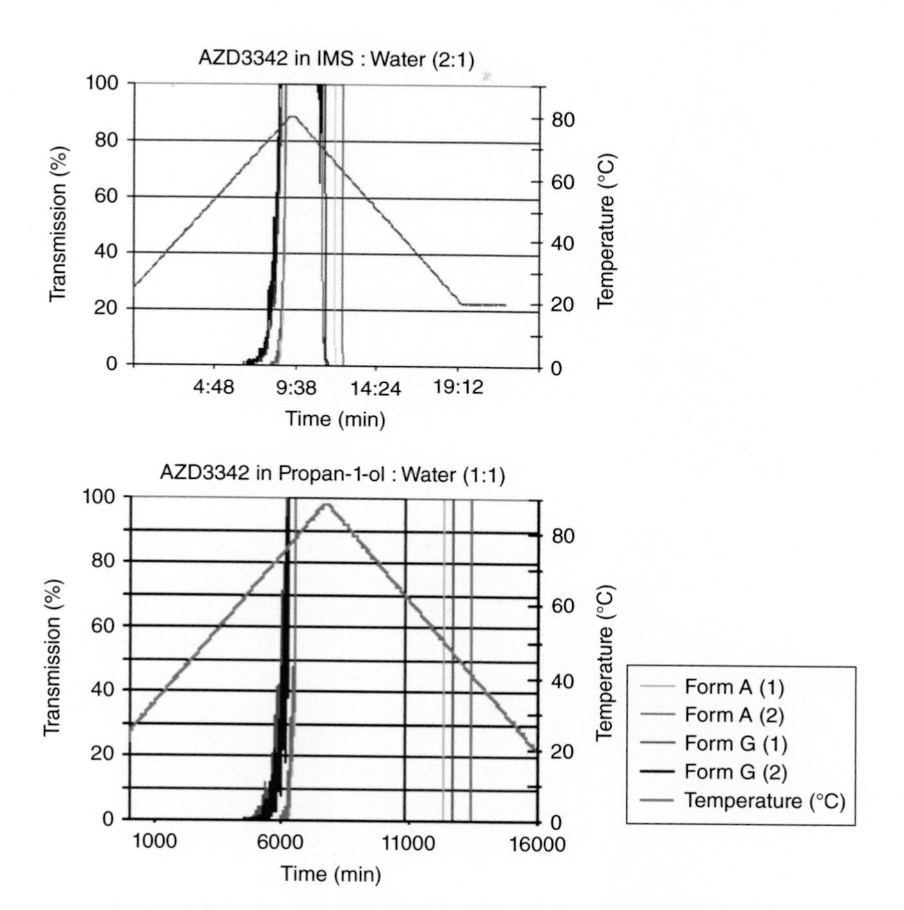

Figure 4.5 *Crystal16 data for AZD3342 polymorphs A and G.*

Pharmaceutical Quality by Design: A Practical Approach, First Edition.
Edited by Walkiria S. Schlindwein and Mark Gibson.

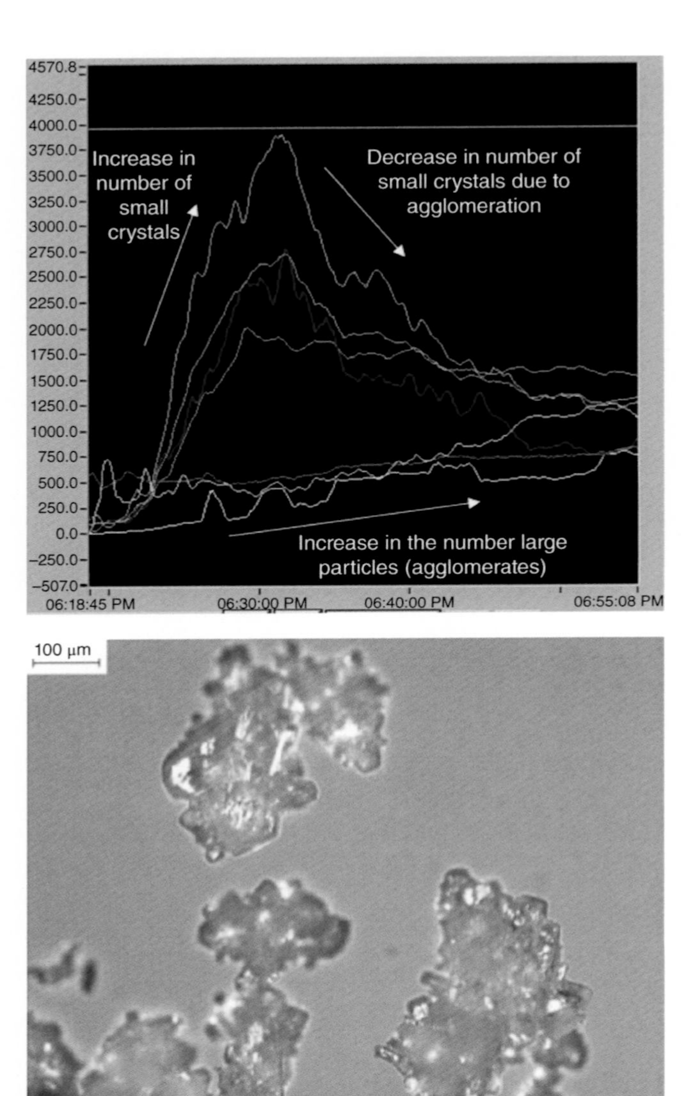

Figure 4.9 *FBRM Lasentec data and Photomicrograph for an unseeded, linear cool crystallization of AZD3342.*

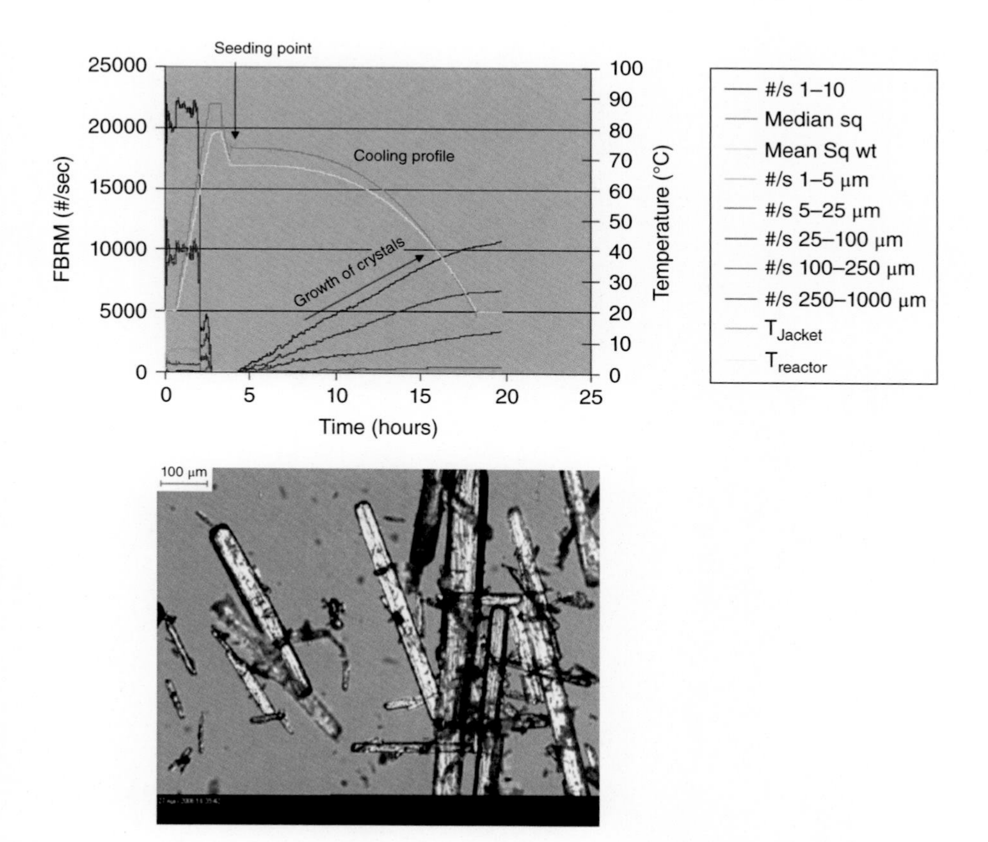

Figure 4.11 *FBRM Lasentec data and optical microscopy of AZD3342 crystals using seeding and a cubic cooling profile (controlled cooling).*

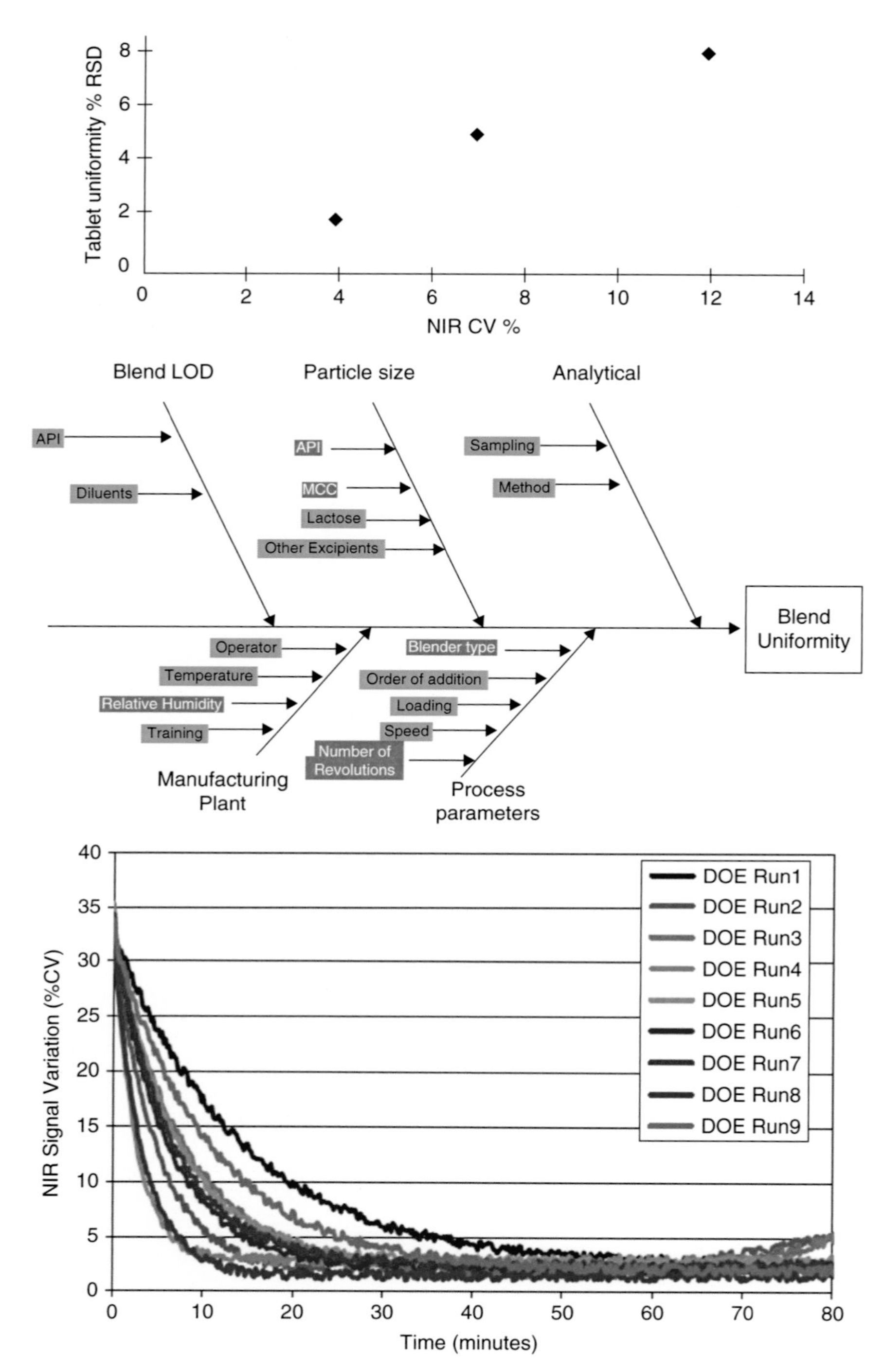

Figure 6.8 *Critical parameters affecting blend uniformity for ACE tablets. (From Pharmaceutical Development Case Study: 'ACE Tablets', prepared by CMC-IM Working Group (March 2008), http://www.ispe.org/pqli-resources.)*

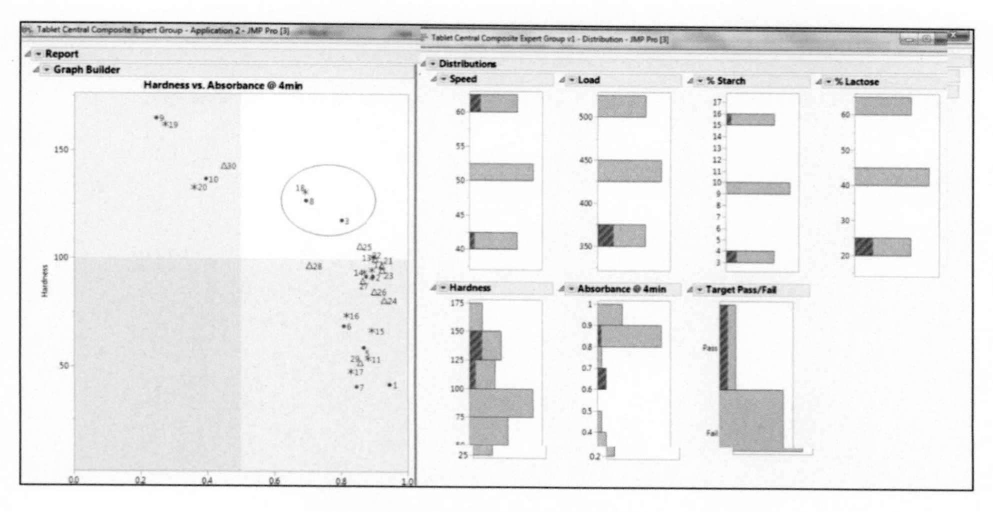

Figure 7.11 Data visualization.

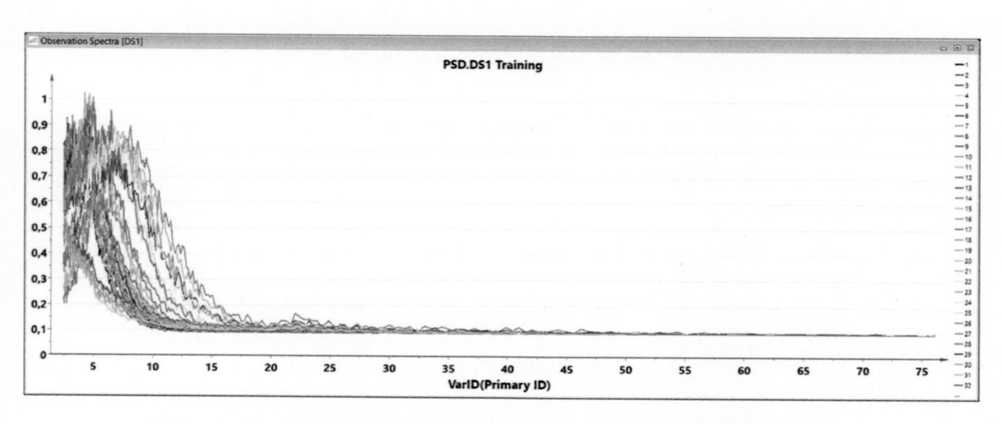

Figure 8.3 The raw data curves of the 45 batches of the training set.

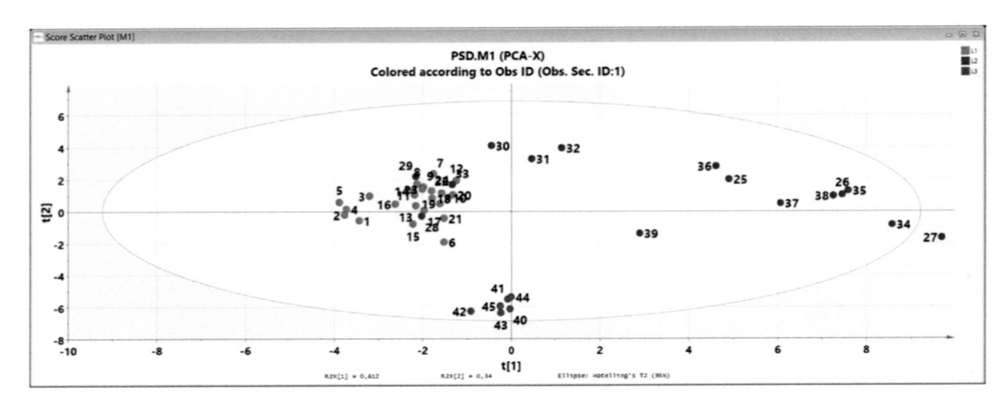

Figure 8.4 Scatter plot of the two principal components. Each point represents one batch of raw material. The plot is color-coded according to supplier.

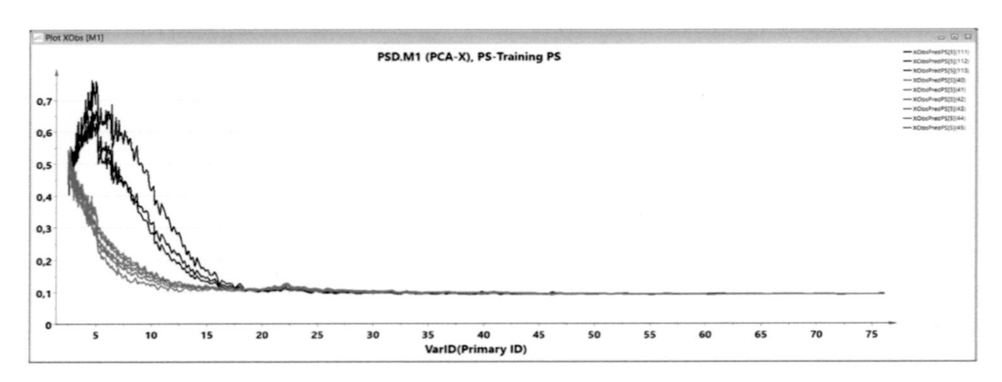

Figure 8.7 Line plot of the power spectral density (PSD) curves of the six early (red color) and three late (black color) batches of supplier L3.

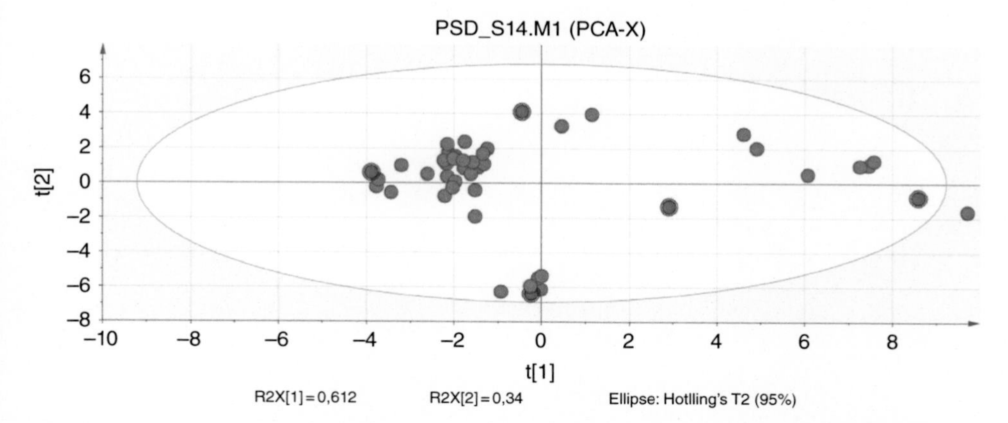

Figure 8.8 *Spanning batches can be selected using DoE. Such spanning batches can be subjected to a thorough investigation in order to ensure production robustness in all part of the PCA score space.*

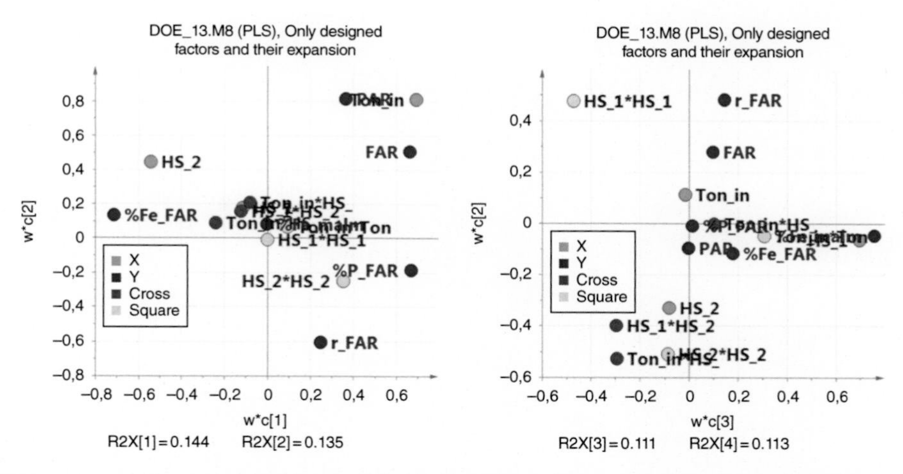

Figure 8.11 *(Left) PLS w^*c_1/w^*c_2 loading weight plot. Ton_In is the most influential factor for PAR and FAR. The model indicates that by increasing Ton_In, the amounts of PAR and FAR increase. There is a significant quadratic influence of HS_2 on the quality responses %Fe_FAR and %P_FAR. This nonlinear dependence is better interpreted in a response contour plot, or a response surface plot. (Right) PLS w^*c_3/w^*c_4 weight plot.*

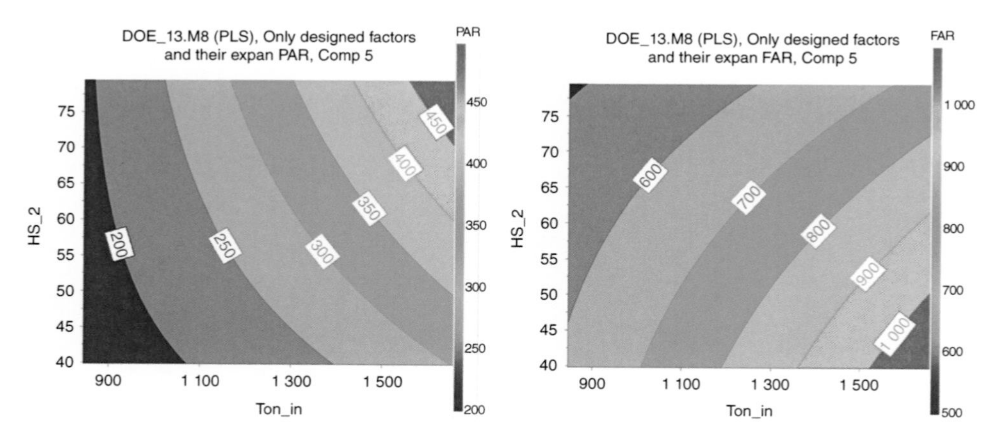

Figure 8.12 *(Left) Response contour plot of PAR, where the influences of Ton_In and HS_2 are seen. (Right) Response contour plot of FAR.*

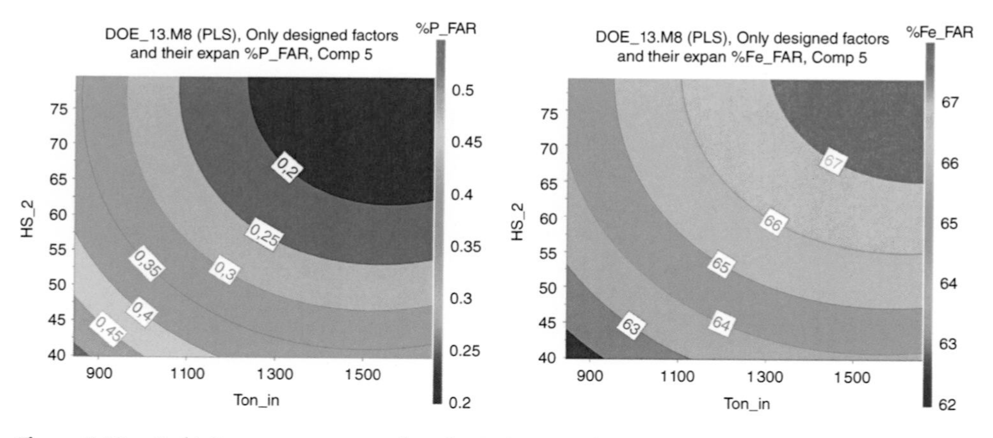

Figure 8.13 *(Left) Response contour plot of %P_FAR. (Right) Response contour plot of %Fe_FAR.*

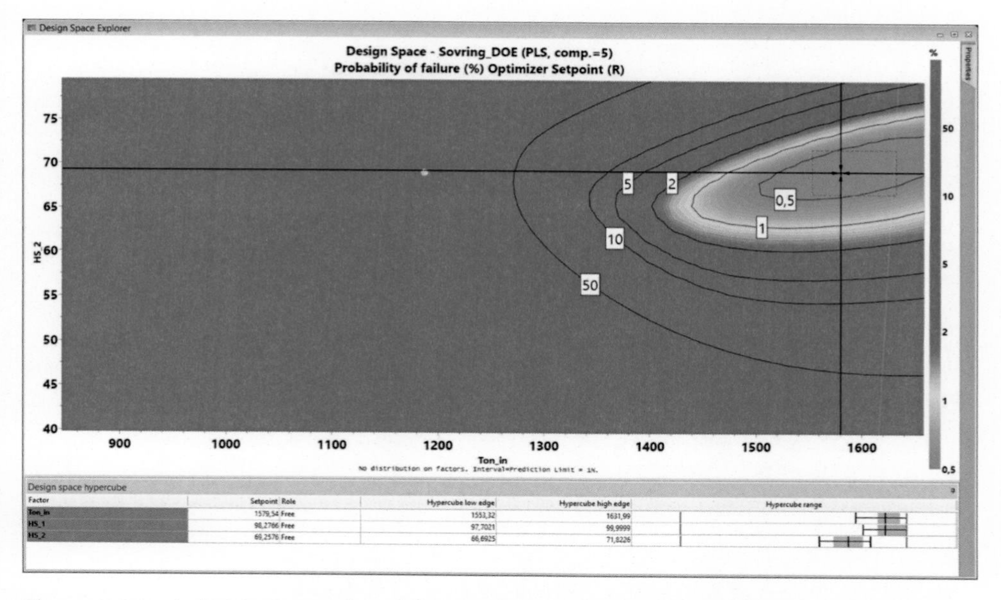

Figure 8.15 *A design space plot of the SOVRING example. The green area corresponds to design space with a low risk, 1%, of failure.*

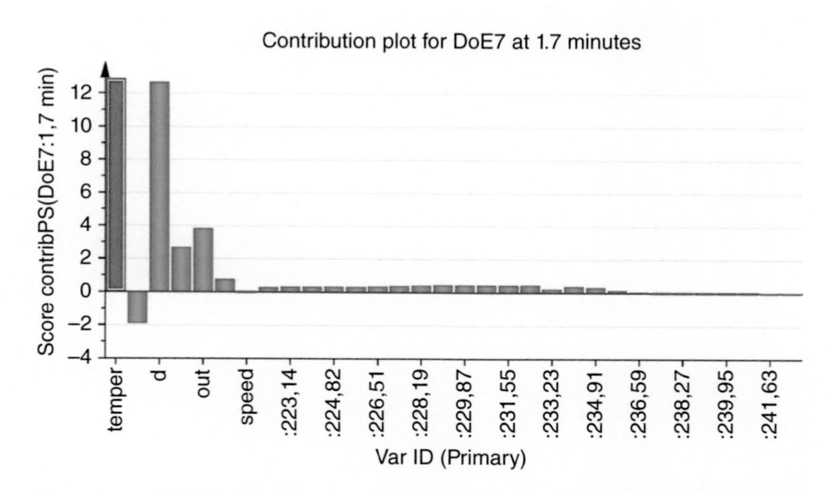

Figure 8.20 *Contribution plot for batch DoE7 at 1.7 min showing which variables are contributing to the process deviation. The process upset is caused by a few of the process parameters. For example, the contribution plot indicates that the reaction temperature (highlighted by red color) is much higher than for a normal batch.*

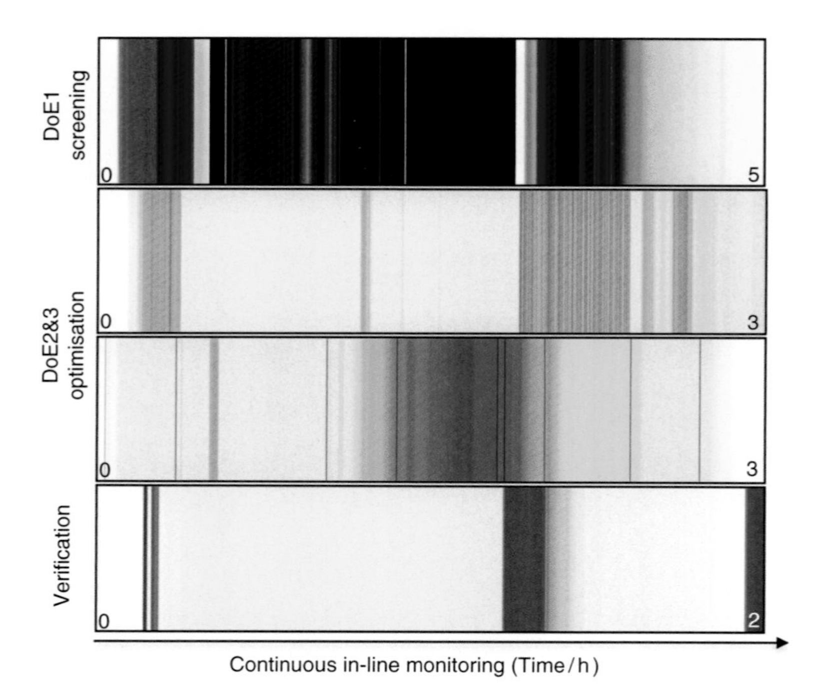

Figure 9.7 *Representation of about 5000 UV-Vis spectra from DoEs 1–3 and verification experiments.*

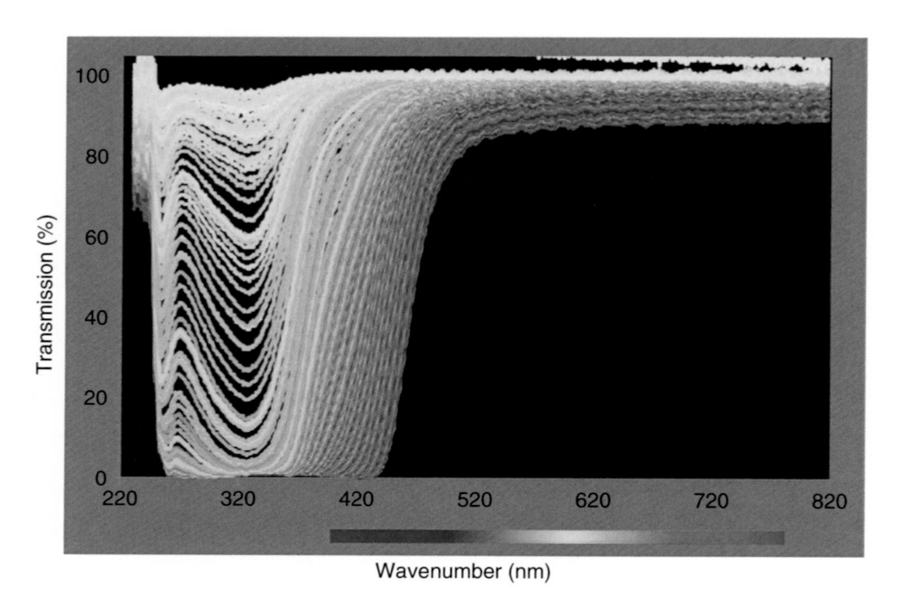

Figure 9.8 *Raw spectra from a DoE set of runs.*

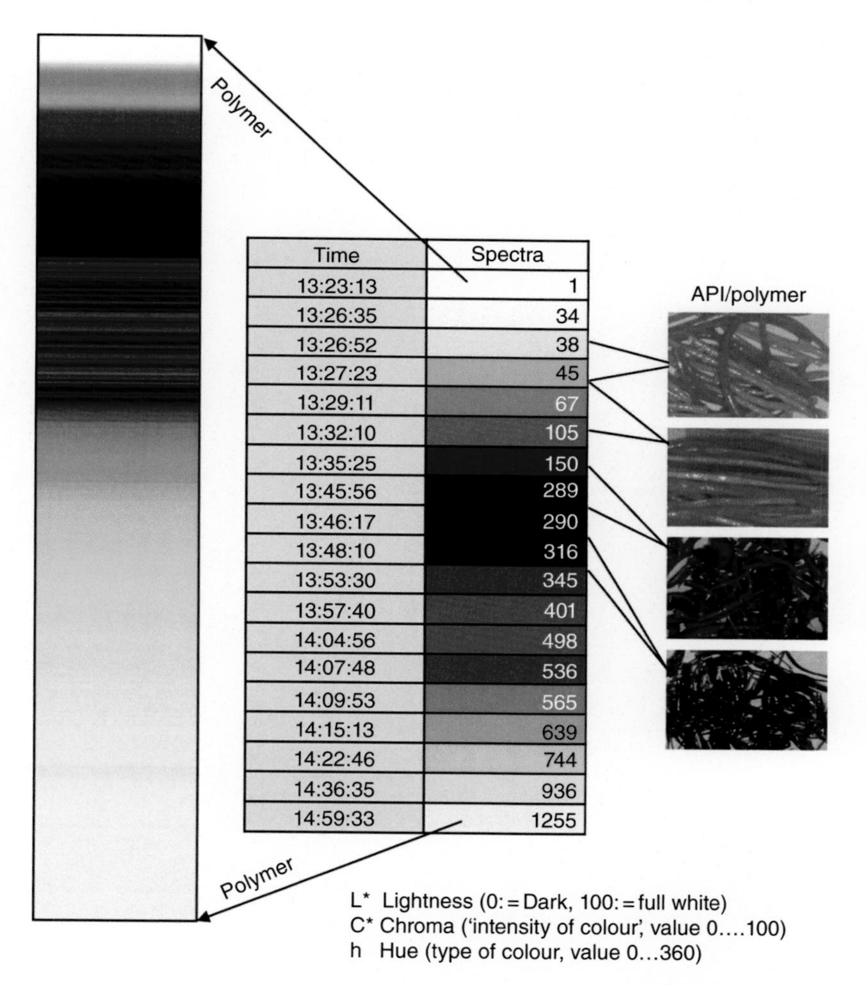

Figure 9.9 *UV-Vis spectra from the first screening design (DoE1) showing a sample of the spectra collected.*

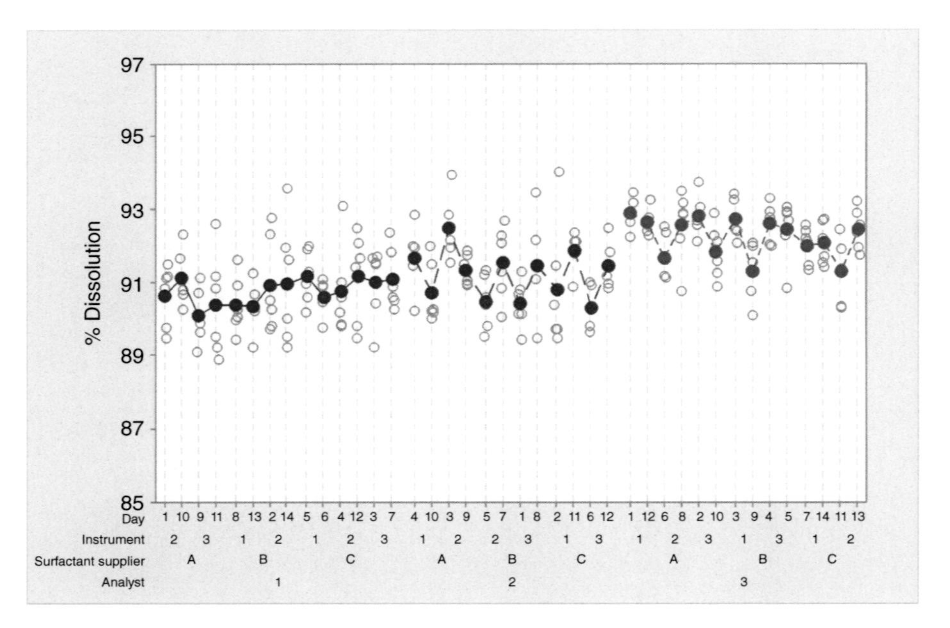

Figure 10.11 *Dissolution data from multivariate experimental study (coloured by analyst) – individual tablet (open circles) and mean data (solid circles) – produced using Minitab® Statistical Software.*

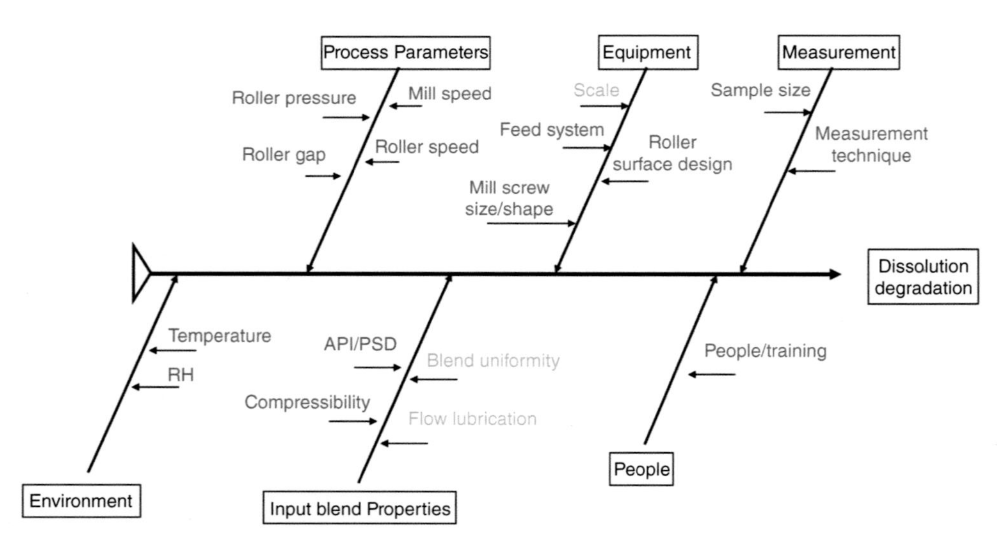

Figure 12.3 *A typical example of an Ishikawa or fishbone diagram. Key: red = potential high impact on CQA; Yellow = potential impact on CQA; green = unlikely to have significant impact on CQA. Example: roller compaction.*

8.5.6 Summary of SOVRING Application

The SOVRING example shows the utility of an optimization design in the process industry. Two main types of responses were distinguished: responses related to *amount of product* (i.e., throughput) and those related to *quality of product*. By increasing the load (Ton_In) to the mineral sorting plant, production of FAR and PAR increased. By regulating the speed of the second magnetic separator (HS_2), it was possible to maintain a balance between the two quality responses %P_FAR and %Fe_FAR.

In summary, the net gain of using DoE and PLS in the SOVRING example was improved process understanding coupled with better product quality. Sometimes, however, DoE cannot be applied at all, or, at least, not to the extent that is desired. Under such circumstances, the PAT approach might be a viable alternative/complement to DoE/QbD.

8.6 Orthogonal PLS (OPLS® Multivariate Software)

Orthogonal PLS (OPLS® multivariate software) is an extension of PLS that makes model interpretation easier [5]. The objective of orthogonal PLS (OPLS® multivariate software) is to divide the systematic variation in the X-block into two model parts, one part which models the correlations between **X** and **Y** and another part which expresses the variation that is not related (orthogonal) to **Y**. Components that are correlated to **Y** are here called *predictive* and are designated by the subscript *P*. Components that are uncorrelated to **Y** are here called *orthogonal* and are designated by the subscript *O*. Note that with a single Y-variable, there is only one predictive component and any number of orthogonal (no relation to **Y**) components.

In the case of a single response variable, we can write the X-part of the orthogonal PLS model as

$$\boldsymbol{X} = 1\overline{\boldsymbol{x}}' + \boldsymbol{t}_p \boldsymbol{p}_p' + \boldsymbol{T}_0 \boldsymbol{P}_0' + \boldsymbol{E} \quad (8.4)$$

and the orthogonal PLS model prediction of y as

$$\boldsymbol{y} = 1\overline{\boldsymbol{y}}' + \boldsymbol{t}_p \boldsymbol{q}_p' + \boldsymbol{f} \quad (8.5)$$

In the expressions above, scores, loadings, and residuals are similarly depicted as in the PLS model, which was outlined in Section 8.4. When the OPLS® multivariate software model encompasses a single Y-variable, only *one* predictive component can result. A consequence of this is that the regression coefficients ($\mathbf{b}_P = \mathbf{w}_P\mathbf{q}'_P$) become redundant, since the prediction vector is given by the loading p of the predictive component. Similarly, a single-Y OPLS® multivariate software model also makes the VIP plot redundant, as this parameter is directly linked to the loading weight w of the predictive component.

It should be noted that the OPLS® multivariate software model for multiple Y-variables involves slightly more complicated matrix expressions; however, the multi-Y OPLS® multivariate software model is beyond the scope of the current chapter.

8.7 Orthogonal PLS (OPLS® Multivariate Software) Case Study – Batch Evolution Modeling of a Chemical Batch Reaction

8.7.1 Dataset Description

The example relates to a chemical reaction run as a batch process [6]. The reaction is a hydrogenation converting nitrobenzene to aniline. Six reaction batches representing normal operating condition (NOC) batches were run as part of an MSPC study. Five additional batches corresponding to various process upsets were available as a test set. Seven process variables were measured: reactor temperature, reactor pressure, gas feed, jacket-in temperature, jacket-out temperature, flow rate of oil, and stirrer speed. UV spectra were recorded in the interval 220–290 nm, and each spectrum was pre-processed using the first derivative. This resulted in 80 spectral variables.

8.7.2 Batch Evolution Modeling

A batch evolution model (BEM) using the methodology described in [7] was fitted using OPLS® multivariate software model and the six NOC batches. The OPLS® multivariate software model had one predictive and two orthogonal components. Figure 8.18 shows the resulting batch control chart for the predictive component. Its ability to diagnose the non-NOC test batches as deviators is obvious. For clarity, results are given for two non-NOC batches only. In a similar manner, Figure 8.19 presents results obtained from an alternative BEM fitted using PLS. It can be concluded that major differences are seen in the first score vectors of the two BEMs, which may have an impact on early fault detection. Classical PLS is not as sharp and capable as orthogonal PLS.

Close inspection of Figure 8.18 shows that the batch DoE7 is predicted within the normal trajectory for the first 1.7 min. Thereafter it deviates sharply from the normal trajectory for

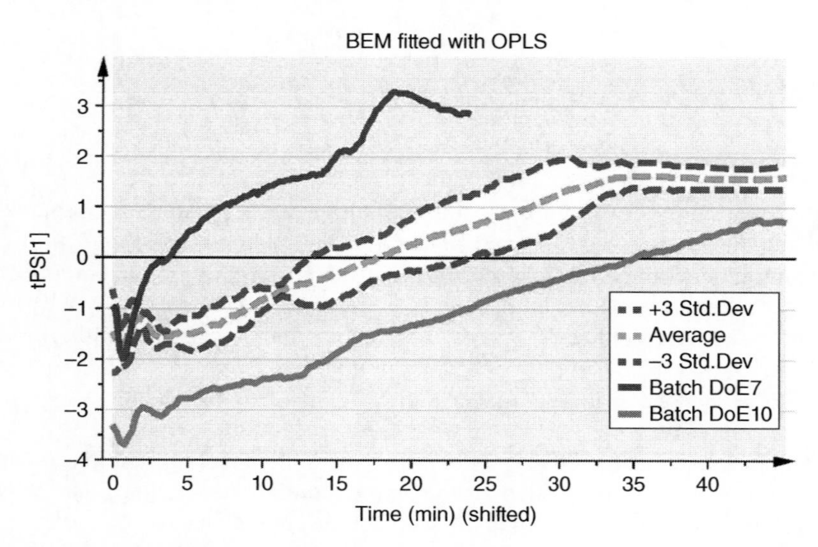

Figure 8.18 *Batch control chart showing predictions for two bad batches. The BEM fitted using orthogonal PLS readily detects the deviating batches.*

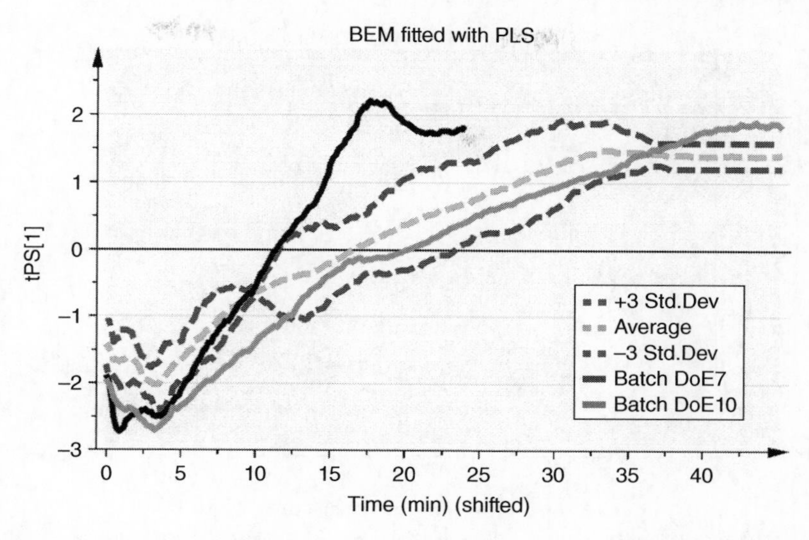

Figure 8.19 *Batch control chart showing predictions for two bad batches. The BEM was fitted using classical PLS.*

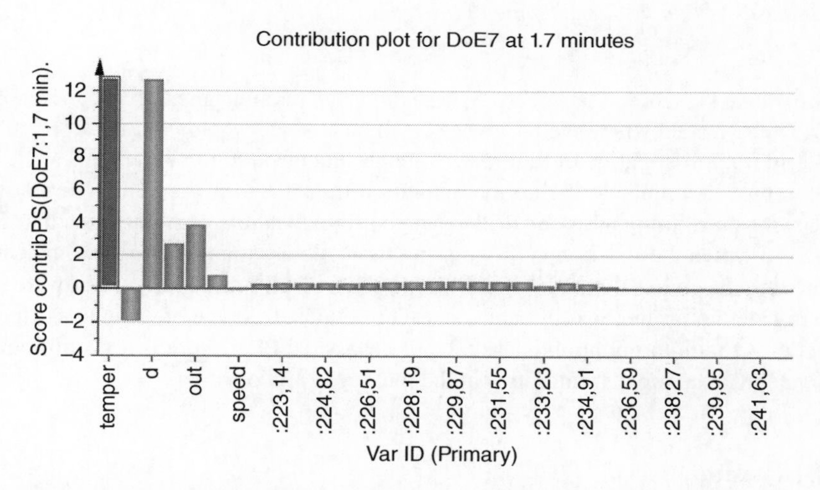

Figure 8.20 *Contribution plot for batch DoE7 at 1.7 min showing which variables are contributing to the process deviation. The process upset is caused by a few of the process parameters. For example, the contribution plot indicates that the reaction temperature (highlighted by red color) is much higher than for a normal batch. (See insert for color representation of the figure.)*

the rest of the batch duration. Figure 8.20 shows a contribution plot at the time point 1.7 min. This plot indicates that the reaction temperature is much higher than anticipated. Using drill-down technology, Figure 8.21 shows the raw data plot of the reaction temperature. It can be concluded that the reaction temperature is almost 15° higher than for the good batches. In an analogous manner, it can be inferred that gas feed is twice the normal

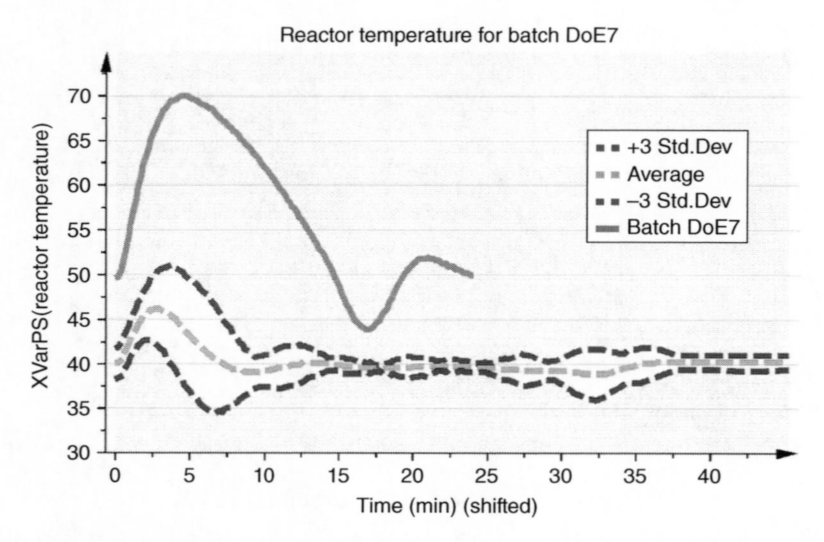

Figure 8.21 *Line plot of the reaction temperature for the non-NOC batch DoE7. At the time point of 1.7 min, the reaction temperature is almost 15° higher than the average temperature across the six NOC batches. The deviation from normality of the reaction temperature for DoE7 increases further into the lifetime of the batch.*

level (no plots shown for gas feed). These unusual circumstances have increased the reaction rate to a state where the progression of batch is very rapid.

Furthermore, investigation of Figure 8.18 shows that the second plotted non-NOC batch, DoE10, is predicted outside the normal evolution trajectory for the entire lifetime of the batch. Using contribution plots and drill-down to original data, it can concluded that DoE10 is a batch in which the reaction is progressing more slowly than for the NOC batches (no plots shown). In conclusion, this study shows that OPLS® multivariate software model works very well in the analysis of batch data. The added advantage of using this orthogonal PLS model is that interpretation is easy. Unlike classical PLS, there is no confounding of predictive and orthogonal variabilities in the measured process data.

8.8 Discussion

We end this chapter by providing an overview discussion on MVDA as applied within the PAT and QbD arena, and how this methodology may assist in CPV.

8.8.1 The PAT Initiative

PAT has its origins in an initiative of the US Food and Drug Administration, FDA, first launched in mid-2002. The goal of PAT is to improve the understanding and control of the entire pharmaceutical and biopharmaceutical manufacturing process. One way of achieving this is through timely measurements of critical quality and performance attributes of raw and in-process materials and processes, combined with MVDA. This needs to be coupled with DoE to maximize the information content in the measured data. A central concept within the PAT paradigm is that quality should arise as a result of design-based understanding of the processes, rather than merely by aiming to generate products that meet minimum criteria within defined confidence limits, and rejecting those that fail to meet the criteria.

There are many objectives associated with the PAT concept, but the foremost goal is to improve the understanding and control over the entire pharmaceutical or biopharmaceutical manufacturing process. Briefly, PAT can be understood as a framework of tools and technologies for accomplishing this goal. Interestingly, the US FDA defines process understanding as "the identification of critical sources of variability, management of this variability by the manufacturing process, and the ability to accurately and reliably predict quality attributes." MVDA tools are clearly needed to achieve this, together with insight, process knowledge, and relevant measurements.

A common question is, "what can be accomplished with PAT?" The answer to this is multifaceted. In part, this is because PAT means different things to different people. Its answer may also depend on whether one is involved in academia, industry, or a regulatory agency. Moreover, the connotations associated with PAT depend to a great extent on the context, for example, whether it is to be applied to an existing process (continuous or batch) or a process still under development.

MVDA and PAT may be applied to important unit operations in the pharmaceutical industry. In our experience, the most common applications of PAT relate to on-line monitoring of blending, drying, and granulation steps. Here, the measurement and analysis of multivariate spectroscopic data are of central importance. These spectroscopic data form the X-matrix, and if there are response data (Y-data), the former can be related to the latter using classical PLS or orthogonal PLS to establish a multivariate calibration model (a so-called soft sensor model). However, MVDA should not be regarded as the only important approach; DoE should also be considered. DoE is very useful for defining optimal and robust conditions in such applications.

8.8.2 What Are the Benefits of Using DoE?

DoE is complementary to MVDA. In DoE, the findings of a multivariate PCA, PLS, OPLS or O2PLS model are often used as the point of departure because such models highlight which process variables have been important in the past. Systematic and simultaneous changes in such variables may then be induced using an informative DoE protocol. In addition, the controlled changes to these important process factors can be supplemented with measurements of more process variables which cannot be controlled. Hence, when DoE and MVDA are used in conjunction, there is the possibility of analyzing both designed and non-designed process factors at the same time, along with their putative interactions. This allows us to see how they jointly influence key production attributes, such as the cost-effectiveness of the production process, and the amounts and quality of the products obtained.

Generally, DoE is used for three main experimental objectives, regardless of scale. The first of these is screening. Screening is used to identify the most influential factors, and to determine the ranges across which they should be investigated. This is a straightforward aim, so screening designs require relatively few experiments in relation to the number of factors. Sometimes more than one screening design is needed.

The second experimental objective is optimization. Here, the interest lies in defining which approved combination of the important factors will result in optimal operating conditions. Since optimization is more complex than screening, optimization designs demand more experiments per factor. With such an enriched body of data, it is possible to derive quadratic regression models.

The third experimental objective is robustness testing. Here, the aim is to determine how sensitive a product or production procedure is to small changes in the factor settings. Such small changes usually correspond to fluctuations in the factors occurring during a "bad day" in production, or the customer not following the instructions for using the product. In this context, it is appropriate to consider the notion of normal operating ranges (NORs) of the process factors. One way to address the robustness testing objective is to verify robustness within the NOR of each factor, which will constitute a well-documented and scientific test approach to the selected NOR.

The great advantage of using DoE is that it provides an organized approach, with which it is possible to address both simple and tricky experimental and production problems. The experimenter is encouraged to select an appropriate experimental objective, and is then guided through devising and performing an appropriate set of experiments for the selected objective. It does not take long to set up an experimental protocol, even for someone unfamiliar with DoE. What is perhaps counterintuitive is that the user has to complete a whole set of experiments before any conclusions can be drawn.

The rewards of DoE are often immediate and substantial; for example, higher product quality may be achieved at lower cost, and with a more environment-friendly process performance. However, there are also more long-term gains which might not be apparent at first glance. These include a better understanding of the investigated system or process, greater stability and greater preparedness for facing new, hitherto unforeseen challenges.

8.8.3 QbD and Design Space

The importance of DoE to the pharmaceutical and biopharmaceutical industry is underlined by the growing interest in the QbD concept inspired by ICH. ICH, the International Conference on Harmonization of Technical Requirements for Registration of Pharmaceuticals for Human Use, is a collaborative body bringing together the regulatory authorities and pharmaceutical industry of Europe, Japan, and the United States to discuss scientific and technical aspects of drug registration. ICH's mission is to achieve greater harmonization to ensure that safe, effective, and high-quality medicines are developed and registered in the most resource-efficient manner. ICH regularly releases guidance documents relating to quality, safety, and efficacy. A number of guidelines on the quality side, denoted ICH Q8, Q9, and Q11 (Q for Quality), concern methods and documentation principles of how to describe a complex production or analytical system.

The ICH guidance documents position QbD based on DoE as a systematic strategy to address pharmaceutical development. It should be appreciated that it is both a drug product design strategy and a regulatory strategy for continuous improvement. QbD is based on developing a product that meets the needs of the patient and fulfills the stated performance requirements. QbD emphasizes the importance of understanding the influence of starting materials on product quality. Moreover, it also stresses the need to identify all critical sources of process variability and to ascertain that they are properly accounted for in the long run.

As mentioned earlier, the major advantage of DoE is that it enables all potential factors to be explored in a systematic fashion. Let us take, as an example, a formulator, who will be operating in a certain environment, with access to certain sets of equipment and raw materials that can be varied across certain ranges. With DoE, the formulator can investigate the effect of all factors that can be varied and their interactions. This means that the optimal formulation can be identified, consisting of the best mix of all the available excipients in just the right proportions. Furthermore, the manufacturing process itself can be enhanced and optimized in the same manner. When the formulation recipe and the manufacturing process have been screened, optimized, and fine-tuned using a systematic approach based on DoE and MVDA, issues such as scale-up and process validation can be addressed very efficiently because of the comprehensive understanding that will have been acquired of the entire formulation development environment.

Thus, one objective of QbD is to encourage the use of DoE. The view of this approach is that, once appropriately implemented, DoE will aid in defining the design space of the process. The design space of a process is the largest possible volume within which one can vary important process factors without risking violation of the specifications (the demands on the responses). Outside this window of operation, there can be problems with one or more attributes of the product.

8.8.4 MVDA/DoE Is Needed to Accomplish PAT/QbD in Pharma

Because the pharmaceutical and biopharmaceutical industry has complex manufacturing processes, MVDA and DoE have great potential and applicability. Indeed, the pharmaceutical sector has long been at the forefront of applying such technology. However, this has mainly been at the laboratory and pilot plant scale and has historically been much less widespread in manufacturing. The main explanation for this dichotomy is that the pharmaceutical industry has been obliged to operate in a highly regulated environment. This environment has reduced the opportunity for change, which in turn has limited the applications of MVDA and DoE in manufacturing.

Pharmaceutical processes are complex with potential product variability due to both variation in operating conditions and raw materials. Final drug quality is influenced by everything that happens during manufacturing – each process step, each ingredient, the condition of the equipment and even subtle changes in the manufacturing environment itself can lead to variations in product quality. The univariate specifications presently used for characterization of raw materials cannot adequately describe their quality or influence on the final product, and often allow problems to go undetected. Only by developing multivariate models that account for each potential cause of variability, and by applying process analytics, can drug manufacturers establish a foundation for manufacturing quality systems at their facilities. MVDA, MSPC, BPSC, and DoE should therefore be important elements of any PAT program, MVDA being the indispensable cornerstone.

8.8.5 MVDA: A Way to Power up the CPV Application

In the final section, we will comment on CPV. The goal of CPV is to corroborate that the process in question is in a state of control during manufacturing. The guidance documentation recommends that CPV be based on data that "include relevant process trends and quality of incoming raw materials or components, in-process material, and finished product."

Furthermore there is a recommendation that the data "should be statistically trended" and that the information so obtained "should verify that the quality attributes are being appropriately controlled throughout the process."

Our first example involving characterizing raw material properties based on particle size distribution curves is a good example of how MVDA can be used to understand quality variations of incoming raw materials. Both the second and the third examples demonstrate how process data can be plotted in control charts, for example, statistically trended, in a multivariate fashion. And the second example, although not being drawn from the pharmaceutical industrial sector, exemplifies how cause-and-effect relationships between influential process factors and critical quality attributes can be manifested and used to understand what sort of process factor settings it takes in order to appropriately control the critical quality attributes.

In general terms, it can be stated that CPV as described by the guidance documents is still fairly new to the pharmaceutical industry. Presently, many are only monitoring and tracking final product quality, but the guidance is clear that process variation relating to quality should also be included. Trending multiple univariate parameters is common, but in the light of what has been discussed in this chapter, this is not a very efficient approach. The MVDA toolbox consisting of the PCA, PLS and OPLS® multivariate models is flexible and can be applied to continuous and batch processes, different types of process data (e.g. process parameters and spectral data), and can link variability in process inputs to variability in process outputs and qualities. With this toolset, we can handle several critical CPV steps in a very efficient manner; notably we here think of the ability to

- Summarize variability in raw material qualities and composition.
- Summarize process variability relative to correlation to quality variation; the latter is especially powerful with OPLS®, where we can isolate process variability in the direction of quality (enhanced interpretation).
- Summarize process performance in a meaningful way, especially for batch processes using limits arising from multivariate model parameters (e.g., Hotelling T2 and DModX).
- Summarize and visualize a low-risk process factor region within which there will be a low probability to fail to comply with quality specifications.

With the strict quality demands on pharmaceutical products, product variability must be kept acceptably small, and consequently there is a need for *process control techniques* to keep critical process variables on specified trajectories, and *process monitoring* to ensure that all parts of the process remain inside specified "trajectory volumes" formed by the data in multidimensional space. Such trajectories can easily be specified using multivariate parameters, such as scores, Hotelling´s T^2, DModX, contributions, and so on as seen in the case studies reported in this chapter.

8.9 References

[1] Jackson, J.E. (1991), *A User's Guide to Principal Components*, John Wiley, New York. ISBN 0-471-62267-2.

[2] Wold, S., Albano, C., Dunn, W.J. *et al.* (1984), Multivariate data analysis in chemistry, in *Chemometrics: Mathematics and Statistics in Chemistry* (ed. B.R. Kowalski), D. Reidel Publishing Company, Dordrecht, Holland.

[3] Wold, S., Geladi, P., Esbensen, K., and Öhman, J. (1987), Multiway principal components and PLS-analysis, *Journal of Chemometrics*, **1**, 41–56.
[4] Wold, S., Sjöström, M., and Eriksson, L. (2001) PLS-regression: A basic tool of chemometrics, *Chemometrics and Intelligent Laboratory Systems*, **58**, 109–130.
[5] Trygg, J., and Wold, S. (2002), Orthogonal projections to latent structures (OPLS), *Journal of Chemometrics*, **16**, 119–128.
[6] Gabrielsson, J., Jonsson, H., Airiau, C. *et al.* (2006), OPLS methodology for analysing pre-processing of spectroscopic data, *Journal of Chemometrics*, **20**, 362–369.
[7] Souihi, N., Lindegren, A., Eriksson, L., and Trygg, J.(2015), OPLS in batch process monitoring – opens up new opportunities, *Analytica Chimica Acta*, **857**, 28–38.

9

Process Analytical Technology (PAT)

Line Lundsberg-Nielsen,[1] Walkiria S. Schlindwein,[2] and Andreas Berghaus[3]

[1] Lundsberg Consulting Ltd and NNE, London, United Kingdom
[2] De Montfort University, Leicester, United Kingdom
[3] ColVisTec AG, Berlin, Germany

9.1 Introduction

PAT was originally introduced by the US Food and Drug Administration (FDA) [1] and later adopted by the International Council for Harmonisation (ICH Q8) [2] as 'a system for designing, analysing, and controlling manufacturing through timely measurements (i.e. during processing) of critical quality and performance attributes of raw and in-process materials and processes with the goal of ensuring final product quality'. The term 'analytical' in PAT is viewed broadly to include chemical, physical, microbiological, mathematical and risk analysis conducted in an integrated manner. The goal of PAT is to enhance understanding and control the manufacturing process. Quality cannot be tested into products; it should be built in or should be by design. PAT is an enabler for designing the quality into the product when Quality by Design (QbD) principles are applied.

It is important to highlight that PAT is a *system*, not an analyser, that requires *action* to be taken based on *real-time information*. Such information could come from a process analyser, a mathematical model or a laboratory test but must be available in 'real-time' or 'near real-time'. This ensures understanding of the process, allowing adjustment to a process parameter set point, or termination of a process when an endpoint has been

Pharmaceutical Quality by Design: A Practical Approach, First Edition.
Edited by Walkiria S. Schlindwein and Mark Gibson.

reached. Although some PAT systems are being used to monitor a process, the full value of a PAT system is not utilised unless the information from the system is used to learn actively about the process or to control it.

The definition refers to 'timely measurements (i.e. during processing) of critical quality and performance attributes', meaning measurements of the finished product critical quality attributes (CQAs), intermediate CQAs or raw material CQAs, that is, measurements of attributes of critical materials. Because the measurements are performed during processing, in-line process analysers are frequently used and are often called PAT tools, PAT analysers or PAT probes. Use of such terms should be discouraged, as they form just the analytical instrument of the PAT system.

Finally, the definition refers to *'the goal of ensuring final product quality'*, that is, using the real-time measurement system to ensure that a product with the required quality is being produced. Often, the measurements are used in a predictive model to determine a process endpoint or to predict a quality attribute of a product, that is, a value of a raw material, an intermediate material or a finished product CQA.

PAT is therefore an important element of the control strategy as it enables the control of CQAs directly or through critical process parameters (CPPs) and materials CQAs that have a direct impact on specific product CQAs.

In 2004, the FDA published *Guidance for Industry, PAT — A Framework for Innovative Pharmaceutical Development, Manufacturing, and Quality Assurance*[1] after years of discussions on how to encourage the pharmaceutical industry and the healthcare authorities to facilitate innovation and the application of advanced technology for development and manufacturing of pharmaceutical products. The PAT Guidance turned out to be the first of a collection of guidelines that we today refer to as the 'science and risk based guidelines' or 'the QbD guidelines'. These includes the ICH Q8(R2), [2] and Q11, [3] that have both adopted the PAT definition and concept.

The guidance describes the goal of PAT: 'To enhance understanding and control the manufacturing process, which is consistent with the FDA current drug quality system: quality cannot be tested into products; it should be or should be by design. Consequently, the tools and principles described in the PAT Guidance should be used for gaining process understanding and can also be used to meet the regulatory requirements for validating and controlling the manufacturing process'.

To realise products that comply with the regulatory expectations, it is essential to focus on understanding the manufacturing process, understanding sources of variability and being able to manage these variations to ensure consistent product quality, and PAT is therefore becoming more essential that it was in the past.

Despite the age of the PAT Guidance for industry, it is worth reading. It addresses the essential issues for gaining a higher level of process understanding, the use of science- and risk-based approaches, setting specifications based on mechanistic understanding, the use of different PAT tools for gaining understanding or for controlling the process, including the opportunity for real-time release and continual improvement.

Terminology related to PAT is defined in a pharmaceutical standard guidance developed by the PAT community within ASTM [4]. ASTM is a consensus-based global standard organisation with members from the industry, health authorities (FDA and EMA) and academia [4].

9.2 How PAT Enables Quality by Design (QbD)

PAT and QbD share similar goals for pharmaceutical and biopharmaceutical manufacturing as they both focus on process understanding, process control and making science- and risk-based decisions. In the QbD framework, PAT plays the role of the toolbox and therefore the enabler for gaining process understanding, defining the control strategy and supporting continual improvement throughout the product lifecycle.

The tools of PAT can, during development, facilitate the identification of CPPs and relevant material CQAs, identification of interactions between process parameters, attributes and product CQAs, for establishing the design space, the related control strategy, scale-up, technology transfer and product and process specifications. In commercial manufacturing, PAT can be used as an element of the control strategy to monitor, control and improve the process, as well as for Real-Time Release Testing (RTRT) measurements of CQAs in lieu of end-product testing.

9.3 The PAT Toolbox

Different tools are available under the PAT framework for gaining process understanding or for controlling a commercial process:

- Process analysers – multivariate.
- Process sensors – univariate.
- Process modelling tools.
- Design of experiments (DoE).
- Multivariate data analysis (Chemometrics, MVDA or MVA).
- Process control tools.

The various tools can be used extensively throughout the product lifecycle to gain process understanding, to design the product and processes, for technology transfer and to establish the controls that will ensure the product is being produced by a robust process in a consistent manner within specifications. A typical PAT system consists of several elements, for example, as shown in Figure 9.1 and Table 9.1.

Several examples of how to use PAT during development and in manufacturing have been presented at various conferences, published in scientific literature [6, 7], through different PAT societies (e.g. the ISPE Process Analytical Technology and Lifecycle Control Strategy (PAT&LCS) Community of Practice (CoP), www.ispe.org), vendor application notes and YouTube presentations.

9.4 Process Sensors and Process Analysers

Real-time measurements are essential for a PAT system being used in development for gaining process understanding and being used in production for controlling the process. Both process parameters and material or product attributes can be measured in real time. These measurements are typically done by process sensors or process analysers.

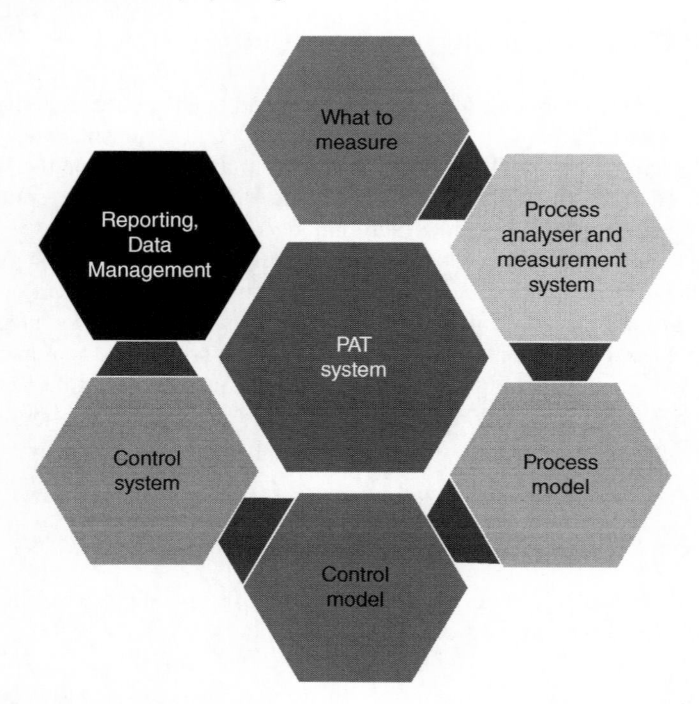

Figure 9.1 *Elements of a PAT system (each element is further described in Table 9.1).*

Process sensors are univariate measurements, and their control systems present values through instruments that only read one variable (parameter or attribute) at the time such as temperature, pressure, conductivity, dissolved oxygen (DO), pH, speed, agitation rate, power consumption, time and frequency.

Process analysers, on the other hand, are multivariate systems. They handle many biological, physical and chemical variables at the same time and are typically used to measure a raw material CQA (e.g. identity or strength of a raw material), an intermediate CQA (e.g. blend endpoint, water content or cell density), a finished product CQA (e.g. assay, uniformity of dosage units or water content) or a process signature (trajectory) following the process in real time from the beginning to the end.

The typical process analyser cannot stand alone but is connected to a more complex system including analytical software, a data management system and a measurement interface to the process.

Process analysers may be operated in different modes:

Off-line: A sample is taken from the process and analysed in a laboratory. It may take several days to get the results.

At-line: A sample is taken from the process and analysed in close proximity. It usually takes minutes to get the results.

On-line: Part of the process stream is diverted from the main process and analysed without interruption. A sample might be collected for reference purpose or the analysed sample returned to the process stream. It usually takes seconds.

Table 9.1 *Elements of a PAT system.*

Element	Description	Example
What to measure	The purpose of the analysis. The raw material, in-process, intermediate or final product CQAs, CPPs or process signature (trajectory) that must be monitored and controlled.	Drying endpoint, characterised by the intermediate CQA of moisture content. The drying process will be characterised by a process signature. The drying endpoint will be defined as the point in time when the moisture content (e.g. measured by NIR (Near Infrared) is below a certain acceptance criteria.
Process Analyser and Measurement System	The preferred process analyser technology and its measurement system. The system of sensors, instruments and/or process analysers collects signals from the process (materials or process equipment) and converts those signals into data.	Drying: NIR instrument, a non-invasive measurement head attached to a window in the dryer that collects NIR spectra. By a multivariate model (established by the NIR instruments analytical software), these are converted to moisture content values.
Process model	A mathematical algorithm, often based on a calibration, e.g. a partial least squares (PLS) model, (see Chapter 8) that uses data from the measurement system as input to calculate the value of the CQAs in real time, i.e. at the time of the measurement. Calibration-free models are possible for some applications.	An NIR model can be developed using an analytical reference method, e.g. the Loss of Drying (LOD) for the calibration. The process model is developed using a PLS model established between collected NIR spectra and LOD reference measurements of samples collected at the same time as the NIR spectra were collected. The model might include spectra from lab, pilot and commercial scale. After implementation and validation of the model, it can be used to predict the moisture content from a spectrum that is not included in the model. If it is sufficient to demonstrate that the moisture content is below a certain threshold and an exact value is not needed, a calibration-free model such as moving block standard deviation (MBSD) can be used to identify the process endpoint.
Control model	A procedure or mathematical algorithm that uses the outputs of the process model combined with any other data inputs required to calculate values for the CPPs for the process. It uses input data from the process to generate an actionable command or commands that are issued to the control system [5].	A control model is used to compare the predicted moisture content against acceptance criteria. If the value is too high, a 'message' is sent to the dryer to continue the drying process. If the predicted value is below the threshold, a 'message' is sent to the dryer, instructing it to stop.

(Continued)

Table 9.1 *(Continued)*

Element	Description	Example
Control system	A system that responds to inputs signals from the process, its associated equipment, other programmable systems or an operator or both, and generates output signals causing the process and its associated equipment to operate in the desired manner [5].	The system that translates the 'messages' from the control model to equipment controls, i.e. that makes the dryer continue or stop.
Reporting, data management	Reporting of the outcome of the control system, e.g. a CQA value, management of data related to the process and control models as well as data collected during the actual control.	Documentation of the final moisture content (e.g. manufacturing execution system (MES), laboratory information management system (LIMS) and the batch records). Management of data related to the model, validation, execution, collection and analysis of NIR spectra, final drying time and other CPPs. How often spectra are collected, storage of raw data, which model is used for analysis of the data, how it is being reviewed, storage of the prediction result of the analysis, and so on.

In-line: A process analyser is located within the process stream in a non-invasive way. This could, as an example, be a measurement through a measuring head (probe) connected directly to a process equipment window (e.g. a blender). In this case, the probe is not in direct contact with the process material. Alternatively, the installation can be invasive. A probe (measuring head) inserted into the process and is in direct contact with the process material. The material analysed (samples) are not removed from the process stream. This can take less than a second to a few seconds and provides often the best live picture of and insight into the process.

Table 9.2 below gives examples of some chemical or physical attributes that can be measured at common unit operations employed typically in pharmaceutical drug substance and drug product manufacture, possible process analyser and the purpose of the application, for gaining understanding or for controlling.

9.4.1 Process Sensors – Univariate

The value of using conventional univariate sensors has not decreased since the introduction of multivariate process analysers, and they still serve their purpose and often combined they can be used for establishing a process model. The issue is 'knowing what to measure', 'when and why', as well as understanding the analysis of the data and making use of the data in development work or in quality decisions.

Process sensors measure one attribute (CQA) or process parameter (CPP) (Table 9.3). Process sensors are well known, simple to use and the data are simple to analyse; they are cheap and should be used whenever it is possible to keep the system as simple as possible.

9.4.2 Process Analysers – Multivariate

Along with the development of chemometrics (see Chapter 8), spectroscopy is becoming more and more important when developing, understanding and controlling complex pharmaceutical or biopharmaceutical processes. The mainstream technologies or analysers will be explained in brief, but it is worth mentioning that new technologies and applications are ongoing development activities. When considering a technology, it is highly recommended to seek out the most up-to-date information.

Many of the most popular process analysers use spectroscopic techniques, analysing transmission, reflection or absorption of light. Depending on the electromagnetic nature of the light, that is, the wavenumber, wavelength or frequency used for an application, information of different chemical and physical characteristics of the sample are obtained.

The laboratory versions are complex and have facilities for analysing many different sample types including solids, liquids and gases. Analysis of discrete samples or continuous flow can often be arranged. Process instruments are robust industrial designs often fitted with optical fibre technology to transmit light between the analyser system and the process itself. This is not trivial. It relies on the equipment design and where the probe can be placed and needs to be placed. Below (Table 9.4) are some examples of typical applications and technologies.

9.4.3 Infrared (IR)

Absorption in the IR or mid-infrared (MIR) region (2.5–50 μm or wavenumber: 4000–200 cm^{-1}) gives structural information about a compound.

Table 9.2 *Typical examples of attribute measurements by process analysers within common unit operations.*

Unit operation	Attribute measurement	Analytical techniques to gain process understanding	Process analysers for process control
		Drug Substance, active pharmaceutical ingredient (API), small molecule	
Dispensing	Identity	NIR (Near Infrared) or NIRS (Near Infrared Spectroscopy) at-line	NIR in-line
Solvent recovery	Assay	NIR	NIR in or on-line
Reaction	Assay, intermediates, process signature polymerisation	IR (Infrared) NIR Raman THz (Terahertz)	NIR in or on-line Raman in-line
Homogenisation	Precipitation, flocculation in suspension and emulsion	FBRM (Focus Beam Reflectance Measurement) laser diffraction	FBRM in-line In-line particle size analyser
Crystallisation	Nucleation and crystal growth Particle size Cord length Particle analysis Turbidity Morphology, growth, suspensions, emulsion sedimentation	Obtain direct information about the morphology and size of co-precipitates FBRM PVM (Particle Vision and Measurement) Video microscope Turbidity analyser NIR THz	FBRM in-line turbidity analyser in-line In-line particle analyser
Solid/liquid separation	Turbidity	Turbidity analyser	Turbidity analyser
Drying	Moisture content Gas analysis	NIR, TDLS (Tunable Diode Laser Spectroscopy) or just TDL	NIR in-line TDL in-line
Milling	Particle size Particle characterisation API, intermediate characterisation	FBRM, PVM, Video microscope (in-line), XRF (X-ray fluorescence), IR	In-line particle analyser

		Drug substance, large molecule	
Fermentation	Process signature	NIR	NIR in-line
Upstream	Cell count	NIR	NIR in-line
Biomass growth, cell cultivation	Cell density	Turbidity analyser	Solid-state refractive index analyser
	Viable Cell Density	Refractive index	
	Cell growth	Raman	Raman in-line
	Feed concentration and rate	Capacitance	Capacitance in-line
Downstream harvest, purification	Protein concentration, impurity	LC (Liquid Chromatography HPLC or UPLC at-line and on-line)	LC on-line
		MS (Mass Spectrometry)	MS on-line
		Conductivity	Conductivity in-line
		UV-Vis (Ultra Violet-Visible)	UV in-line
		Drug product, solids	
Dispensing	Raw material identification, particle size distribution	NIR at-line	NIR in-line (handheld)
		Raman at-line	Raman at-line (handheld)
Granulation	Homogeneity, composition, assay, endpoint, particle size distribution	NIR	NIR in-line, Acoustics, UV-Vis in-line
		Acoustics	
		UV-Vis	
		FBRM	
Drying	% Water content, endpoint detection	NIR	NIR in-line
		MRT (Microwave Resonance Technology)	
Blending/mixing	Powder mixing, Homogeneity, endpoint	NIR	NIR in-line
		Raman	
	% API content uniformity		
Extrusion	Concentration of API in opaque or transparent material	NIR	NIR in-line
		UV-Vis	UV-Vis in-line
Compression	Identification, assay, content uniformity	NIR	NIR in-line, at-line

(Continued)

Table 9.2 (Continued)

Unit operation	Attribute measurement	Analytical techniques to gain process understanding	Process analysers for process control
Coating	Coating thickness, endpoint, colour	NIR, THz UV-Vis	NIR in-line UV-Vis in-line
Product characterisation	Composition, morphology, particle size, particle shape, coating thickness, salt form	THz IR-Imaging NIR-Imaging Raman-Imaging XRF	In-line particle analyser NIR in-line Imaging Vision systems
Filling	Identification, head space analysis (oxygen)	TDLS	NIR, TDLS in-line
Lyophilisation	Moisture content	NIR MS	MS on-line
		Drug product, liquids	
Blending/mixing	Homogeneity, endpoint	NIR Raman	NIR in-line, Raman in-line
Clean process steps	Bioburden	LIF (Light Induced Fluorescence)	LIF
Lyophilisation	Moisture content	NIR MS TDLS	MS TDLS in-line

Table 9.3 Typical attribute measurements by process sensors.

Process sensor	In-process or intermediate CQA	Application
pH in-line	pH	pH sensors are widely used in aqueous processes.
Conductivity in-line	Purity	Measurement on total ionic concentrations (e.g. dissolved compounds) in aqueous chemical and biologics processes. Widely used applications are water and product purification, cleaning in place (CIP)/SIP (clean in place/ steam in place) control and the measurement of solution concentration.
Capacitance in-line	Viable Cell Density	Bioreactor, cell growth
Total organic carbon (TOC)	TOC	Measurement of ppb-level of organic contamination using UV oxidation and conductivity measurement technology. Used for water purification processes.
Refractive index (RI)	Concentration of dissolved solids in a liquid (or liquid mixtures)	Measurements based on refracted and reflected light Used, e.g. for feed control in a bioreactor
Dissolved oxygen sensor	DO	Bioreactor, fermentation processes
Dissolved carbon dioxide sensor	CO_2	Fermentation processes
Turbidity analysers	Colour, cloudiness	Bioreactors

One of the advantages of IR is that it can provide a snapshot or even a video of the process in real time without requiring multivariate modelling.

IR has found a niche in Active Pharmaceutical Ingredient (API) development and manufacturing, especially in synthesis monitoring, because of its ability to detect structural information quantitatively. One of the weaknesses of IR is its short penetration depth in material due to strong absorption. It is therefore often difficult to justify that an IR measurement is representative of the entire process.

Mid-IR is a versatile tool in pharmaceutics and biopharmaceutics, with a wide field of applications ranging from characterisation of drug formulations to elucidation of kinetic processes in drug delivery. FTIR-ATR (Fourier Transform Infrared - Attenuated Total Reflection) is a well-established standard method used to study drug release in semi-solid formulations, drug penetration and the influence of penetration modifiers. It has also been demonstrated as a useful technology in various clinical in vivo studies [8].

Table 9.4 *Typical attribute measurements by process analysers.*

Process sensor	In-process or intermediate CQA	Application
NIR In-line	% API content uniformity	Powder mixing, tablet manufacturing
NIR In-line	% Water content	Drying, wet granulation processes
Raman In-line	Polymorphism	Monitoring of isothermal phase transformation
Chemical Imaging (CI)	Coating thickness	Monitoring of thickness during coating process
Particle Vision and Measurement (PVM)	Nucleation and crystal growth	Obtain direct information about the morphology and size of co-precipitates
NIR In-line	Powder blend bulk density	Real time prediction based on NIR calibration models for blend density and drug concentration
Solid State Nuclear Magnetic Resonance (NMR)	Interactions, drug-drug or drug-excipients	Measurement nuclei and chemical environment within a molecule.
Powder X-Ray Diffraction (XRD)	Polymorphism	Structural information from 5 to 90^0 2θ
Terahertz Pulsed Imaging (TPI)	Film thickness	Three dimensional space-time data analysis.

9.4.4 Near Infrared (NIR)

NIR spectroscopy is the most popular technology among the modern process analysers because it is not only applicable to liquids, but also to powders and solids. The NIR region of the electromagnetic spectrum covers 750 nm to 2500 nm (wavenumbers: 13,300 to 4000 cm^{-1}).

NIR spectra provide information on overtones and combinations of fundamental vibrations. They convey structural information, although this is not as specific as with IR. NIR spectra are often analysed using qualitative or quantitative multivariate data analysis such as principal component analysis (PCA) or partial least squares regression (PLS) (see Chapter 8) to provide a process signature, qualify or quantify the presence of an analyte (chemical).

The following are the typical advantages of using NIR:

- The absorption is 10–100 times weaker than in IR (mid-infrared) resulting in deeper material penetration.
- High NIR throughput is possible, even when employing low-cost fibre optics.
- NIR light penetrates PE (polyethylene) materials and glass, making non-invasive analysis possible.
- Low-cost instruments with high signal-to-noise ratio (SNR).
- Endpoint determination is possible, for example, reaction, drying, granulation, blending and coating.
- Material characterisation is possible, for example, identity, polymorphs, moisture content, bulk density, homogeneity and concentration.
- Robust industrial designs possible.

Near infrared (NIR) has been increasingly used for real-time measurements of critical process and product attributes, as this technique allows rapid and non-destructive measurements without sample preparations [9–14].

9.4.5 Tunable Diode Laser Spectroscopy (TDLS)

An NIR diode laser tunable, in a very narrow wavelength region, can be used for measuring several physical or chemical analytes known to have absorption in the range of the diode laser. The laser frequency is tuned to match the internal vibration frequency of a target substance, thereby making it a dedicated application. TDLS is used widely from headspace analysis to lyophilisation.

9.4.6 Ultraviolet-Visible (UV-Vis)

UV-Vis is another spectral technique that can be used to study how radiation interacts with samples. UV-Vis spectroscopy is used for quantitative analysis of solution concentrations where straightforward detection is needed (though there are examples of mathematical models being used to gain much more information). UV-Vis is robust and can be used for detection of impurities in cleaning in place (CIP) of cycles, measurements of API concentrations in hot-melt extrusion and controlling a dosing process (see case study example in this chapter). It can be used to characterise the behaviour of an extruder in general, for example, cleaning, residence time distribution, homogeneity of mixture, deterioration of polymer and or API and amorphous versus crystalline phases.

9.4.7 Raman

Raman spectroscopy is used for routine qualitative and quantitative measurements of both inorganic and organic materials, and it is successfully employed to solve complex analytical problems such as determining chemical structures and polymorphic identity. Gases, vapours, aerosols, liquids and solids can be analysed. Raman spectroscopy does not detect water, making it attractive for analysing aqueous solutions.

Raman spectroscopy is popular for raw material identification, as it does not require the use of multivariate modelling and has been demonstrated successfully in a number of drug substance and drug product processes in an industrial environment [9, 15].

9.4.8 Focused Beam Reflectance Measurements (FBRM) and Laser Diffraction

A laser beam is focused on the sample spot and is reflected by the particles in the light beam. The reflected or diffracted light is detected and gives a measurement of the core length of a particle which when measured multiple times can be used as a measure of the particle size distribution. This in-line technology is widely used in the pharmaceutical industry and is applied for particle characterisation both during development and for full-scale manufacturing in both API and drug product manufacturing. Many FBRM analysers allow invasive in-line monitoring of particle growth, size, and surface area and for capability for process end-point detection [16–19].

9.4.9 Particle Vision and Measurement (PVM)

A probe-based video microscope can be used to gain an understanding of particle characteristics such as morphology, shape, size and surface area.

9.4.10 X-Ray Fluorescence (XRF)

XRF is an elemental, non-destructive technique and is typically used to assay a material of interest. Liquids, semi-solids and solids can be analysed in situ, making the technique simple but highly sample dependent.XRF is normally used when there is a specific element of interest, for example, a chloro group used to measure a salt concentration in a finished product or heavy metal concentrations of an API.

9.4.11 Imaging Technologies

Many different chemical imaging systems are available that are based on IR, NIR, Raman or THz spectroscopy and can be used to characterise materials and products in 2D or 3D depending on the technology. These systems combine advanced imaging technology with a relevant light source. Using a light source that can penetrate the sample makes investigation of the 3D structure possible.

These techniques are typically used during development for product design or for product characterisation, including revealing counterfeiting products. [20]

9.5 Analyser Selection

Analyser selection should be based on the principles of simplicity and relevance. Simple and 'commercially off-the-shelf' solutions can often do the job. For complex operations, where a simple solution is not available, breakthrough technologies might be investigated. Technologies should be selected for specific purposes. Selection of the right technology depends on the proposed control strategy and requires a cross-functional, science- and risk-based approach including project management, Quality Risk Management (QRM), analytical development (multivariate models), process engineering, QA and IT.

Some of the typical issues that must be addressed come from the control strategy such as: which attribute (CQA) or parameter (CPP) or process to investigate (understand, monitor or control), where and how to measure, who should be able to use the technology, how to integrate the technology in the facility, data management and process control system. Many of these elements can be included in a risk analysis, focusing on both risk related to installing the technology, but also the related business risk by investing in the technology. It might be very difficult during early-stage development to conduct a proper business analysis, but some qualitative and possibly quantitative implications are important to qualify to support the decision. Many examples of appealing return of investment implementing PAT have been reported, and the benefits of various technologies shown in Figure 9.2.

9.6 Regulatory Requirements Related to PAT Applications

When PAT applications are to be used for process control or measurements of product CQAs, health authorities must approve the use of the technology and the method. Both EMA and FDA have published guidelines clarifying the information needed for approval.

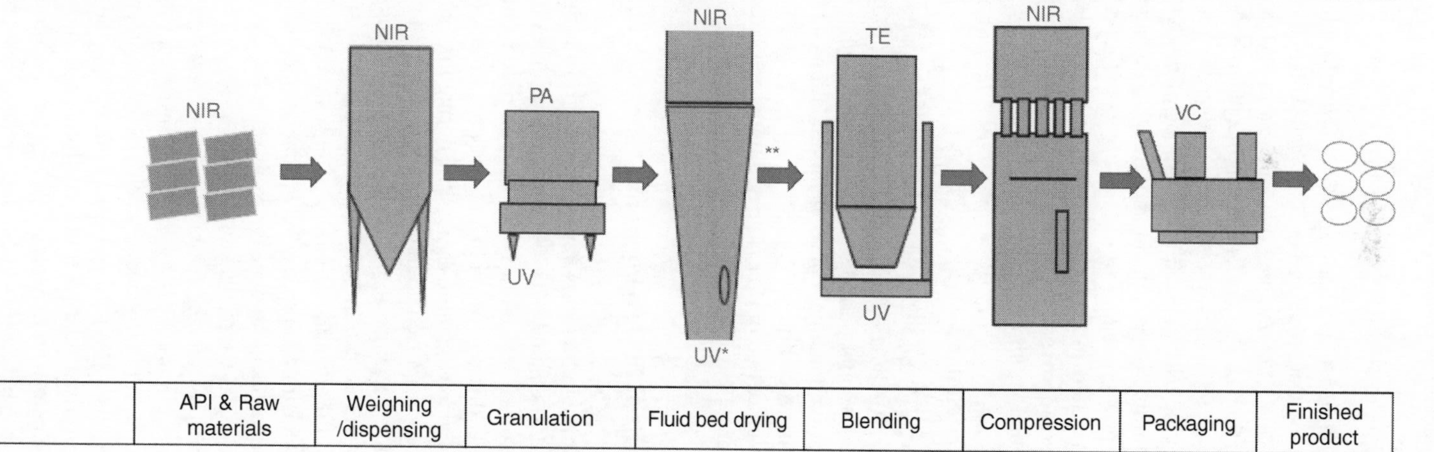

Spec	Identity	Water content & particle size	Granulate size (end point)	Water content & fines (end point)	Blend homogeneity (end point)	Assay & physical properties	Correct label & pack	
PAT	NIR	NIR	Passive acoustics (PA)	NIR	Thermal effusivity (TE)	NIR	Video capture (VC)	Real time release
Benefit	Release time ⇨ 99%	Cycle time ⇨ 10%	Cycle time ⇨ 30% (+ ⇨ 40% with CIP) Scrap & rework ⇨ 2%	Cycle time ⇨ 40% (+ ⇨ 40% with CIP) Scrap & rework ⇨ 2%	Cycle time ⇨ 70% (+ ⇨ 40% with CIP)	Scrap & rework ⇨ 2%	Scrap & rework ⇨ 2% Cycle time ⇨ 30%	Release time ⇨ 99%

	API & Raw materials	Weighing /dispensing	Granulation	Fluid bed drying	Blending	Compression	Packaging	Finished product

*CIP linked with PAT to reduce cycle time 40% and release 90%
**Perform HPLC for impurity release here

Figure 9.2 *Examples of PAT applications used for an oral solid dosage process and benefits.*

9.6.1 Europe

EMA has published 'Guideline on the Use of Near Infrared Spectroscopy by the Pharmaceutical Industry and the Data Requirements for New Submissions and Variations' [21], in operations from 2014. This NIR guideline is applicable to PAT applications that employ NIR or other similar technologies. The guideline is applicable to both quantitative and qualitative measurements used for process control or quality decisions in commercial manufacturing.

It requires the user to establish the scope of the NIR or PAT application (procedure) to facilitate continual improvement and lifecycle management. Changes within the approved scope of the NIR procedure would be subject to Good Manufacturing Practice (GMP) only and can be handled under the users own Pharmaceutical Quality System, whereas changes outside of the approved scope of the procedure would be subject to variation application, i.e. need regulatory approval.

9.6.2 United States

Similarly, in 2015 the FDA published a draft guideline: 'Development and Submission of Near Infrared Analytical Procedures Guidance for Industry' [22].

This guidance only pertains to the development and validation of NIR analytical procedures, but the fundamental concepts of validation can be applied to other PAT technologies including other techniques.

The draft guideline outlines the additional information that should be submitted to the regulators for a PAT application and how to manage changes after approved application.

9.7 PAT Used in Development

The time available for development of the product, the process and the control strategy as outlined in ICH Q8 and ICH Q11 will in most cases be directed from the clinical development programme, and there is not enough time available for developing a high level of understanding that will result in a robust process design and control strategy. PAT can help gain as much understanding as possible during this narrow time window. Process analysers can be used to visualise and investigate the process in real time, enabling understanding to be gained and decisions to be made. The PAT community has developed several ASTM standard guidances useful for gaining process understanding [23–30].

Similarly, PAT can be used to speedup scale-up from lab to pilot or to commercial scale and for transferring the process between instruments, lines or sites. Process analysers measure scale-independent product attributes such as CQAs, rather than scale-dependent process parameters such as CPPs, that are typically measured by process sensors. The understanding gained at the lab scale and the knowledge of how to measure relevant quality attributes by PAT tools are used to verify the scale-up or transfer. A process signature or CQA measurement should be similar between batch sizes, equipment sizes and site locations. Additional advantages of using PAT for process verification is the fast-real-time in situ process analysis and that the process can be followed from beginning to end, which is not possible in the same fast manner with conventional sampling and laboratory analysis. In other cases, sampling is not feasible due to safety issues or lack of sampling access, and the process signature will then provide information that is not possible to achieve with other means.

In summary, the use of PAT during development is to gain sufficient understanding of what is critical during processing, that is, which process parameters and material attributes have a direct impact on the CQAs and therefore must be controlled and become elements of the control strategy. PAT tools are used to establish a more efficient and robust control strategy.

There are many examples of PAT tools used during development for gaining understanding and defining the control strategy available in the scientific literature, including the public domain [6, 7].

Process analyser manufacturers can normally help in proposing an appropriate technology for a specific application, including offering theoretical or practical feasibilities studies (see Section 9.12)

9.8 PAT Used in Manufacturing

The purpose of PAT for commercial manufacturing is mainly process control, but it is also used for process improvement and troubleshooting.

The control strategy defines what should be controlled during processing, how it is done and where in the process. The controls could be measurement of an intermediate CQA, final product CQA or a process signature. The PAT applications explored during development can be developed further for control purposes and implemented in manufacturing.

Examples include the determination of a process endpoint for a batch process, for example, a reaction or a drying endpoint or controlling a continuous process step such as a tablet compression stage or a hot-melt extrusion stage, as will be discussed in more detail below.

A typical example of how process analysers are used for a tableting process including the manufacture of the API is shown in Figure 9.3.

The boxes on the right-hand side show the analysers or models (mostly NIR) and the attributes that can be measured. The boxes on the left-hand side show the conventional testing done during end-product testing of the API, and drug product, respectively. As illustrated, most of the conventional Quality Control (QC) tests can be replaced by process analysers or a soft sensor, which is a model based on data from specific intermediate CQAs and CPPs. The results of the analysis performed during processing can be used in lieu of end-product testing or as in-process controls, depending on how the final control strategy will be composed.

When implementing process analysers in a commercial facility, the entire PAT system must be considered and implemented as appropriate and in alignment with current GMP requirements.

The development and implementation aspects are further discussed in Section 9.10 [21, 22]. The PAT community has developed a set of practical standards that can be used for implementation [23–30]. PAT applications used for quality decisions will need regulatory approval, and the requirements relate to model development, implementation, validation and maintenance (see Section 9.6).

The advantage of implementing a PAT-based control strategy is a higher degree of process control of what is critical, reduced risk of failing a batch and manufacturability issues. Despite these benefits, only a relatively low number of applications have been filed and approved. PAT systems might still be in place but only for gaining

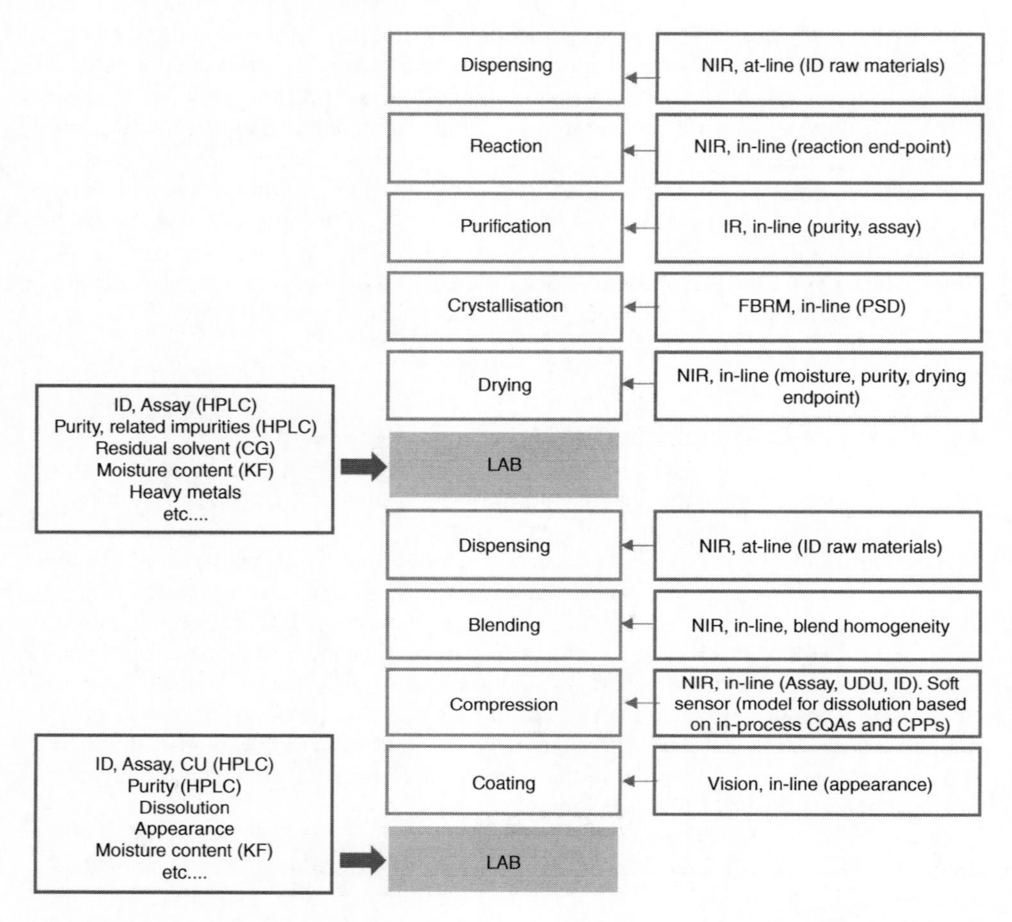

Figure 9.3 *PAT applications used during processing in lieu of conventional end-product QC testing.*

understanding and troubleshooting not for quality decisions. There are probably several reasons for the lack of a high number of submitted and approved applications. It is a change in mindset for both the industry and regulatory authorities. We are used to making quality decisions based on end product testing rather than moving the process control and measurements upstream. The increased process understanding might be a challenge to share with authorities. When QbD and PAT principles are used during development, the developed process and control strategy are much more robust compared to a more conventional development approach. It is understood how to manage the variability within the process to get a consistent product without having to put PAT controls in place. Another obstacle is the lack of confidence, capabilities and skills in applying PAT and the fear of any regulatory impact it may have.

The good news is that this trend is changing in line with the availability of more universities being involved, publicised success stories, cost-efficient equipment, the move towards continuous manufacturing (that cannot be done without QbD and PAT) and a regulatory community that encourage innovation, increased level of process understanding and the use of PAT for process control.

9.9 PAT and Real Time Release Testing (RTRT)

When a PAT system is used for process control, it typically measures a CQA in real time at a process step that is particularly critical for that CQA. If it can be demonstrated that the risk of affecting this CQA further during the remaining process, this predicted value can be used as an In-Process Control (IPC) verifying that the process is in control or to replace conventional end-product testing performed in the quality control (QC) laboratory, with the potential of significant savings in terms of time and cost. This feature is called RTRT and is defined in ICH Q8(R2), [2] as 'the ability to evaluate and ensure the quality of in-process and/or final product based on process data, which typically include a valid combination of measured material attributes and process controls'.

Replacing end-product testing by RTRT is performing the analysis in real time during processing rather than after the entire process has been completed. In many cases, a traditional IPC could be used as RTRT if it is developed on the basis of science and risk principles.

Although RTRT is the goal for both the pharmaceutical industry and the regulators, not many companies have currently achieved it in practice, although successful cases have been published [31–35].

The concept of RTRT is indispensable when establishing continuous manufacturing. A trend in pharmaceutical manufacturing is moving from batch production to continuous manufacturing (see Chapter 11). When raw materials are continuously fed into the beginning of a process at the same time as acceptable product is continuously removed from the process at the end, the process is called 'continuous' [31]. Such a process is an example of a process that is nearly impossible to develop and run without PAT-enabled controls. The PAT system will in real time control the different process steps to ensure that consistent quality is being produced. The output from one unit operation serves as the input to the next unit operation and must be consistent if at the end of the process, consistent quality of the finished products is to be ensured at any given point in time. The first regulatory approval of a new product made by continuous manufacturing, including PAT and RTRT, was by the FDA in July 2015, and in April 2016, FDA approved a legacy product being moved from batch to continuous processing.

RTRT can therefore be applied to any process: batch, continuous, existing or new and for any type of drug molecule or product type. RTRT as an element of the control strategy and a replacement of conventional end-product testing must be approved by the relevant regulatory authority. This is described in greater detail in [31–35].

9.10 PAT Implementation

To facilitate a successful implementation of PAT, it is important to establish the task as a cross-functional project with appropriate sponsors to ensure that the project is supported and keeps moving in the right directions. Relevant stakeholders must be involved and 'buy in' right from the beginning.

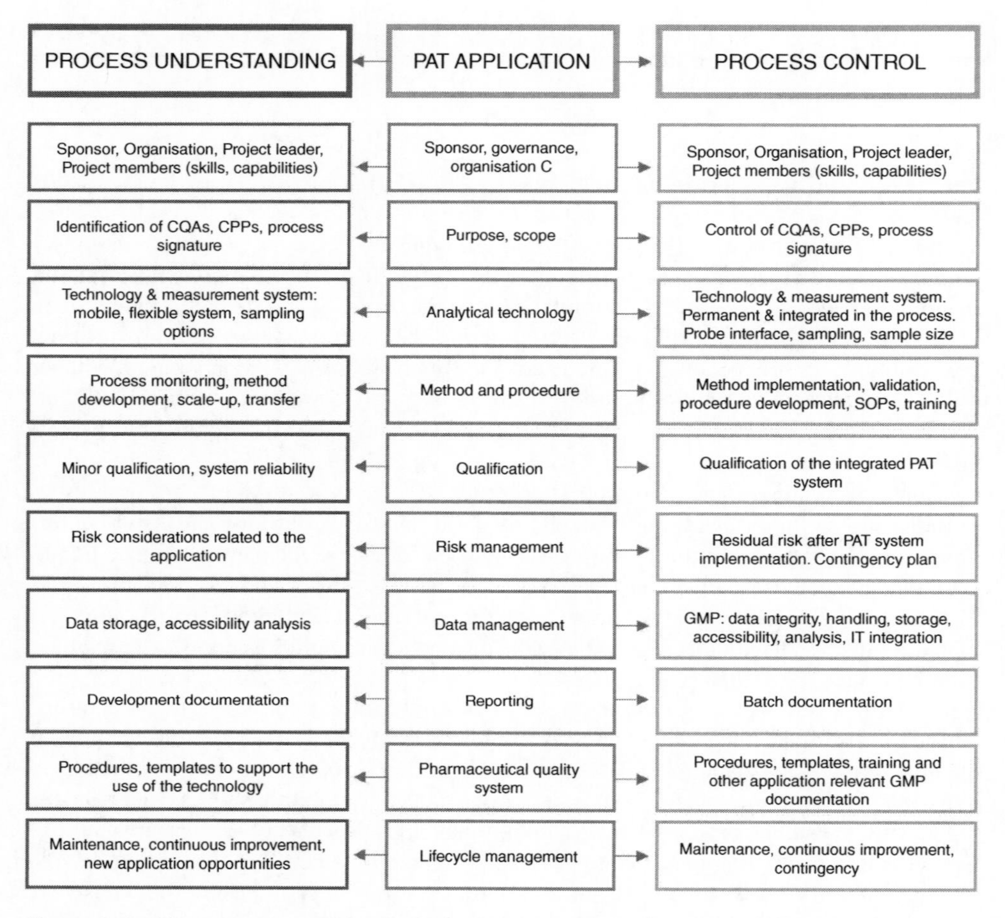

Figure 9.4 *Elements to consider for inclusion in a PAT application, whether it is for gaining process understanding or for process control.*

There are several elements that need to be considered for inclusion in the project plan, such as elements related to cGMP. Figure 9.4 shows some of these elements for consideration, whether for 'gaining process understanding' in development or 'controlling the process' in commercial manufacturing.

9.11 Data Management

The concept of data management is about making sure the right data can be acquired from the analyser at the right point in time, and simultaneously transmitted to the right people for analysis and actions. Data management is the practice of organising and using data resources for the maximum benefit of the organisation, and it ranges from a high-level strategic discipline to very low-level technical practices. It includes data identification, acquisition, synchronisation, modelling, coordination, organisation, distribution, architecting and storage (Figure 9.4). Without careful data management considerations, handling and analysis of data can be very painful.

Both for development and manufacturing purposes, data should be handled with care and follow common standards and practices for data management and data integrity.

A PAT system will be able to collect much data, so it is recommended to use a risk-based approach to plan for data handling early on. The PAT system must be qualified and verified in accordance with the respective application. If it will be used as an element of the control strategy, it needs full qualification, whereas less can be done to a development system if it is not used for making quality decisions. ASTM has published a standard guidance on verification of a PAT system [7].

9.12 In-Line Process Monitoring with UV-Vis Spectroscopy: Case Study Example

The example described here is taken from the first practical QbD workshop: 'Integrating QbD Principles in Continuous Pharmaceutical Product Development and Manufacture' developed in collaboration between De Montfort University, GlaxoSmithKline, Leistritz, ColVisTec and MKS Data Analytics Solutions. A team of 20 people including pharmacists, chemists, physicists, chemical engineers and pharmaceutical scientists were tasked to develop and manufacture a 'real' medicine (Oral Solid Dose, OSD) in four days. The main requirement was to demonstrate continuous development and manufacture concepts including in-line process monitoring. The main objectives were to

- Integrate *QbD principles* in development and manufacture.
- Practice the use of *QRM including risk assessment tools.*
- Understand basics of *DoE.*
- Gain *process understanding* by using in-line monitoring.
- Design *control strategies* to minimise risks.
- Understand *design space* and considerations for operation.
- Understand monitoring *product and process performance* and *continual improvement.*
- Understand *regulatory issues.*

The Quality Target Product Profile (QTPP) requirements were identified on the basis of patients, business and regulatory needs. The OSD selected was an immediate release tablet containing an active pharmaceutical ingredient (API) with poor aqueous solubility, but high permeability (Biopharmaceutical Classification System, BCS Class II). The manufacturing process was based on Hot-Melt Extrusion (HME) [36–38] with in-line process monitoring using a UV-Vis spectrophotometer. The spectral range of the UV-Vis system was from 220 to 820 nm with a resolution of 1 nm. The manufacturing process, showing the main unit operations, is illustrated in Figure 9.5.

Figure 9.6 shows the preliminary hazardous assessment (PHA) carried out to identify the potential critical material attributes (material and intermediate CQAs, sometimes referred to as CMAs) and potential CPPs that could impact the CQAs of the final product.

The selection of factors and ranges for the first experimental design of the extrusion process (DoE1, screening design) was based on the risk analysis shown in Figure 9.6. Four factors (temperature, screw speed, powder flow rate and API concentration) were

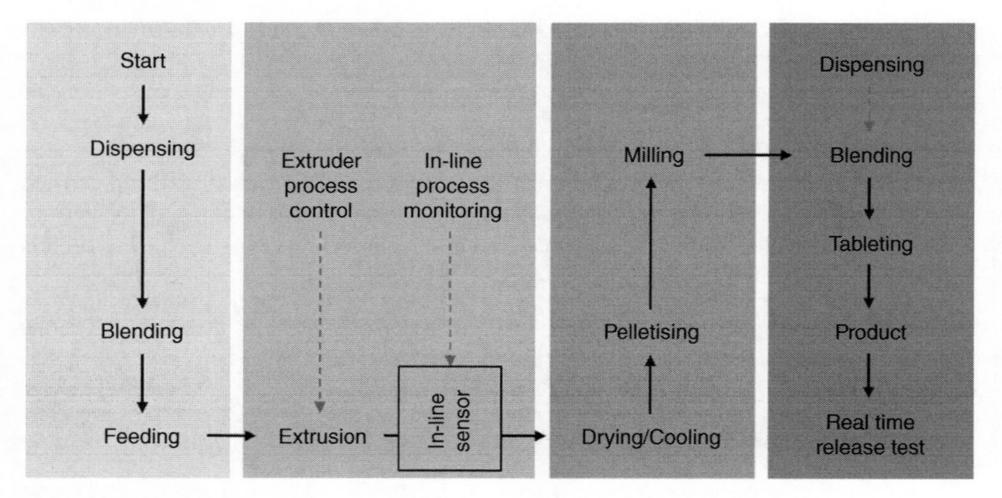

Figure 9.5 *Manufacturing process based on hot-melt extrusion showing the main unit operations from powders to final product, tablets.*

potentially critical to the CQAs of the intermediate product, in this case the extrudate API/polymer mixture. Two subsequent experimental designs (DoE2 and DoE3, optimisation designs) were performed based on the knowledge previously gained from DoE1. The final experiment was performed to verify the 'best' process conditions identified in DoEs2&3.

The focus of this case study is to show the potential of in-line process monitoring using UV-Vis spectroscopy. Figure 9.7 represents about 5000 UV-Vis spectra that were collected during DoEs 1-3 and the spectra from the verification experiments (note that the time scales are not the same).

The spectra for each DoE set of runs were recorded continuously and are shown from left to right in Figure 9.7. The '*white*' colour spectra represent pure polymer, no API. Rapid changes in colour from '*yellow-grey*' shows the formation of bubbles within the extrudated API/polymer sample. This could be due to the presence of moisture. The colour '*red/orange*' indicates deterioration of API due to excess temperature. The '*yellow*' colour suggests optimum operational conditions producing a homogenous API/polymer mixture.

The extended grey stripe is due to undissolved API embedded in the saturated polymer matrix. The finely dispersed particles scatter the light evenly over all wavelengths.

Each row in Figure 9.7 shows the sequence of spectra of each DoE runs (typically 11 runs for each DoE). The colours represent the visible part of the spectra, measured in transmission, of the mixture of an API with a compatible polymer. In the last set of spectra, highlighted as 'verification' experiments, the '*dark-grey*' colour is from the cleaning polymer material (not transparent).

The spectrophotometer recorded spectra from 220 to 820 nm. The visible part of the spectrum extends from 380 to 780 nm. For this range, colour values were calculated. Their red-green-blue (rgb)-equivalent is shown in these plots. Figure 9.8 shows raw spectra from a DoE set of runs. The temperature and API concentration were varied. The rainbow indicates the visible part of the spectrum, accessible to the human eye.

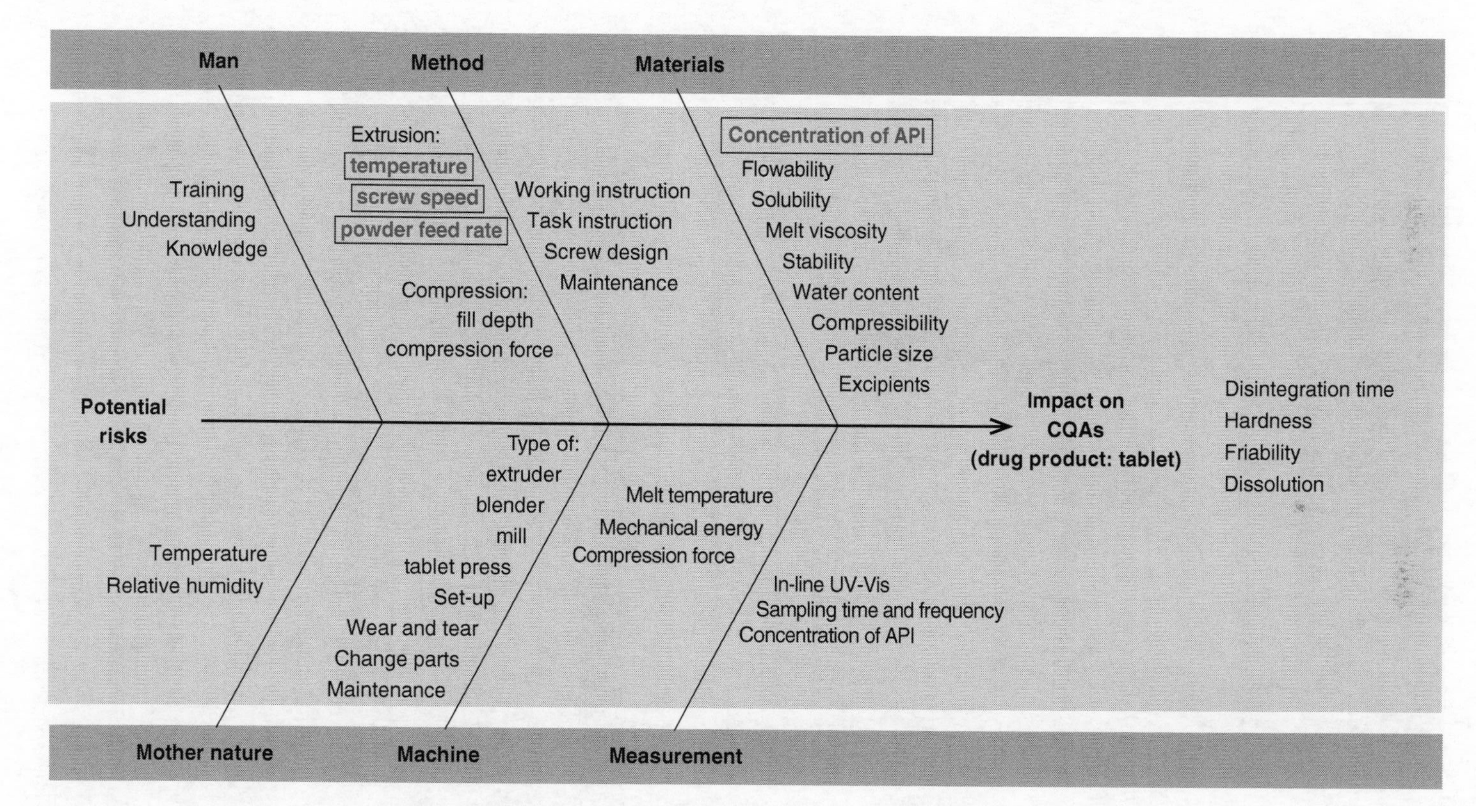

Figure 9.6 *Preliminary risk assessment (RA) analysis (Ishikawa diagram). The factors highlighted were considered critical for the extrusion process.*

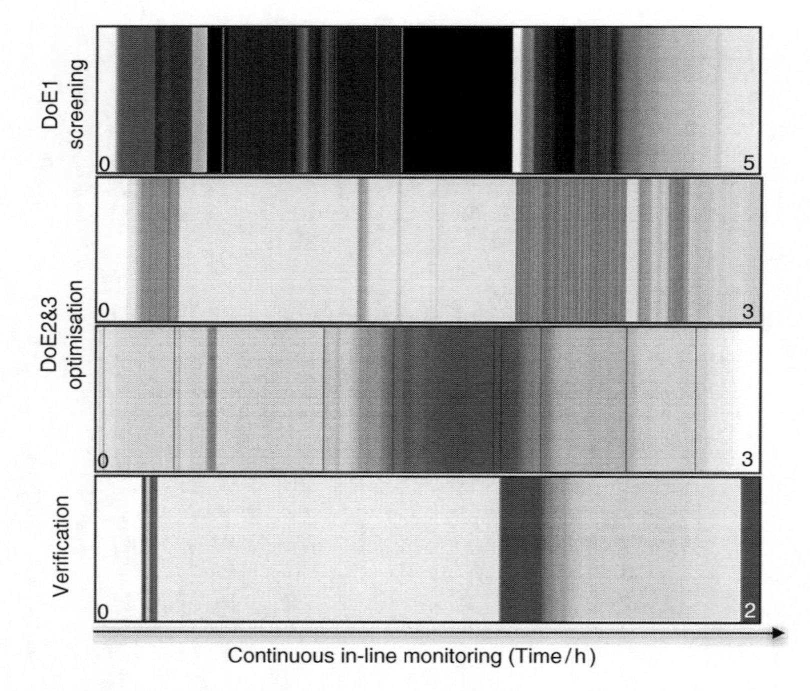

Figure 9.7 *Representation of about 5000 UV-Vis spectra from DoEs 1–3 and verification experiments. (See insert for color representation of the figure.)*

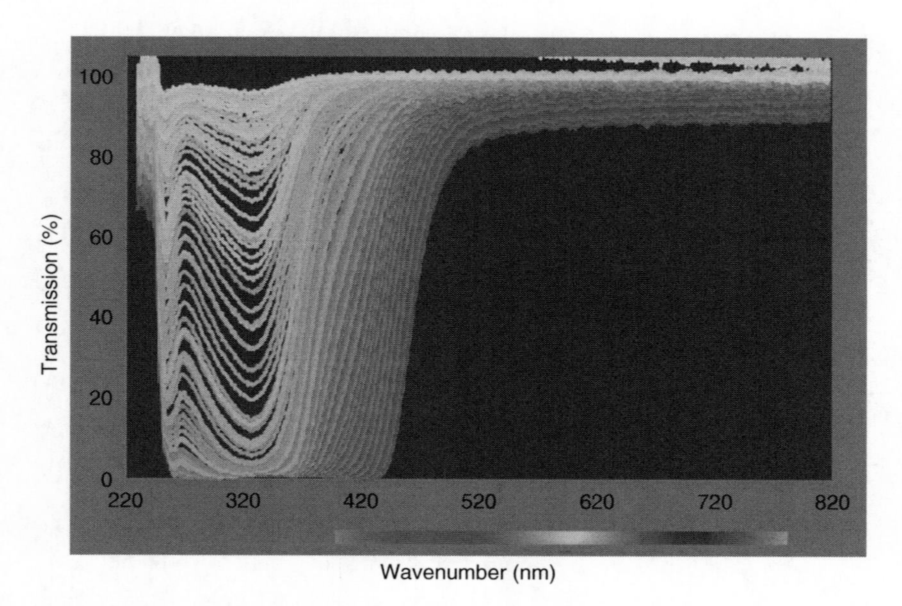

Figure 9.8 *Raw spectra from a DoE set of runs. (See insert for color representation of the figure.)*

Typically, many APIs and all the common pharmaceutical polymers show distinct absorption edges or even full peaks within the mentioned wavelength range. In this case study, various process conditions were tested (e.g. temperature, API concentration, screw speed, solid feed rate) using different types of DoE (e.g. screening, optimisation and verification designs)

The first row in Figure 9.7 shows initial experiments to determine the outer boundaries of the process parameter windows in the various dimensions. The optimum ranges for temperature, API concentration, screw speed and solid feed rate were determined from only 22 experiments and with only 500 g of API. It was clear from this design that temperature and concentration of API are both critical to product CQAs (e.g. solubility and stability of the API). Figure 9.9 shows a selection of spectra and pictures of 4 of the 11 samples collected.

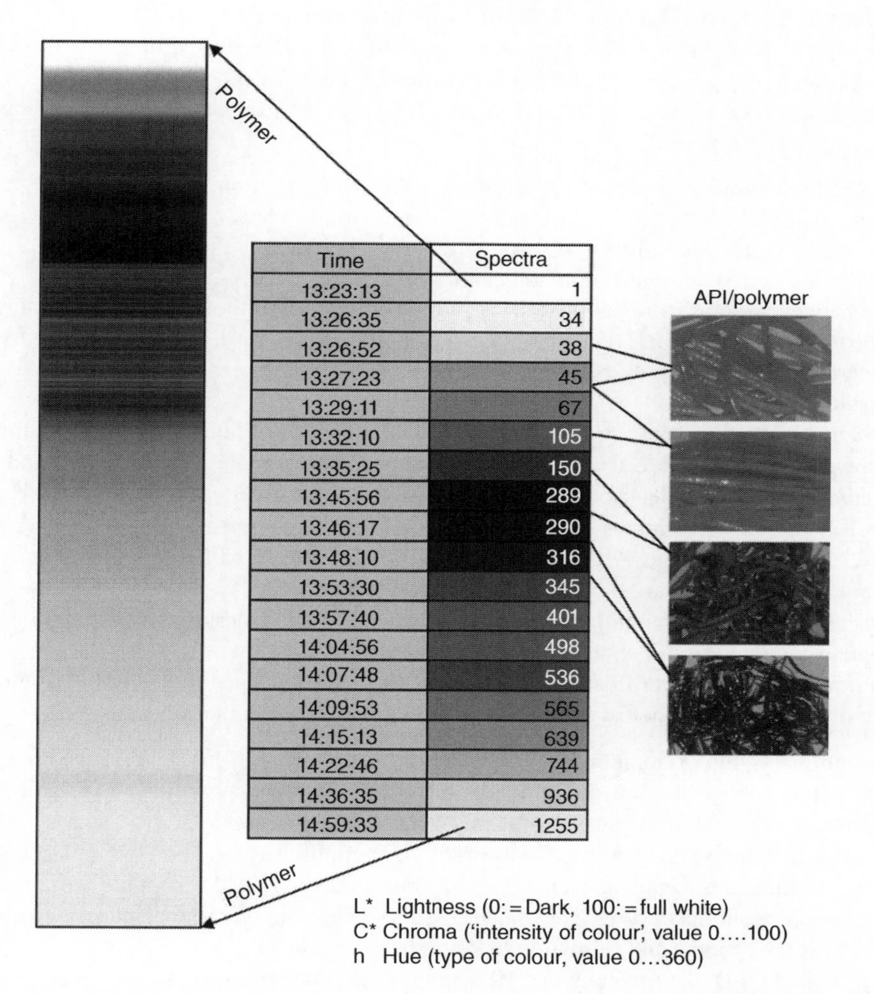

Figure 9.9 *UV-Vis spectra from the first screening design (DoE1) showing a sample of the spectra collected. (See insert for color representation of the figure.)*

The dark-red colour clearly indicates decomposition of the sample. Temperature and API concentration were both found to impact product CQAs (e.g. solubility of the API in the polymer and stability) significantly more than powder flow rate and screw speed.

The third row in Figure 9.7 shows spectra from experiments where only temperature and API concentration were varied using narrow ranges of temperature and API concentrations.

Even without any further detailed knowledge, the progress made during those four experiments towards a stable (production) operation is quite apparent.

Based on the way the spectroscopic system was calibrated the 'intensity' of the colour (Chroma) in these rows in Figure 9.7 represents the API concentration if the API is in solution. If it is not fully dissolved, then small particles of undissolved API will scatter the light out of the transmission path, which reduces the intensity (light vs. dark). It is quite common for the maximum concentration of API that will dissolve in the polymer to be dependent on the temperature at which both components are being mixed. A change in colour from yellow to orange/red is the result of the deterioration of the API due to excess heat.

Using the knowledge acquired during these experiments, one can make the following qualitative statements:

- A change in chroma represents a variation in API concentration.
- A decrease in Lightness L* ('Intensity') is the result of a lack of dissolution of the API or caused by another effect that scatters light (e.g. bubbles).
- A rapid change in colour/lightness is undesirable.
- A change in colour from yellow to orange/red is the result of the deterioration of the API due to excess heat (don't take that tablet!).
- Operating close to conditions that are difficult to control increases the risk of unstable results.
- *DoE1*: Evaluation of four potential critical factors (temperature, API concentration, screw speed and solid feed rate) and their edge of failure boundaries (operating window). Red/orange denotes deterioration of API possibly due to excess heating.
- *DoE2*: Higher temperature limit was decreased, and this prevented the deterioration of the API (no red/orange). Bubble formation and undissolved API still apparent at times. Rapid *yellow-grey* changes indicate bubble formation.
- *DoE3*: Higher concentration limit was decreased. There is less bubble formation, but undissolved API is still apparent.
- *Verification*: Model verification run. Data from DoE3 was used to determine a suitable set of operating parameters for the extruder and API/polymer concentration ratio.

In summary, this example nicely illustrates the potential of in-line optical spectroscopy for real-time data generation, measurement of API stability, concentration and solubility. In addition, it is possible to measure feeding accuracy, residence time distribution (RTD), extruder cleaning behaviour (washout phase), temperature/shear stability of polymer/API against thermal and mechanical energy behaviour, batch-to-batch or supplier variations of raw materials (old material vs. new material), crystalline and amorphous phases, mapping of the process window and formation of degradation products.

The use of QbD methodology and PAT tools for product design and continuous process monitoring in pharmaceutical product manufacture can offer clear advantages to ensure product quality [39]. A structured framework for education, training and continued professional development to cover fundamental principles of modern pharmaceutical development and manufacture is needed [40–42]. The case study described here shows how QbD principles and PAT can effectively transform the way pharmaceutical products can be developed and manufactured.

9.13 References

[1] FDA PAT Guideline (2004) *Guidance for Industry, PAT – A Framework for Innovative Pharmaceutical Development, Manufacturing, and Quality Assurance*, (http://www.fda.gov/downloads/Drugs/GuidanceComplianceRegulatoryInformation/Guidances/ucm070305.pdf) (accessed 30 August 2017).

[2] ICH Q8 (R2) (2009) *Pharmaceutical Development*, http://www.ich.org/fileadmin/Public_Web_Site/ICH_Products/Guidelines/Quality/Q8_R1/Step4/Q8_R2_Guideline.pdf (accessed 30 August 2017).

[3] ICH Q11 (2012) *Development and Manufacture of Drug Substances (Chemical Entities and Biotechnological/Biological Entities)*, http://www.ich.org/fileadmin/Public_Web_Site/ICH_Products/Guidelines/Quality/Q11/Q11_Step_4.pdf (accessed 30 August 2017).

[4] ASTM E2363-14 (2014) *Standard Terminology Relating to Process Analytical Technology in the Pharmaceutical Industry*.

[5] ASTM E2629-11 (2011) *Standard Guide for Verification of Process Analytical Technology (PAT) Enabled Control Systems*.

[6] Levente, L., Simon, L.L. *et al.* (2015) Assessment of recent process analytical technology (PAT) trends: *A Multi-Author Review, ACS Publications*, dx.doi.org/10.1021/op500261y | *Org. Process Res. Dev.*, **19**, 3–62.

[7] Bakeev, K.A. (2010) *Process Analytical Technology: Spectroscopic Tools and Implementation Strategies for the Chemical and Pharmaceutical Industries*, 2nd edition, Katherine A. Bakeev (Editor), Wiley, ISBN: 978-0-470-72207-7.

[8] Wartewig, S. and Neubert, R.H.H. (2005) Pharmaceutical applications of Mid-IR and Raman spectroscopy. *Adv Drug Deliv Rev.*, 6/15, **57**(8), 1144–1170.

[9] De Beer, T. *et al.* (2011) Near infrared and Raman spectroscopy for the in-process monitoring of pharmaceutical production processes. *Int J Pharm.*, 9/30, **417**(1–2), 32–47.

[10] Cárdenas, V., Cordobés, M., Blanco, M. and Alcalà, M. (2015) Strategy for design NIR calibration sets based on process spectrum and model space: An innovative approach for process analytical technology. *J Pharm Biomed Anal.*, 10/10, **114**, 28–33.

[11] Chavez, P. *et al.* (2015) Active content determination of pharmaceutical tablets using near infrared spectroscopy as Process Analytical Technology tool. *Talanta* 11/1, **144**, 1352–1359.

[12] Fonteyne, M. *et al.* (2015) Process Analytical Technology for continuous manufacturing of solid-dosage forms. *TrAC Trends AnalChem.*, 4, **67**, 159–166.

[13] Schaefer, C. *et al.* (2013) On-line near infrared spectroscopy as a Process Analytical Technology (PAT) tool to control an industrial seeded API crystallization. *J Pharm Biomed Anal.*, **9**, 83,194–201.

[14] Wood, C. *et al.* (2016) Near infra-red spectroscopy as a multivariate process analytical tool for predicting pharmaceutical co-crystal concentration. *J Pharm Biomed Anal.*, 9/10, **129**, 172–181.

[15] De Beer, T.R.M. *et al.* (2007) Spectroscopic method for the determination of medroxyprogesterone acetate in a pharmaceutical suspension: Validation of quantifying abilities, uncertainty assessment and comparison with the high-performance liquid chromatography reference method. *Anal Chim Acta.*, 4/25, **589**(2), 192–199.

[16] Arellano, M., Benkhelifa, H., Flick, D. and Alvarez, G. (2012) Online ice crystal size measurements during sorbet freezing by means of the focused beam reflectance measurement (FBRM) technology. Influence of operating conditions. *J Food Eng.*,11, **113**(2), 351–359.

[17] Arellano, M. *et al.* (2011) Online ice crystal size measurements by the focused beam reflectance method (FBRM) during sorbet freezing. *Procedia Food Sci.*,**1**, 1256–1264.

[18] Clain, P. *et al.* (2015) Particle size distribution of TBPB hydrates by focused beam reflectance measurement (FBRM) for secondary refrigeration application. *Int J Refrig.*, **2**,50, 19–31.
[19] Narang, A.S. *et al.* (2017) Application of in-line focused beam reflectance measurement to Brivanib alaninate wet granulation process to enable scale-up and attribute-based monitoring and control strategies. *J Pharm Sci.*, **106**(1), 224–233.
[20] Gendrin, C., Roggo, Y. and Collet, C. (2008) Pharmaceutical applications of vibrational chemical imaging and chemometrics: A review. *J Pharm Biomed Anal.*, 11/4, **48**(3), 533–553.
[21] EMA NIRS *Guideline on the Use of Near Infrared Spectroscopy by the Pharmaceutical Industry and the Data Requirements for New Submissions and Variations*(2014). http://www.ema.europa.eu/docs/en_GB/document_library/Scientific_guideline/2014/06/WC500167967.pdf (accessed 30 August 2017).
[22] FDA draft NIR (2015). *Development and Submission of Near Infrared Analytical Procedures Guidance for Industry*. http://www.fda.gov/downloads/Drugs/GuidanceComplianceRegulatoryInformation/Guidances/UCM440247.pdf (accessed 30 August 2017).
[23] ASTM E2474-14 (2014) *Standard Practice for Pharmaceutical Process Design Utilizing Process Analytical Technology*.
[24] ASTM E2475-10 (2010) *Standard Guide for Process Understanding Related to Pharmaceutical Manufacture and Control*.
[25] ASTM E2898-14 (2014) *Standard Guide for Risk-Based Validation of Analytical Methods for PAT Applications*.
[26] ASTM E2476-09 (2009) *Standard Guide for Risk Assessment and Risk Control as it Impacts the Design, Development, and Operation of PAT Processes for Pharmaceutical Manufacture*.
[27] ASTM E2537-08 (2008) *Standard Guide for Application of Continuous Quality Verification to Pharmaceutical and Biopharmaceutical Manufacturing*.
[28] ASTM E2500-13 (2013) *Standard Guide for Specification, Design, and Verification of Pharmaceutical and Biopharmaceutical Manufacturing Systems and Equipment*.
[29] ASTM E2891-13 (2013) *Standard Guide for Multivariate Data Analysis in Pharmaceutical Development and Manufacturing Applications*.
[30] ASTM E2968-14 (2014) *Standard Guide for Application of Continuous Processing in the Pharmaceutical Industry*.
[31] EMA PDA QbD Workshop London (2014) *Case Study 4: Challenges in the Implementation of Model Based and PAT based RTRT for a New Product*. Gorm Herlev Jørgensen, Danish Health and Medicines Authority & Theodora Kourti, GSK http://www.ema.europa.eu/docs/en_GB/document_library/Presentation/2014/02/WC500162149.pdf (accessed 30 August 2017).
[32] Hiroshi Nakagawa, Daiichi Sankyo Co., Ltd. (2016) *QbD/PAT/RTRT Application and Regulatory Interaction in Japan*. IFPAC, Washington. http://www.infoscience.com/JPAC/ManScDB/JPACDBEntries/1457368191.pdf (accessed 15 January 2017).
[33] EudraLex (2015) *Volume 4 EU GMP, Draft Annex 17: Real Time Release Testing*. http://ec.europa.eu/health//sites/health/files/human-use/quality/pc_quality/consultation_document_annex_17.pdf (accessed 30 August 2017).
[34] EMA/CHMP/QWP/811210/2009-Rev1 (2012) *Guideline on Real Time Release Testing*. http://www.ema.europa.eu/docs/en_GB/document_library/Scientific_guideline/2012/04/WC500125401.pdf (accessed 30 August 2017).
[35] EMA PDA QbD Workshop London (2014) *Case Study 3: Applying QbD for a Legacy Product and Achieving RTRT by a Design Space Approach with Supportive PAT and Soft Sensor Based Models: Challenges in the Implementation*. Lorenz Liesum, Novartis & Lama Sargi, ANSM http://www.ema.europa.eu/docs/en_GB/document_library/Presentation/2014/03/WC500162194.pdf (accessed 30 August 2017).
[36] Kolter, K., Karl, M. and Gryczke, A. (2012) *Hot Melt Extrusion with BASF Pharma Polymers, Extrusion Compendium*, 2nd revised and enlarged edition, ISBN 978-3-00-039415-7.
[37] Crowley, M.M. *et al.* (2007) *Pharmaceutical Applications of Hot-Melt Extrusion: Part I, Drug Development and Industrial Pharmacy*. 10.1080/03639040701498759
[38] Maniruzzaman, M. *et al.* (2012) A review of Hot-Melt Extrusion: Process technology to pharmaceutical products, *International Scholarly Research Network, ISRN Pharmaceutics*, 10.5402/2012/436763

[39] Aksu, B. *et al.* (2012) Strategic funding priorities in the pharmaceutical sciences allied to Quality by Design (QbD) and Process Analytical Technology (PAT). *Eur J PharmSci.*, 9/29, **47**(2), 402–405.
[40] de Matas, M. *et al.* (2016) Strategic framework for education and training in Quality by Design (QbD) and Process Analytical Technology (PAT). *Eur J PharmSci.*, 7/30, **90**, 2–7.
[41] Fisher, A.C. *et al.* (2016) Advancing pharmaceutical quality: An overview of science and research in the U.S. FDA's Office of Pharmaceutical Quality. *Int J Pharm.*, 12/30, **515**(1–2), 390–402.
[42] Hertrampf, A. *et al.* (2015) A PAT-based qualification of pharmaceutical excipients produced by batch or continuous processing. *J Pharm Biomed Anal.*, 10/10, **114**, 208–215.

10

Analytical Method Design, Development, and Lifecycle Management

Joe de Sousa, David Holt, and Paul A. Butterworth

AstraZeneca, Macclesfield, United Kingdom

10.1 Introduction

Analytical methodology is an intrinsic part of any pharmaceutical manufacturing process.

The nature of medicinal products means that analytical methodology plays a fundamental part in ensuring that patients receive the right amount of the therapeutic agent in the manner that has been demonstrated through clinical trials to have the intended effect. Analytical testing also ensures that the patient is not exposed to impurities and other undesirable contaminants that may arise during the manufacturing process or on storage of the medicinal product prior to use.

The implementation of Quality by Design (QbD), advocated in ICH Q8(R2) 'Pharmaceutical Development' and applied by the pharmaceutical industry to product development and manufacture, is well established, and the regulatory authorities are increasingly encouraging its adoption [1]. There is an opportunity to apply the same principles of QbD to the design, development and lifecycle management of analytical methods that are required to monitor the performance of the processes used to manufacture medicinal products, and to control the quality of the medicinal product that is produced [2–5].

Analytical QbD enables a systematic approach to the development of analytical methods that begins with predefined method objectives and emphasises product and process

Pharmaceutical Quality by Design: A Practical Approach, First Edition.
Edited by Walkiria S. Schlindwein and Mark Gibson.

understanding and process control, based on sound science and quality risk management [1]. Recently, the US Food and Drug Administration (FDA) has approved some new drug applications (NDAs) with regulatory flexibility for QbD-based analytical approaches [6]. This flexibility allows for improvements to be made to the analytical method control strategy. Analytical methods developed using a QbD approach reduce the number of out-of-trend (OOT) and out-of-specification (OOS) results due to improved method robustness and ruggedness. Thus, it is considered best practice that the pharmaceutical industry implement QbD in method development processes [7].

In this chapter, the concepts of QbD: design, development, implementation and lifecycle management and how they can be applied to analytical methods are described.

10.2 Comparison of the Traditional Approach and the Enhanced QbD Approach

An overview of the process steps during the lifecycle of analytical methods using the traditional approach compared to the QbD lifecycle approach has been described in the literature [8]. The US FDA has also discussed the enhanced QbD lifecycle approach to analytical methods, as illustrated in Figure 10.1 [6].

The traditional approach in the pharmaceutical industry to analytical method validation adheres to the procedures documented in, for example, [9]. These approaches are sound in many ways and provide confidence to regulators by demonstrating that an analytical procedure is suitable for its intended use.

However, the traditional approach focuses on compliance, as the ICH Guidance is often used in a check-box manner so that validation of methods will be acceptable to the

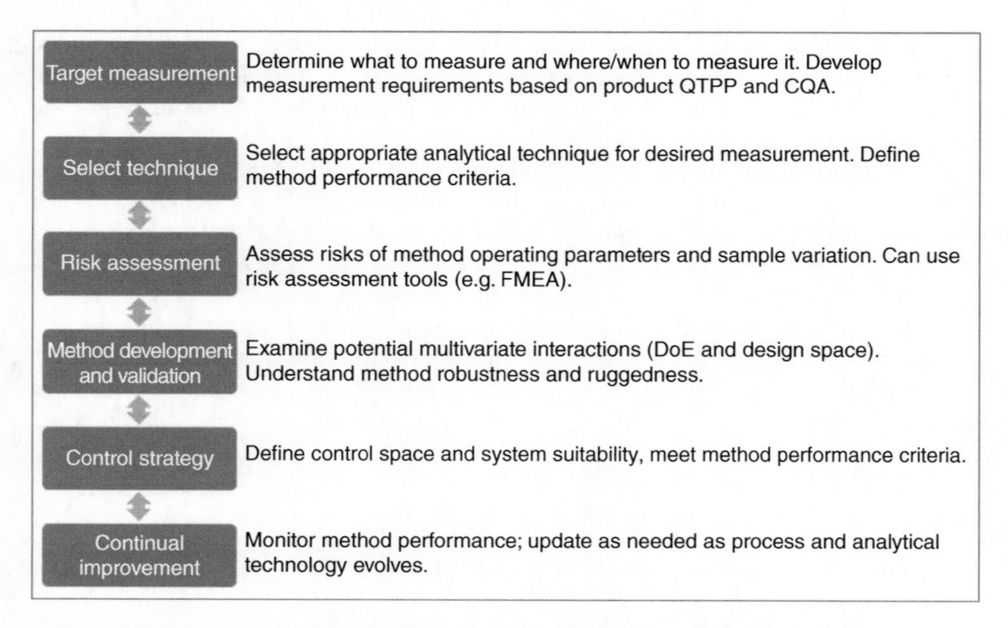

Figure 10.1 *Enhanced QbD lifecycle approach to analytical methods (adapted from [6]).*

regulatory authorities. With this in mind, there may have been little consideration given to how the methods are likely to perform outside the R&D laboratory that they were developed in, or how method improvements can be incorporated throughout the lifecycle of the medicinal product. Often there is a lack of knowledge or incomplete understanding of the critical variables in an analytical method, and therefore these may not be well controlled.

Traditional approaches are centred around a one-off validation exercise in a controlled set of conditions, which is then expected to provide reassurance of future method performance during the lifecycle of the product. Robustness studies are typically performed at the end of the validation process using a univariate approach without consideration of the interactive effects.

Traditional approaches have little focus on the end users' needs and do not help ensure ongoing monitoring of method performance during routine use. This snapshot in time approach fails to apply risk management concepts that fully consider the many changes that inevitably occur during the lifecycle of an analytical method. Making improvements to a method developed and validated by these traditional approaches is more difficult, since any changes will require a variation approval to the registered methodology and so can hinder continuous improvement.

These changes to analytical procedures can arise for various reasons, for example:

- Changes to supply chains or the manufacturing process.
- New, improved analytical technology.
- Obsolescence of original analytical technology.
- New product understanding.

The nature of science and technology means that improved methodologies will emerge for the testing and control of medicinal products during the lifetime of the product. These improved methodologies all have the potential to either directly or indirectly benefit patients. An analytical method that is more discriminating will help control future product quality. An analytical method that enables analysis to be performed more accurately and precisely will reduce product wastage by avoiding rejection of materials that were in fact suitable for use, and also reducing the amount of safety stock carried to ensure that supplies to patients are not interrupted. An analytical method that enables testing to be performed more rapidly enables supply chains to be shortened and inventory to be reduced. All of these improvements in analytical methodology ultimately lead to patients, care providers, and society being better served through safer products, uninterrupted supply or through reductions in the cost of the supply of medicines.

The traditional approach to method validation and the associated regulatory expectations present a significant barrier to bringing improvements to analytical procedures following initial licensing approval. The QbD approach, on the other hand, is an enabler for improvement, ensures that product quality is maintained, and leads to advantages at all stages of the analytical method lifecycle. The QbD approach when applied to analytical procedures brings the following key advantages:

- Methods that are designed with the end user in mind.
- More robust and rugged methods.
- A science- and risk-based approach to method validation, method transfer, and change control.
- Increased regulatory flexibility that enables methods to be improved throughout the lifecycle while continuing to ensure product quality.

With regard to method development, by using a QbD approach, methods are developed with the end user in mind rather than being based solely on the method developer's extensive experience and expertise. All factors that could impinge on the method's performance are considered, and a clear understanding of the performance requirements of the method are taken into consideration and fully explored.

With method validation, a QbD approach ensures a thorough investigation of all factors that might influence the performance of the method. Risk assessment is used to identify those variables with a potential to impact method performance, rather than assuming that nothing changes following a one-time validation exercise.

When it comes to introducing method improvements, a QbD approach brings many advantages over a traditional approach: Method changes can be made where equivalent performance is demonstrated using internal change control, rather than a much more onerous country-by-country regulatory interaction. A QbD approach also facilitates sharing of knowledge and experience.

A QbD approach enables enhanced method understanding and a more structured approach to development, so that better controls can be implemented that lead to method performance being much more closely related to the needs of the process.

QbD represents a conceptual shift in the way methods are developed, validated, transferred, and controlled throughout the product lifecycle, and is underpinned by robust science.

As described in ICH Q11, traditional and enhanced approaches are not mutually exclusive. A company can use either a traditional approach or an enhanced approach, or a combination of both [10].

10.3 Details of the Enhanced QbD Approach

Enhanced pharmaceutical development approaches, which apply QbD concepts, rely heavily on the development of fit-for-purpose analytical methods to deliver product and process understanding, and a well-defined overall control strategy. Quality cannot be 'tested into' a product, it should be 'designed into' the process and product. Analytical methodologies play a key role in bringing product and process understanding to modern-day pharmaceutical development across the entire control strategy, as outlined in the Table 10.1 [11].

As outlined in the FDA process validation guidance [12], any commercial manufacturing process should be capable of consistently producing acceptable quality product within commercial manufacturing conditions. Knowledge and understanding from product and process development is the basis for establishing an approach to control of the manufacturing process that results in products with the desired quality attributes. Manufacturers should

- Understand the sources of variation.
- Detect the presence and degree of variation.
- Understand the impact of variation on the process and ultimately on product attributes.
- Control the variation in a manner commensurate with the risk it represents to the process and product.

Analytical testing is key in providing quality assurance, and as a consequence there is a need to better understand measurement variability if we are to truly understand the

Table 10.1 *The role of analytical testing in pharmaceutical development across the entire control strategy.*

Analytical testing	Enhanced pharmaceutical development approach
Raw material testing	• Specification based on product and process needs • Effect of variability is understood
In-process testing	• Timely on-line or in-line measurements • Active control of process to minimise product variation • Criteria based on multivariate process understanding
Release testing	• Quality attributes predictable from process inputs • Specification are only part of quality control strategy • Specification based on patient needs (quality, safety, efficacy)
Stability testing	• Predictive models at release minimise stability failures • Specification set on desired product performance over time

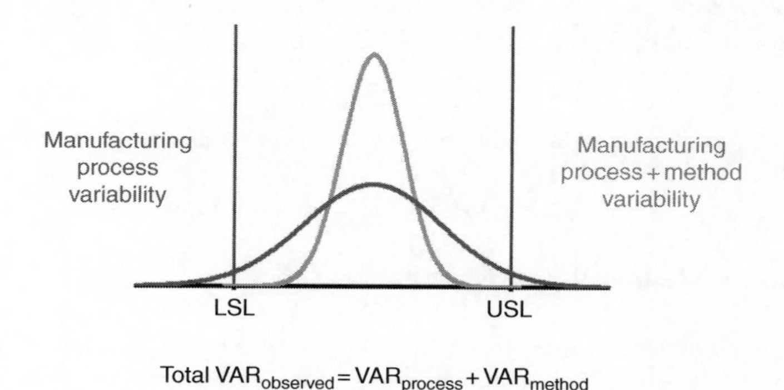

Total $VAR_{observed} = VAR_{process} + VAR_{method}$

Figure 10.2 *Example of the impact of method variability on overall variability for drug product assay (LSL=95%LC and USL=105%LC).*

performance of manufacturing processes. Understanding and minimising the contribution of analytical measurement variation, predominantly focused on accuracy and precision, to the overall variability observed can have a significant impact on improving the quality and manufacturing process capability of pharmaceutical products, as illustrated in Figure 10.2.

The concepts of QbD, which apply a systematic approach to development that begins with predefined objectives and emphasises product and process understanding and process control, based on sound science and quality risk management, can be extended to analytical methods. An enhanced development approach provides greater opportunities for understanding and improving analytical method performance, and in facilitating continuous improvement through the lifecycle of the analytical method.

Appropriate application of QbD principles will lead to robustness and ruggedness being built into analytical methods, providing a high degree of confidence in our ability to monitor and control the quality of pharmaceutical products. Each stage of the systematic science- and risk-based approach to analytical method design, development, and lifecycle management is depicted in Figure 10.3 and described in detail in the following.

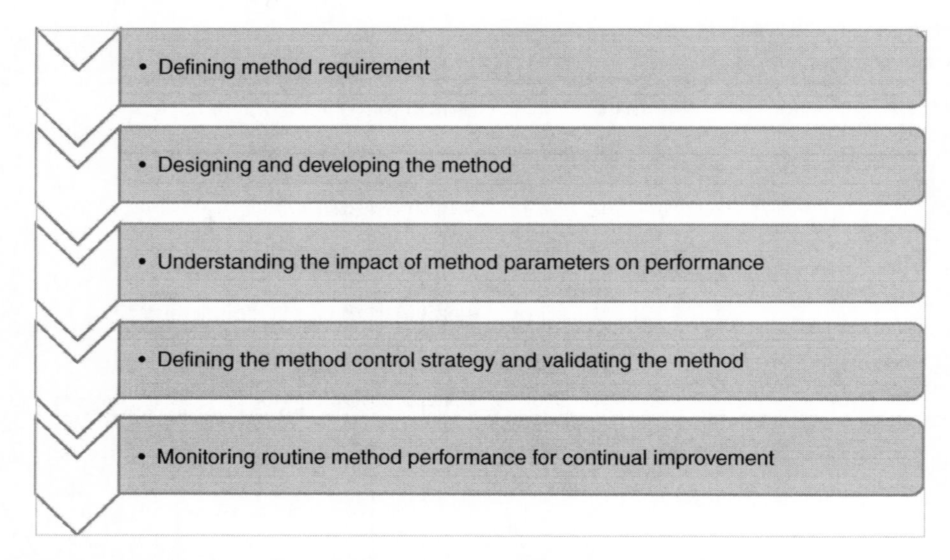

Figure 10.3 *Science- and risk-based approach to analytical method design, development and lifecycle management.*

10.4 Defining Method Requirements

The first stage of the enhanced approach should focus on a holistic evaluation of the likely analytical control strategy requirements, followed by an analytical quality risk assessment, which will focus on analytical activities to meet the product and manufacturing process requirements. An understanding of potential analytical method criticality and risks will allow development efforts to be focused most appropriately, in order to understand and control the greatest risks to product quality.

As defined by ICH Q8, the quality target product profile (QTPP) will provide a prospective summary of the quality characteristics, including critical quality attributes (CQAs), of a drug product that ideally will be achieved to ensure the desired quality, taking into account safety and efficacy.

The holistic analytical control strategy should identify the potential CQAs that require analytical measurement, for both drug substance and drug product, along with an assessment of potential criticality for each CQA. A high-level definition of the analytical control strategy for each potential CQA should be defined, at key stages of the product lifecycle, describing opportunities where control could be achieved by end-product testing, upstream in-process control(s), in-line/at-line testing using process analytical technology (PAT), or by control of processing parameters. The initial analytical control strategy will be based upon limited data and assumptions derived from the available information, and should answer the following types of questions:

- What analytical measurements are required to develop product and process understanding?
- Which CQAs require in-process control testing within the manufacturing process?
- Which CQAs require specification control across the entire manufacturing process (e.g. starting materials, intermediates, final drug substance, excipients, drug product intermediate materials and final drug product)?

- What end-product specification limits are likely to be applied to CQAs during development and at the point of marketing submission?
- Where are the opportunities to *not include* certain CQAs on end-product specifications, based on product and process understanding and/or parametric controls?
- Which CQAs will require monitoring during stability studies?

Elaboration and refinement of the analytical control strategy should be performed as the knowledge of the product and process increases.

Following definition of the analytical control strategy, the analytical quality risk assessment should focus on the ability to develop analytical methods in order to monitor and control potential CQAs. The risk assessment should consider the following areas related to the development of analytical methodology:

- Drug substance physicochemical properties and solid state characteristics.
- Proposed formulation and process development strategy.
- Analytical complexity (e.g. ability to measure CQAs).
- Ability to monitor chemical and physical stability changes.
- Need for specialist skills and equipment.
- Understanding manufacturing process risks associated with achieving the quality target for a given CQA. The analytical measurement becomes more critical for a situation where the manufacturing process is producing material that close to the proposed specification limit(s).
- The importance placed upon the analytical method within overall control strategy in order to monitor and control quality.

In general, there is likely to be a greater emphasis placed on end-product test methods, and any in-process controls or real-time release test methods employed. An example risk assessment output for a drug substance manufacturing process is presented in Table 10.2. Analytical risks are visualised through simple traffic light colour coding,

Table 10.2 *Summary analytical control strategy and risk assessment for a drug substance manufacturing process. Note: red = dark grey; green = light grey and amber = medium grey*

API CQAs	Starting materials			Synthetic stages				
	SM1	SM2	SM3	Inter 1	Inter 2	Crude	Pure unmilled	Pure milled
Physical state	L	L	L	L	L	L	L	M
Identification	L	L	L	L	L	L	L	M
Assay	L	L	L	L	L	L	L	H
Organic impurities	L	L	L	L	L	M	L	H
Genotoxic impurities	L	L	H	L	H	L	L	L
Residual solvents	L	L	L	L	L	L	M	H
Water content	L	L	L	L	L	L	H	L
Inorganic impurities	L	L	L	L	L	L	L	L
Metal content	M	L	M	L	L	L	L	L
Particle size distribution	L	L	L	L	L	L	L	H
Polymorphic form	L	L	L	L	L	L	H	L
Microbial purity	L	L	L	L	L	L	L	L

L = low; M = medium; H = high; SM = starting material; Inter = intermediate.

summarising which analytical measures are most critical to monitoring and confirming product quality. 'High' and 'Medium' risks identify the critical analytical measures, with a combination of upstream controls for starting materials and intermediates, along with in-process controls for pure unmilled drug substance and end-product tests for pure milled drug substance.

For analytical methods which have been identified as being potentially critical to the overall analytical control strategy, the method performance characteristics should be defined, and should focus on the general analytical requirements needed to adequately measure the respective CQA (independent of analytical technique employed). For example:

- A chromatographic assay and organic impurities method may focus on performance characteristics which correspond to the well-known characteristics described in the ICHQ2(R1) [9] validation guidelines, such as specificity, linearity, accuracy, precision and limits of detection and quantification.
- A tablet dissolution method may place more emphasis on the discriminating ability, comparing the dissolution profiles of the reference (target) product to test products that are intentionally manufactured with meaningful variations for the most relevant critical manufacturing variables. Additional considerations may also be given to the *in vivo* relevance of the dissolution method.

10.5 Designing and Developing the Method

Once method requirements have been defined, this should be used to drive the design and development of the analytical method that will be used to measure the CQA. Method design and development should deliver selection of appropriate analytical technique(s) and the method operating conditions to meet the method performance characteristics for a given CQA. Screening of generic method options should be the starting point for any new method development, if they are available. Only when generic method options are unavailable, or screening is unsuccessful, should bespoke method options be investigated.

Having identified suitable lead analytical method(s), the factors which can influence the performance of an analytical method should be evaluated. To understand method risks, potential causes which can impact the method performance characteristics should be prioritised in order to identify potentially critical method parameters. As discussed in ICH Q8(R2) [1], an Ishikawa (fishbone) diagram can be used to brainstorm and visualise the potential causes which can impact method performance. A generic Ishikawa (fishbone) diagram for a drug product chromatographic assay method is presented in Figure 10.4.

Risk assessment (e.g. FMEA or multi-voting) can be utilised to prioritise method parameters worthy of further investigation. One important consideration, prior to the next stage of method development, is the updating of method definition within the analytical procedure to reflect any additional controls that have been identified during the processes described earlier. This will ensure that appropriate method controls are in place prior to undertaking any experimental work so that the method performance can be understood.

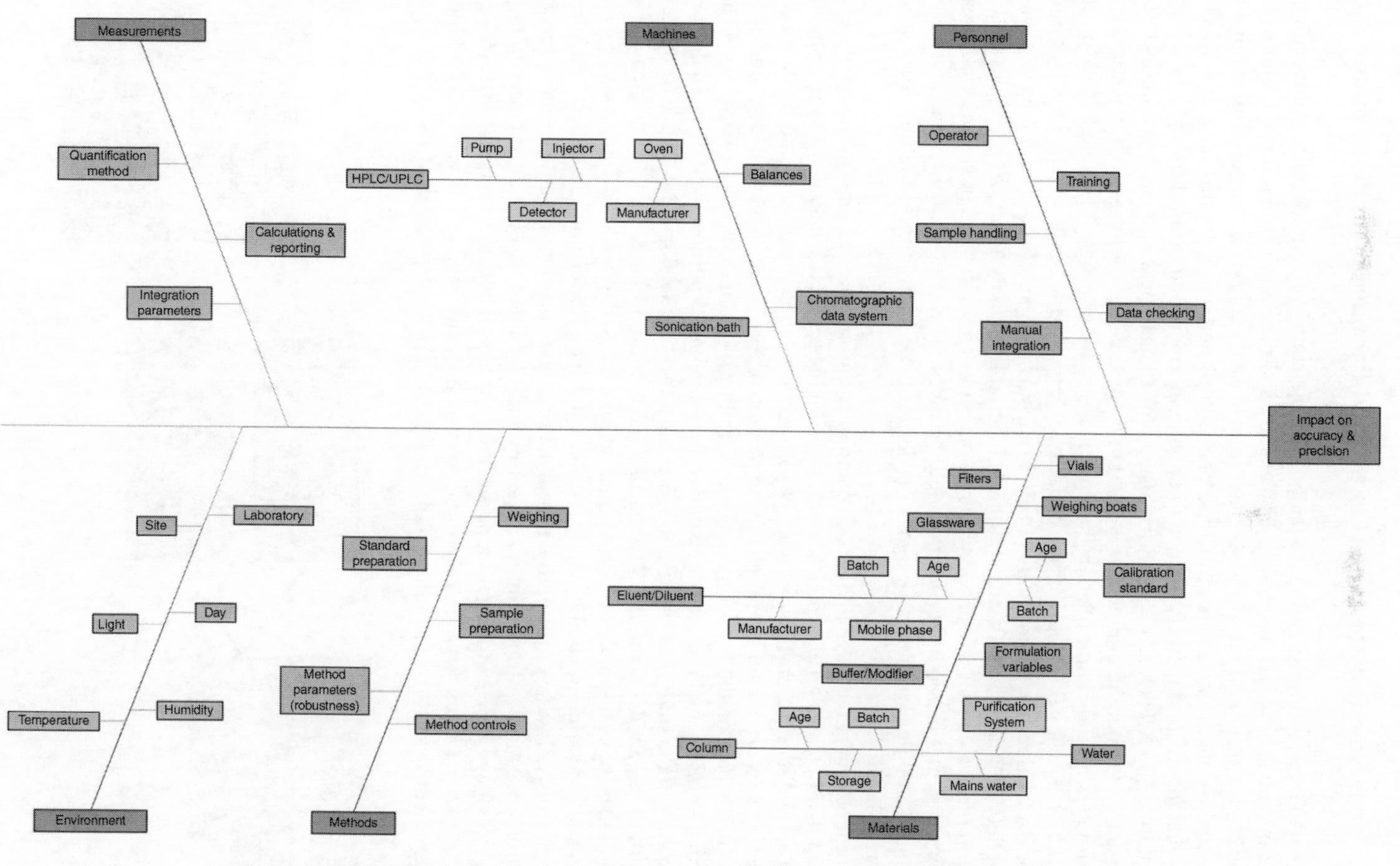

Figure 10.4 *Ishikawa (fishbone) diagram for a drug product chromatographic assay method.*

10.6 Understanding the Impact of Method Parameters on Performance

The goal of this experimentation is to understand and control potential sources of variation, so that the robustness and ruggedness of the analytical method can be evaluated, improved and verified. This provides confidence that the analytical method has the capacity to remain unaffected by small variations in method parameters, and unintentional variations due to noise, that the method is likely to encounter under normal conditions of use.

As defined by ICH Q2(R1) [9], the robustness of an analytical procedure is a measure of its capacity to remain unaffected by small, but deliberate variations in method parameters and provides an indication of its reliability during normal usage. The ruggedness of an analytical method is a measure of the capacity of a method to remain unaffected by unintentional variation that the method is likely to encounter under normal conditions (e.g. due to a variety of noise factors such as different laboratories, analysts, instruments, days, LC/GC columns, batches of reagents, sample preparations and so forth).The highest risk noise and experimental method parameters should be investigated through structured and iterative multivariate experimental design, in order evaluate and optimise method robustness and ruggedness.

In order to get to the stage in method development where method robustness can be verified, multivariate design of experiments (DoE) should be used to systematically investigate cause-and-effect relationships between experimental method parameters and method performance. This can be done through the application of a series of sequential DoE studies to arrive at a robust method (e.g. scoping → screening → optimisation → robustness). The choice of experimental designs should be fit for its intended purpose, and this will depend on the aims of the study, knowledge from previous work carried out and available resources.

Method ruggedness should also be assessed through structured experimentation. Measurement systems analysis (MSA) methods can be used to evaluate your measurement systems and determine whether you can trust your data. For example, Minitab Statistical Software provides several approaches to help determine how much of your process variation arises from variation in your measurement system [13]. The most common type of MSA is gage repeatability and reproducibility (Gage R&R; see Figure 10.5). A Gage R&R study evaluates your measurement system precision and estimates the combined measurement system repeatability and reproducibility. As defined in ICH Q2(R1) [9],

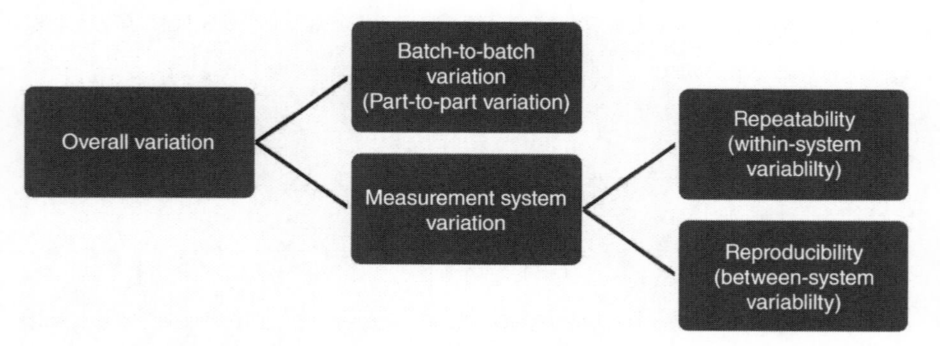

Figure 10.5 Gage repeatability and reproducibility study. (Illustration based on Minitab® Help printed with permission of Minitab Inc. All such material remains the exclusive property and copyright of Minitab Inc. All rights reserved.)

repeatability expresses the precision under the same operating conditions over a short interval of time. Repeatability in a gage study represents the variation that occurs when the same operator measures the same batch (part) with the same measuring device. Reproducibility is described in ICHQ2(R1) [9] as the precision between laboratories. Reproducibility in a gage study represents the variation that occurs when different operators measure the same batch (part) with one or more devices.

Gage R&R studies usually employ crossed and/or nested hierarchical experimental designs. Following experimentation, statistical analysis of variance (ANOVA) can be performed on the data to quantify all the variance components. The results from these studies provide method understanding and key information about the magnitude of variation coming from different sources, and whether or not the variability observed is acceptable for routine operation of the method. These studies can also be used to assess inter-laboratory equivalence during method technology transfer processes.

The experimental approaches discussed earlier are exemplified in the case studies presented at the end of the chapter.

10.7 Defining the Method Control Strategy and Validating the Method

As defined in ICH 10 [14], a control strategy is a planned set of controls derived from current product and process understanding that ensures process performance and product quality. The controls can include parameters and attributes related to drug substance and drug product materials and components, facility and equipment operating conditions, in-process controls, finished product specifications and the associated methods and frequency of monitoring and control.

With respect to analytical methods, in order to ensure the long-term performance of the analytical method, it is imperative to have a well-defined method control strategy. This control strategy should be defined to deliver the predefined method performance characteristics for a given CQA. For a given analytical method, the operating parameters, parameter ranges and set points required to ensure performance of the analytical method should be specified, and form the basis of the method control strategy. Additional aspects of the method control strategy should also be specified, such as performance monitoring system suitability checks and limits. It is important to recognise that if elements of the written analytical procedure are ambiguous or not documented at all, then invariably this leads to different interpretations by different analysts, and ultimately may lead to an increase in observed method variability, predominantly associated with analytical method reproducibility.

Once the method control strategy has been finalised, the method performance should also be tested and verified through formal method validation. The objective of analytical method validation is to demonstrate that it is suitable for its intended purpose. The extent of the method validation, and acceptance criteria employed, will also be dependent on the phase of development, the purpose of the method and the type of method being employed. Typical validation characteristics, based on ICH Q2(R1) [9] requirements, along with other aspects that should be considered, are as follows:

- Specificity.
- Linearity and range.
- Accuracy.
- Precision (repeatability, intermediate precision).

- Quantitation limit and detection limit.
- Method robustness.
- Analytical solution stability.
- Establishment of system suitability test parameters.

10.8 Monitoring Routine Method Performance for Continual Improvement

As outlined in 'Guidance for Industry – Process Validation: General Principles and Practices (FDA 2011)', continued process verification (CPV) [12] is defined as providing assurance that during routine commercial production the process remains in a state of control (the validated state). In relation to analytical methods, the principles of CPV can be applied to monitor routine method performance, to ensure that they are working as anticipated to deliver quality data. An ongoing monitoring programme could collect and analyse data relating to method performance, for example:

- Set criteria for concordance of samples or standards during the analysis.
- Trend system suitability data.
- Trend the data from the regular analysis of a reference standard lot.
- Monitor OOT and OOS results, and analytical laboratory investigations.

In addition, visualisation and trending of output CQA data for a given product can also provide insight into analytical method performance. Statistical process control (SPC) charts can be used to provide a visual assessment of the stability and capability of a process. A process is said to be stable when predominantly only common causes, not special causes, affect the process output.

For example, in relation to analytical methods, a shift in the spread of batch-to-batch variability in drug product assay, as illustrated after batch 25 on the individual-moving range (I-MR) control chart (Figure 10.6), could be an indicator that analytical method precision has deteriorated.

Monitoring of performance-indicating method attributes may lead to a desire or a need to make modifications during the lifecycle of the analytical procedure. Improvements in method performance may be desirable to take advantage of new technology, for example, new instrumentation leading to higher throughput. On the other hand, method performance may become compromised over time, for example, as a result of incremental changes to chromatographic column chemistry/performance.

Management of substantial method changes post approval is impacted by the number of international markets in which a method is registered, and how post-approval changes are handled by regulatory authorities. The multitude of markets and varying timelines, typically taking months or years, associated with post-approval changes challenge the practical viability of continuous improvement.

Employing a science- and risk-based method development approach to define a well-understood method control strategy, built on relevant and justified criteria for system suitability and performance verification (indicative of method failure), may facilitate

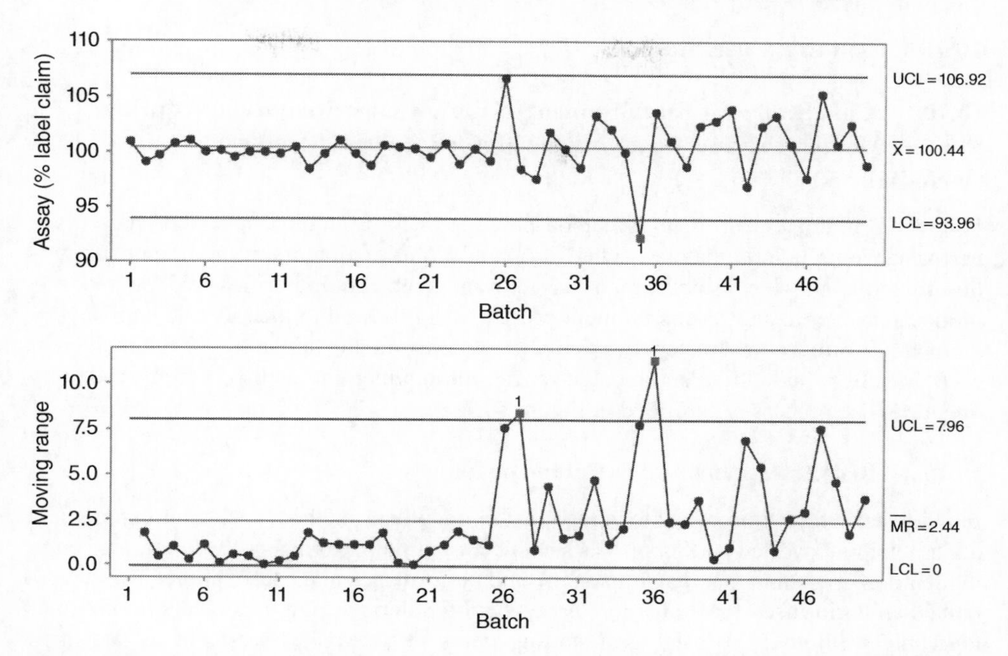

Figure 10.6 *I-MR chart of drug product assay in manufacturing order – produced using Minitab® Statistical Software.*

continuous improvement of methods in routine operation. This in turn could reduce the burden on industry and regulators alike while ensuring continued method performance. This philosophy and approach is exemplified in an article by Åsberg *et al.* [15], who presented a case study describing the development of a pharmaceutical quality control method using high-performance liquid chromatography and the use of ultrahigh-performance liquid chromatography, which allows the method to be improved throughout its lifecycle.

10.9 Summary

The enhanced QbD approach to analytical method design, development and life cycle management has many advantages over the traditional approach, not least that it minimises variability and delivers robust and rugged analytical methods for routine use. This is essential to any enhanced overall pharmaceutical QbD approach, where there is an intrinsic reliance on the analytical data to bring about process knowledge and understanding, and ultimately to support the definition of an overall control strategy. This in turn provides batch-to-batch consistency and ensures the quality, safety and efficacy of the medicinal product for reliable supply to patients.

10.10 Example Case Studies

10.10.1 Case Study 1 – Establishment of Robust Operating Ranges during Routine Method Use and Justifying the Method Control Strategy (Including SST Criteria)

Following establishment of final method parameters for a gradient reversed phase ultra-performance liquid chromatography (UPLC) and UV detection assay and organic impurities method, and the establishment of relevant synthetic impurities and degradation products, a science and risk assessment process was followed to identify and confirm key method performance characteristics that will demonstrate the suitability of the method at each operation, and define an operable range within which the method may be routinely operated. The process is described in Figure 10.7.

10.10.2 Risk Assessment and Definition of Ranges

Individuals with a thorough knowledge of the analytical method, technical experts and other potential invested parties such as statisticians participated in a pre-validation meeting to identify parameters that may impact method performance and cause failure. These were ranked and prioritised for evaluation. A number of generic design parameters for chromatographic methods were used as a starting point for the risk assessment as shown in Table 10.3. Some parameters were considered to have a low likelihood of impact on method performance or for evaluation outside the design to verify that system suitability test (SST) criteria can be fulfilled under routine operating conditions, for example, columns of different types.

For each parameter included in the DoE, ranges were defined on the basis of prior knowledge of the method and an understanding of chromatographic theory. A series of feasibility studies were performed to establish operational constraints, for example, flow limitations on the back pressure of the chromatographic system.

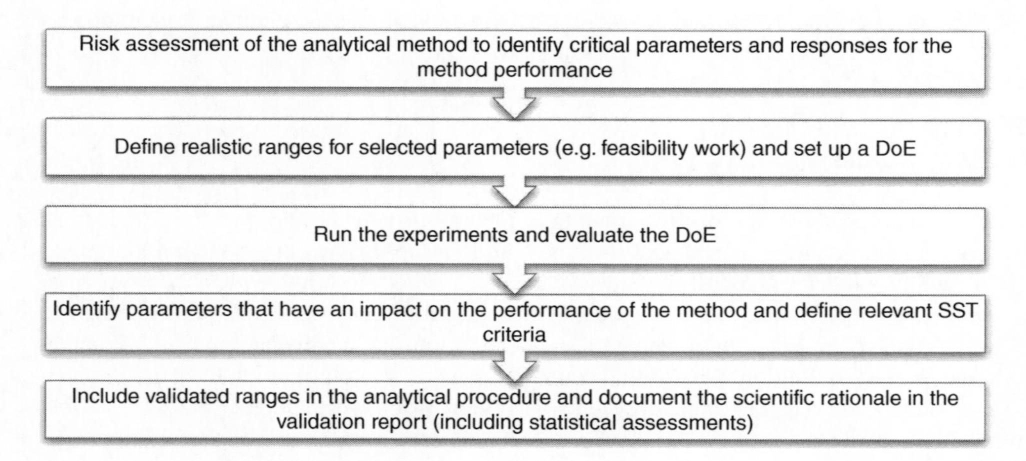

Figure 10.7 Science and risk assessment process.

Table 10.3 Generic design parameters for chromatographic methods.

Gradient method	Isocratic method	Potential response effected/ failure mode
Column type (vendor and batch)	Column type (vendor and batch)	Retention factor, resolution factor and retention order
Column temperature	Column temperature	Retention factor, resolution factor and retention order
Colum size	Colum size	Resolution factor
Detector wavelength	Detector wavelength	Relative response, LOD and LOQ
Injection volume	Injection volume	LOD and LOQ
Injection needle type	Needle type	Precision, LOD, LOQ and S/N
Velocity/flow rate	Velocity/flow rate	Resolution
Run time	Run time	No detection of highly retained samples
Mobile phase (gradient steepness)	N/A	Retention factor and resolution factor
Mobile phase composition, e.g. buffer pH and organic solvent	Mobile phase composition, e.g. buffer pH and organic solvent	Retention factor, resolution factor and retention order
Gradient/mixer size	N/A	Baseline noise, retention factor and resolution factor
Sample preparation[1]	Sample preparation[1]	Peak symmetry, accuracy and specificity
Particle size (stationary phase)	Particle size (stationary phase)	Retention factor, resolution factor and retention order
Column pressure	Column pressure	Retention factor, resolution factor and retention order
Column dead volume	N/A	Retention factor, resolution factor and retention order

[1] It is recommended to perform a separate robustness study (DoE) for the sample preparation.

10.10.3 Experimental Design

Statistical software, such as MODDE or Minitab, was used to create an experimental design with a sufficient number of experiments to gain a thorough understanding of the method experimental space. A fractional factorial, resolution III design was performed in this package of work. The extent to which main effects were confounded with other effects was considered against the likelihood of factor interactions and the need to estimate the effects separately from one another. No main effects were confounded with other main effects, but main effects were confound with two-factor interactions. This type of design obtains relevant information using very few experiments. On the basis of the risk analysis, the following factors were varied according to Table 10.4; injection volume, column temperature, v/v% TFA in the mobile phase, gradient time 1 and gradient time 2. Experiments were performed in a random order defined by the statistical software package to minimise systematic errors.

Table 10.4 *Generic design parameters for chromatographic methods.*

Experiment name	Run order	Injection volume (μL)	Column temperature (°C)	TFA (%)	Gradient[a] time 1 (minutes)	Gradient[b] time 2 (minutes)
N1	1	4	35	0.05	8	4
N2	4	6	35	0.05	6	2
N3	3	4	55	0.05	6	4
N4	11	6	55	0.05	8	2
N5	7	4	35	0.15	8	2
N6	8	6	35	0.15	6	4
N7	10	4	55	0.15	6	2
N8	6	6	55	0.15	8	4
N9	2	5	45	0.1	7	3
N10	5	5	45	0.1	7	3
N11	9	5	45	0.1	7	3
Validated ranges		4–6	35–55	0.05–0.15	6–8	2–4

[a] Length in minutes from 90% A/10% B to 78% A/22% B, mobile phase compositions according to method.
[b] Length in minutes from 78% A/22% B to 75% A/25% B, mobile phase compositions according to method.

Responses studied across all experiments in the DoE were defined on the basis of failure modes. Normal system suitability parameters for separation methods were considered and selected to detect, control and prevent method failure and highlight potential system suitability criteria. Investigated responses for the chromatographic system were retention time (tr), capacity factor (k´), resolution (Rs), separation factor (α), tailing factor, number of theoretical plates (N), response factor and reporting limit (RL).

Samples were prepared containing relevant impurities for the method, identified as those routinely detected in the drug substance during development and known degradation products observed from stability and forced degradation studies.

10.10.4 Evaluate the DoE

Responses for all organic impurities and degradation products were evaluated using multivariate statistical software [16]. Most peaks and peak pairs showed good separation, with responses not significantly affected by the factor ranges studied. However, for selectivity, the Impurity E and Drug Substance (DS) peak pair was considered to be critical within the investigated ranges to the analysis of drug substance and verification of method performance.

The design was evaluated for model fit, significance and prediction properties and factor effects identifying main factors and any interactions. Although not required in this instance, it may be necessary to complement the original design with additional experiments to resolve interactions or add higher-order terms to the model to improve prediction models (Figure 10.8).

The following centred and scaled coefficients for the resolution and separation factor models highlight the magnitude of the impact of each factor. Factors are significant when

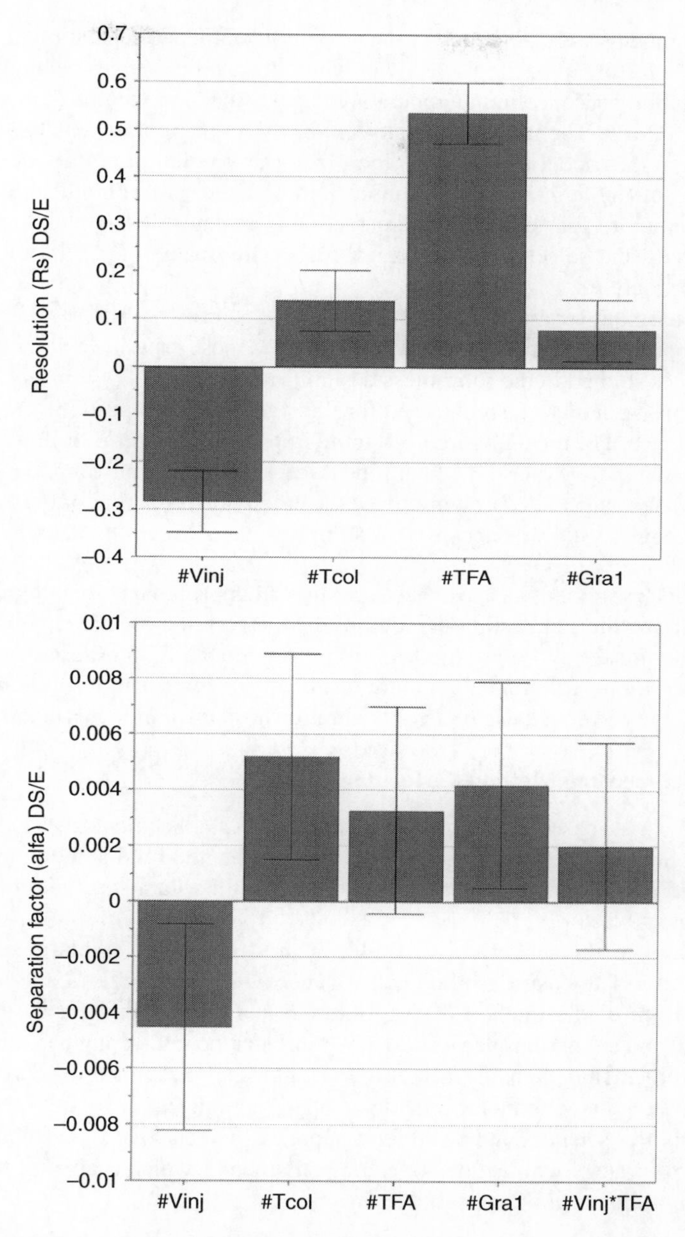

Figure 10.8 *Centred and scaled coefficients for the resolution and separation factor models.*

confidence intervals do not include zero. In addition, an alternative statistical approach using main effect plots, which show the predicted response for the selected main factor with all other factors set to their average, is exemplified for the amount of trifluoroacetic acid (TFA) in the mobile phase.

Significant models were obtained for the resolution factor and separation factor, with the most important factors impacting selectivity within the studied ranges being the amount of TFA in the mobile phase, column temperature (Tcol), injection volume (Vinj) and the first gradient time (Gra1). The interaction between the injection volume and TFA (Vinj*TFA) was included in the model for the separation factor as prediction properties of the model improved. Although the interaction is confounded with the second gradient time (Gra2), it was considered more probable that the interaction is responsible for the effect. Within the validated ranges, the selectivity, Rs, achieved a desired value of 1.5 in all experiments except in N2, where Rs was 1.16.

To ensure baseline separation of the critical peak pair, a resolution value of ≥1.5 was considered appropriate. The experiments confirm that these criteria can be met within the validated ranges studied in the robustness design (Figure 10.9).

A significant model was also obtained for the signal-to-noise ratio (S/N) at the RL for the drug substance. The most important factor to impact S/N was TFA, followed by a minor impact due to the mobile phase gradient time (Gra1). The injection volume (Vinj) was not significant, but it was included as it improved the prediction properties of the model. Within the validated ranges, S/N was above 10 in all experiments except in N5, where it was 7.9 (Figure 10.10).

From the factors studied, it is possible to predict the impact of a chromatographic parameter, thus enabling the successful operation of the method in a routine environment where minor modifications may be required to ensure acceptable SST criteria, for example, a short-term increase in column temperature results in an increase in resolution of a critical pair, allowing the continued use of an old column whose resolution has deteriorated.

10.10.5 Documenting Method Performance

A science- and risk-based approach has enabled the establishment of a method control strategy with defined operating parameters, which enable the routine monitoring and control of method performance. The specific elements of the established control strategy are shown in Table 10.5.

The criterion for the sensitivity, S/N ≥10, at the RL and the resolution, Rs ≥1.5, are adopted for the SST to ensure method performance within the method experimental space in each operation of the method (Table 10.6). In addition, a limit for system precision (repeatability) and an interference test, to make sure that no interfering peaks are present in the blank, were included in the SST criteria, and analytical method chromatographic parameters were updated to include validated operating ranges.

In conclusion, a science- and risk-based approach has enabled the establishment of a method control strategy with defined operating parameters, which enable the routine monitoring and control of method performance.

10.10.6 Case Study 2 – Evaluation of the Ruggedness of a Dissolution Method for a Commercial Immediate Release Tablet Product

As part of dissolution method establishment at the commercial manufacturing site for an immediate release tablet product, method ruggedness was assessed to provide assurance of the likely method performance during routine quality control (QC) testing of batches. This immediate release tablet product has a single point lower-limit dissolution specification Q value at the 45 minute time point for complete release (USP 2 apparatus).

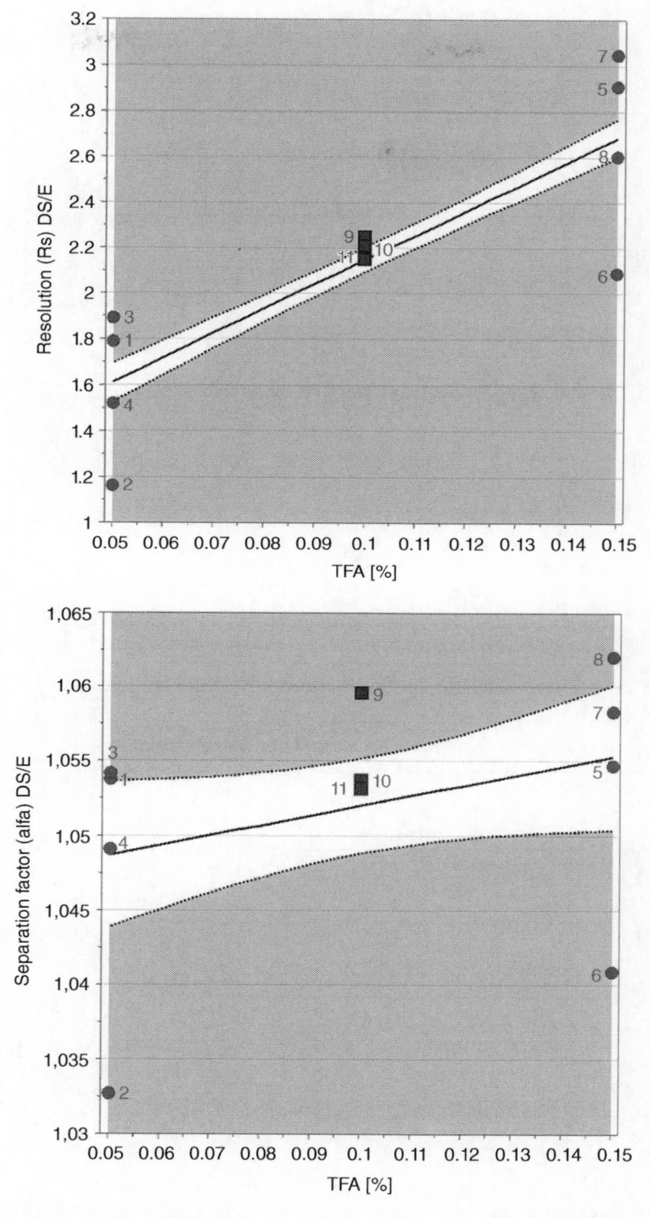

Figure 10.9 *Alternative statistical approach – main effect plot amount of TFA in the mobile phase.*

The dissolution method uses an aqueous surfactant media, and was validated during pharmaceutical development, showing appropriate specificity, linearity, accuracy, precision and range for the determination of dissolution. Typically, validation data are generated in a short time period, with a limited number of analysts and dissolution instruments. As such, validation data are not always wholly representative of the expected long-term

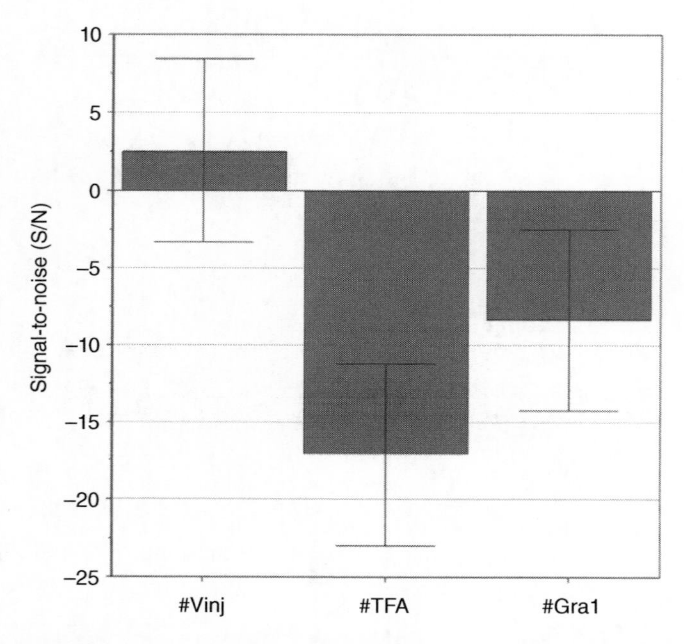

Figure 10.10 *Centred and scaled coefficients for the signal-to-noise model.*

Table 10.5 *Elements of an established analytical control strategy.*

Parameter	Routine settings			Validated ranges
% TFA in mobile phase	0.1%			0.05%–0.15%
Injection volume	5 μL			4–6 μL
Column temperature	45°C			35°C–55°C
Flow	0.6 mL/min			NA
Gradient	Time (min)	% Mobile Phase A	% Mobile Phase B	Time (min)
	0	90	10	0
	7	78	22	6–8
	10	75	25	8–12
	15	50	50	12–18
	16	10	90	13–19
	17	10	90	14–20
	17.1	90	10	14.1–20.1
	20	90	10	17–23

Table 10.6 *System suitability criteria.*

System suitability test	Acceptance criteria
Interference	Confirm that the blank injection is free from interfering peaks.
Resolution factor	Confirm that the resolution between the DS and Impurity E in the system suitability test solution is ≥1.5
Repeatability	Confirm that the relative standard deviation of the DS peak of five injections of the SST solution is ≤1.5%.
Sensitivity	Confirm that the signal-to-noise ratio of the peak due to DS in the sensitivity test solution is ≥10.

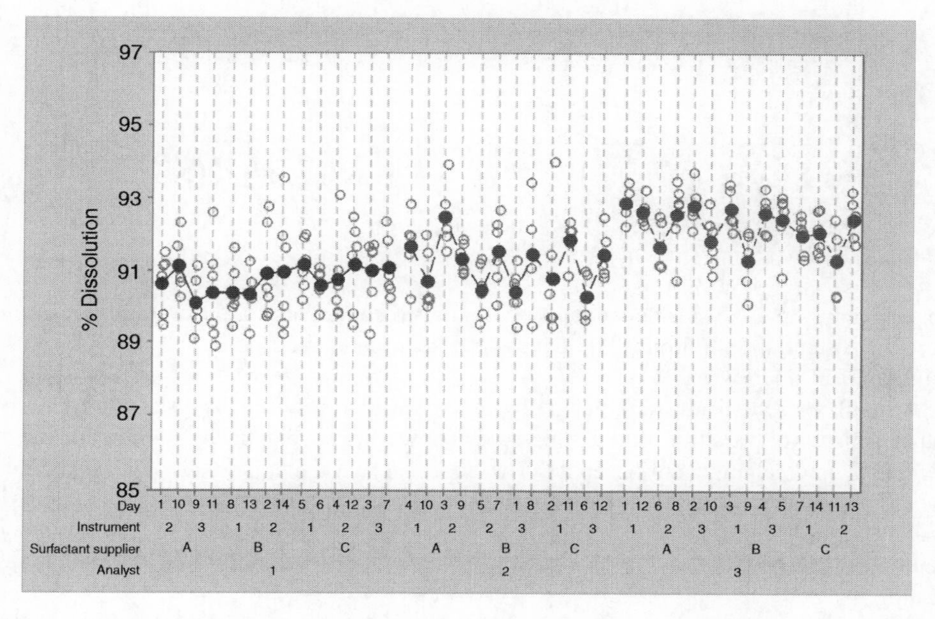

Figure 10.11 *Dissolution data from multivariate experimental study (coloured by analyst) – individual tablet (open circles) and mean data (solid circles) – produced using Minitab® Statistical Software. (See insert for color representation of the figure.)*

analytical variability likely to be experienced in a commercial production setting. The routine performance of the dissolution method was therefore assessed in the commercial QC testing laboratories.

A multivariate experimental study was undertaken to understand the ruggedness of the dissolution method, and the impact of the most likely potential noise factors on method performance. The factors investigated were different analysts, dissolution instruments and surfactant suppliers. An optimal (custom) factorial experimental design was employed to evaluate the factors outlined. In all, 40 dissolution experiments were undertaken using a single commercial batch (n=6 tablets were tested per experiment). The day of analysis and experiment run order was also randomised. For each dissolution

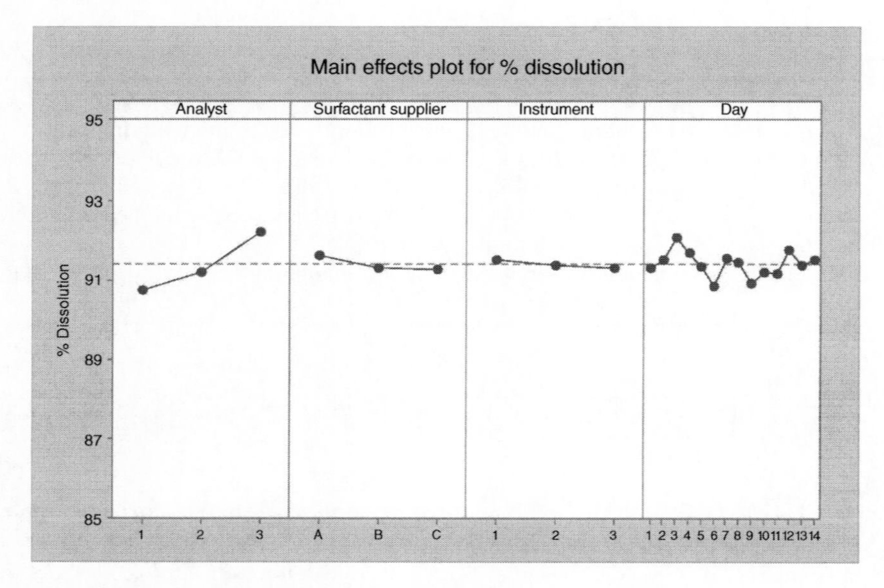

Figure 10.12 *Main effects plot from multivariate experimental study – produced using Minitab® Statistical Software.*

experiment, the analyst independently prepared fresh surfactant dissolution media and reference standard solutions. The data at the 45 minute time point are summarised in Figure 10.11 and Figure 10.12.

A small, yet statistically significant difference was seen between analysts, relative to the other sources of variation during this study; however, no practically significant factor effects or interactions were observed. The expected range of reported mean results that would be obtained from the analysis of the same batch on different occasions is approximately ±2%. Overall, these data support that the dissolution method is rugged, has good reproducibility and is capable of assessing batch-to-batch variability during routine QC testing.

10.10.7 Case Study Acknowledgements

Case Study 1: Olof Svensson, Marie-Louise Ulvinge, Anette Skoog and Charlotta Bergh (Pharmaceutical Technology & Development, AstraZeneca)

Case Study 2: Claire Elliot, Steven Mount (Pharmaceutical Technology & Development, AstraZeneca) and Paul Nelson (Technical Director, PRISM Training and Consultancy)

10.11 References

[1] ICH (2009) *Pharmaceutical Development Q8(R2) Step 4*, http://www.ich.org/fileadmin/Public_Web_Site/ICH_Products/Guidelines/Quality/Q8_R1/Step4/Q8_R2_Guideline.pdf (accessed 30 August 2017).

[2] Schweitzer, M., Pohl, M., Hanna-Brown, M. *et al.* (2010) Implications and opportunities of applying QbD principles to analytical measurements, *Pharmaceutical Technology*, **34**(2), 52–59.

[3] Vogt, F.G. and Kord, A.S (2011) Development of quality by design analytical methods, *Journal of Pharmaceutical Sciences*, **100**(3), 797–812.
[4] Monks, K., Molnar, I., Rieger, H.J., *et al.* (2012) Quality by design: Multidimensional exploration of the design space in high performance liquid chromatography method development for better robustness before validation, *Journal of Chromatography A*, **1232**, 218–230.
[5] Asberg, D., Nilsson, M., Olsson, S. *et al.* (2016) A quality control method enhancement concept – Continual improvement of regulatory approved QC methods, *Journal of Pharmaceutical and Biomedical Analysis*, **129**, 273–281.
[6] Chatterjee, S. (2013) QbD considerations for analytical methods – FDA perspective, *IFPAC Annual Meeting*; January, Baltimore, USA.
[7] Peraman, R., Bhadraya, K. and Reddy, Y.P. (2015) Analytical quality by design: A tool for regulatory flexibility and robust analytics, *International Journal of Analytical Chemistry*, **2015**, 1–9.
[8] Nethercote, P., Borman, P., Bennett, T. *et al.* (2010) QbD for better method validation and transfer, *Pharmaceutical Manufacturing*, 13th April.
[9] ICH Q2(R1) (1994) *Validation of Analytical Procedures: Text and Methodology Q2(1), step 4*, http://somatek.com/content/uploads/2014/06/sk140605h.pdf (accessed 30 August 2017).
[10] ICH (2012) *Development and Manufacture of Drug Substances (Chemical Entities and Biotechnological/Biological Entities) Q11, Step 4*, http://www.ich.org/fileadmin/Public_Web_Site/ICH_Products/Guidelines/Quality/Q11/Q11_Step_4.pdf (accessed 30 August 2017)
[11] Moheb, N. (2008) QbD: Analytical Aspects, *32nd International Symposium on High Performance Liquid Phase Separations and Related Techniques*.
[12] FDA (2011) *Guidance for Industry, Process Validation: General Principles and Practices. US Food and Drug Administration*, http://www.fda.gov/downloads/Drugs/…/Guidances/UCM070336.pdf (accessed 30 August 2017).
[13] Minitab® 17.1.0 Statistical Software. MINITAB® and all other trademarks and logos for the Company's products and services are the exclusive property of Minitab Inc. All other marks referenced remain the property of their respective owners. See http://www.minitab.com for more information (accessed 30 August 2017).
[14] ICH (2008) *Pharmaceutical Quality System Q10, Step 4*, http://www.ich.org/fileadmin/Public_Web_Site/ICH_Products/Guidelines/Quality/Q10/Step4/Q10_Guideline.pdf (accessed 30 August 2017).
[15] Åsberg, D., Karlsson, A., Samuelsson, J. *et al.* (2014)Analytical method development in the quality by design framework, *American Laboratory,* 16 December.
[16] MODDE® Pro 11, MKS data analytics solutions, Umeå, Sweden, http://mksdataanalytics.com/product/modde-pro (accessed 30 August 2017).

11

Manufacturing and Process Controls

Mark Gibson

AM PharmaServices Ltd, Congleton, United Kingdom

11.1 Introduction to Manufacturing and Facilities

In this chapter, the application of Quality by Design (QbD) principles and approach to drug product manufacturing aspects will be discussed. The site selection for either manufacture of clinical trials supplies or commercial manufacture will require a business decision: do it internally or use a third party contract manufacturer. A pharmaceutical company will have to consider several technical aspects before commencing manufacture, such as the product type (e.g. tablets, sterile injection, or pressurized metered dose inhaler (pMDI)), the intended manufacturing process, the scale of manufacture required to meet the demand, the people skills required, and so on and whether these can be accommodated or not. Other important considerations are validation of the chosen facilities and equipment for their intended use, deployment of an effective pharmaceutical quality system and linkage of the product requirements (and control strategy) to the facilities, equipment, and process validation.

So, to enable manufacture, sufficient development work should have been undertaken to establish the product design and the manufacturing process design. The critical process parameters (CPPs) and material attributes should be well understood and linked to the drug product critical quality attributes (CQAs) and the control strategy. Hence, manufacturing is a key output of the QbD process.

Pharmaceutical Quality by Design: A Practical Approach, First Edition.
Edited by Walkiria S. Schlindwein and Mark Gibson.

In the subsequent sections, the important links between QbD and facilities, process equipment, manufacturing, and process controls will be covered. The paradigm shift between the traditional approaches and the current QbD approach is also discussed.

11.2 Validation of Facilities and Equipment

Historically, the legal requirement for validation of pharmaceutical manufacturing facilities, equipment and processes originated in the United States when the Food and Drug Administration (FDA) introduced the Current Good Manufacturing Practice (cGMP) regulation in 1979 to improve the quality of pharmaceuticals. The main purpose of these regulations was to gain assurance of drug quality by adopting cGMP, so that the equipment and processes used in manufacturing, handling, packaging, and the testing operation is suitable for its intended use. In response to this new FDA regulation, pharmaceutical companies endeavored to quickly establish and implement validation procedures to meet their own company requirements based on their philosophy and financial constraints. The FDA regulations contained some ambiguities pertaining to the scope and extent of the requirements that were often misinterpreted by the pharmaceutical industry. In spite of becoming a familiar part of GMP compliance, validation and qualification have frequently been cited in FDA inspection 483s and warning letters over the years for non-compliance.

In hindsight, the reasons for validation/qualification of a new or upgraded facility would seem obvious, since most pharmaceutical companies realize that sound commissioning and facility validation are fundamental for assuring the future success of manufacturing process validation and for maintaining consistency and control during routine operation. There is really no point validating a manufacturing process, until the facility, utilities, and equipment to support the manufacturing process are in place and functioning correctly. Facility qualification is an important part of validation that documents evidence that equipment or ancillary systems are properly installed, are working correctly, and actually lead to the expected results. In summary, validation is now well established within the pharmaceutical industry, with the FDA definition being

> a process of establishing documented evidence that provides a high degree of assurance that the manufacturing processes, including buildings, systems, and equipment, will consistently produce a product that meets the desired results according to predetermined specifications and quality attributes.

11.2.1 The International Society for Pharmaceutical Engineering (ISPE) Baseline® Guide: Commissioning and Qualification

Over the decades, the cost of validating a new GMP facility has escalated with the increased time and resource required for establishing protocols, standard operating procedures (SOPs), collecting data and analyzing it, and for writing the final validation reports. Consequently, engineering representatives from the pharmaceutical industry met in the 1990s, with the International Society for Pharmaceutical Industry (ISPE) and the FDA to establish a set of Baseline® Engineering Guides intended to assist pharmaceutical manufacturers in the design, construction, and commissioning of facilities in the most efficient and compliant way. This was a more focused approach to validation incorporating clearly defined stages of activity to establish and provide documentary evidence that satisfied the requirements of the FDA. The ISPE Baseline® guide, Volume 5 [1], defines the following terms that are still referred to today; for example, they are quoted in EU GMP regulations (Annex 15):

- *Commissioning*: "A well planned, documented, and managed engineering approach to the start-up and turnover of facilities, systems, and equipment to the end user that results in a safe and functional environment that meets established design requirements and stakeholders' expectations."
- *Design Qualification (DQ)*: "The documented verification that the proposed design of the facilities, systems and equipment is suitable for intended purpose."
- *Installation Qualification (IQ)*: "The documented verification that the facilities, systems and equipment, as installed or modified, comply with the approved design and the manufacturer's recommendations."
- *Operational Qualification (OQ)*: "The documented verification that the facilities, systems and equipment, as installed or modified, perform as intended throughout the anticipated operating ranges."
- *Performance Qualification(PQ)*: "The documented verification that the facilities, systems and equipment, as connected together, can perform effectively and reproducibly, based on the approved process method and product specification." Once the facility has been validated (IQ + OQ + Performance Qualification [PQ]), then process validation can commence.
- *Process Validation (PV) or Process Performance Qualification (PPQ)*: "The documented evidence that a specific process will consistently produce a product meeting predetermined specifications and quality attributes."

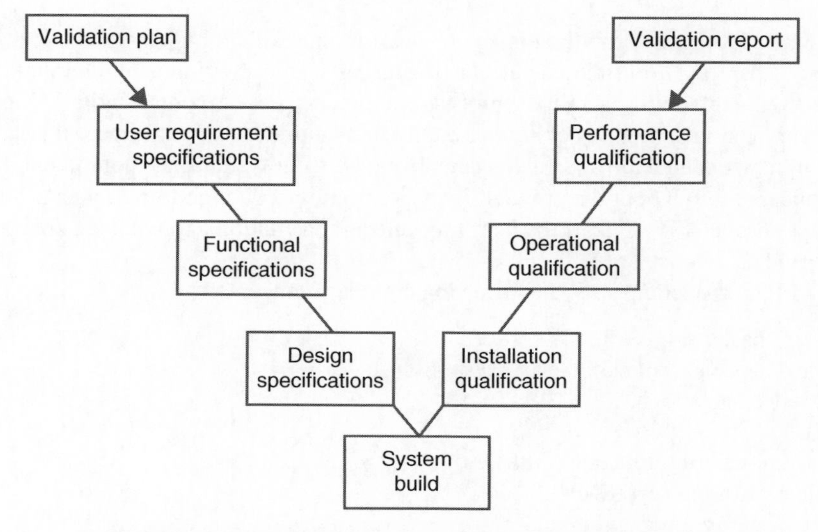

Figure 11.1 *The "V" model concept of validation.*

The ISPE Baseline® Guide Volume 5 was universally accepted at the time, and the process was commonly referred to as the "V" model, as described in Figure 11.1. The left arm of the V deals with defining the requirements and the changes, and the right arm of the V ensures that for each item in the left arm, there is a corresponding activity that verifies that the design and changes have been met. This traditional approach to validation achieved adequate compliance, but did not incorporate the latest QbD and risk-based concepts. In addition, the approach involved inspecting and testing against all engineering specifications, which was very time-consuming and often resulted in a multitude of thick protocols and reports.

11.2.2 ASTM E2500-07: Standard Guide for Specification, Design, and Verification of Pharmaceutical and Biopharmaceutical Manufacturing Systems and Equipment

More recently, a team was formed through ISPE, which worked with the American Society of Testing and Materials (ASTM) committee E55.03 to further develop a new consensus-based commissioning and qualification standard to help pharmaceutical companies focus and prioritize their validation efforts based on risk assessment. Note that consensus standards are voluntary and not legal or regulatory. However, they do tend to add more weight than documents from professional societies.

The main driver for a new standard was to try and simplify the validation process and hence reduce the associated time and cost. The resultant standard, ASTM E2500, which is only five pages long, was finally approved at the end of May 2007 and published in July 2007 [2]. It is applicable worldwide and employs a lean approach to validation that is completely in line with the FDA's "Pharmaceutical cGMPs for the 21st Century" and the latest PV guidance; EMA guidance; International Conference on Harmonization (ICH) Q8, Q9, and Q10; and the QbD concept. Although a standard approach is provided, organizations need to translate this into their own procedures and apply tools and templates suited to their operation.

ASTM E2500 applies to all elements of pharmaceutical and biopharmaceutical manufacturing capability; from the actual manufacturing systems, equipment, and automation systems, to manufacturing facilities, process equipment, process control, utilities, and supporting services such as gases or liquid systems and laboratory and information systems. It applies to new manufacturing systems and also to existing systems and equipment. In fact, it is applicable throughout the product lifecycle, from introduction to retirement. The process steps for the ASTM E2500 system lifecycle and validation approach are summarized in Figure 11.2.

ASTM E2500 encompasses the following eight key concepts:

1. Science-based approach.
2. Critical aspects of manufacturing systems.
3. Risk-based approach.
4. QbD.
5. Good Engineering Practice (GEP).
6. Subject matter expert (SME).
7. Use of vendor documentation.
8. Continuous improvement.

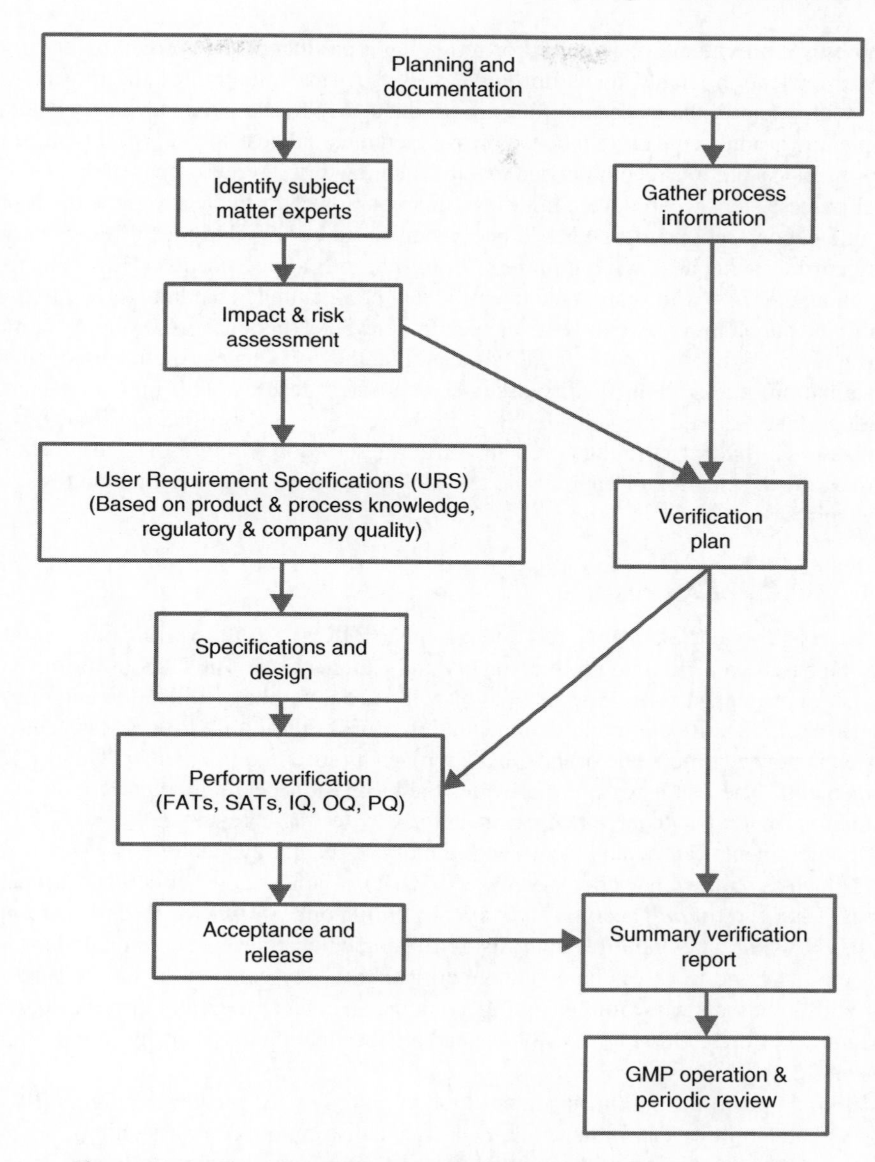

Figure 11.2 *The ASTM E2500 system lifecycle and validation approach.*

11.2.3 Science-Based Approach and Critical Aspects

Product and process information related to product quality and patient safety should be used to ensure that manufacturing systems are designed and verified to be fit for their intended use. Qualification of the facility, process equipment, and systems used in the manufacture of drug product should be based on a *QbD approach* to gain a detailed understanding of the CQAs of the product and the critical aspects of the manufacturing systems and process (CPPs) that are essential for providing a product of the correct quality to the patient. Verification should focus on these aspects. At the initiation of the project, it is important to identify the product or range of products the process is intended to handle; for example, the product could be a tablet, a pressurized metered dose inhaler (pMDI), a sterile injection, and so on. It is recommended that the manufacturing process be then divided into logical process steps to produce a high-level map or flowchart to clearly show the design logic and process boundaries. Each process step should add value by fulfilling product quality attributes that are captured in the User Requirement Specification (URS) step from the appointed *SMEs,* who share their scientific and product understanding. SMEs are individuals who should have several years of specific relevant experience (>10 years), training, and responsibility in a particular field. For example, they may have expertise in engineering, validation, automation, quality assurance, product development, or manufacturing operations. They should take a leading role in the planning and verification strategies for manufacturing systems, defining acceptance criteria, selection of appropriate test methods, risk assessment, selection of appropriate test methods, execution of verification tests, and reviewing results.

11.2.4 Risk-Based Approach

The scope and extent of quality risk management (QRM) for the verification activities should be based on the risk to product quality and patient safety. The URS forms the basis for a risk management process whose output will determine the qualification activities to be undertaken. Identification and documentation of risks during the risk assessment process step is now a cornerstone of the QbD approach as specified in ICH Q9 "Quality Risk Management." The risk assessment provides the basis for accepting the current design in addition to providing a documented rationale for why testing is required.

Risk assessments can be conducted with the aid of readily available tools described in ICH Q9 such as failure mode effects analysis (FMEA) or failure mode, effects and criticality analysis (FMECA), hazard analysis and critical control points (HACCP), hazard operability analysis (HAZOP), and fault tree analysis, for example, to achieve a common understanding of the risks and to assess their relative criticality. Every company is unique and will require different tools when implementing a risk management program. Further discussion on the many different QRM tools available and an introduction to risk analysis are covered in Chapter 2 of this book.

FMEA is a powerful risk management tool commonly used for developing a URS for equipment and utilities and for new facility design, construction, commissioning, and qualification. The benefits are that it can reduce design and manufacturing costs and improve product reliability. FMEA is a bottom-up approach to failure mode analysis used to evaluate a design or process for possible failure modes. It considers each mode of failure of every component of a system, and ascertains the effects on system operation of each failure mode in turn to help eliminate poor design and process weaknesses that could adversely affect safety and performance. FMEA may also be used to analyze the intended manufacturing processes for the drug product. It is best to apply this in the early stages of process design, before any new equipment of facilities has been actually purchased.

A failure mode is a physical description of a defect, condition, or characteristic related to the specific function, component, or operation identified. Potential effects are any conditions that can occur, brought about by a failure mode, if it were present in the product used by the customer(s) or end user(s). This might be during manufacturing or in the clinical setting, for example. For each failure mode, the team of subject experts should list all the possible causes that could bring about the failure. There may be more than one cause for each failure mode.

In most FMEAs, the risk priority ratings for each possible failure mode are calculated by multiplying the severity (S) if a failure occurred, the probability or occurrence (P) of a failure occurring, and the detectability (D) of the failure if it should occur (S × P × D) to give a risk priority number (RPN). The RPN gives a measure of the overall risk associated with the failure mode; the higher the number, the more serious the failure mode will be. Typically, a 10-point scale is used for each of S, P, and D, with (1) being the lowest and (10) being the highest as shown in Table 11.1.

A greater chance of a failure occurring will result in a higher score for P. The more severe the consequence of failure, the higher will be the score for S. The greater the chance that the failure will not be detected, the higher will be the score for D. The RPN can range from 1 to 1000 and can be used for ranking purposes and for identifying the most critical aspects of the system that should receive the most attention. Scoring is based upon factors such as prior knowledge, previous experience and expertise, and known system capability assessed by an assembled team of SMEs representing various disciplines, for example, engineering, product development, manufacturing operations, and quality assurance.

Following the risk evaluation step, the SMEs should consider risk mitigation steps to reduce risks to an acceptable level, or provide a rationale for accepting the risk as it is. For risks considered to be unacceptably high, corrective actions taken to reduce the RPN could be to change the design to reduce the severity, to change the manufacturing process and/or

Table 11.1 *A typical ranking system for severity, probability, and detectability.*

Rating	Severity	Probability	Detectability
10	Dangerously High: Failure affects safety and involves major non-compliance and government regulations. Product could potentially lead to death	Very High: Failure almost inevitable	Absolute Uncertainty: Product is not inspected or defect is not detectable
6	Moderate: Failure causes some dissatisfaction. Patient is made uncomfortable or is upset by the failure, resulting in a patient complaint and product return.	Moderate: Occasional failures.	Low: Controls may detect the existence of a defect. Moderate risk that the product will be delivered with the defect. Some statistical controls employed.
1	Negligible/none.	Remote: Failure is unlikely.	Almost certain: Defect is obvious or there is 100% automatic inspection with regular calibration or preventative maintenance of the inspection equipment.

product design to reduce the probability, or to improve the controls to improve the detectability. Very low-risk items may have their requirements satisfied with commissioning tests and documentation, thus avoiding repeat testing during qualification. For commissioning tests to be accepted, the team must be satisfied that the testing method was appropriate and was conducted with an acceptable quality, including a comparison of the test environment with the final environment. Once action has been taken to improve the design, a new rating for the RPN should be evaluated until the failure mode is deemed to be acceptable.

11.2.5 System and Component Impact Assessments

Before a detailed URS document is developed, a system impact assessment should be performed to identify potential product defects caused by the URS requirement not being fulfilled. The consequences (severity) of the product defect should be considered if it reaches the patient, and requirements to mitigate the failure or its effect should be sought, for example, eliminating the failure by understanding the root cause, or preventing the product defect from reaching the patient by detection and rejection. So, before a new system or process equipment is purchased, installed, commissioned, and validated, an impact assessment should be performed. "Direct Impact" systems are deemed to have an impact on product quality, whereas "No Impact" or "Indirect Impact" systems do not. Although all systems will require commissioning, the "direct impact" system will also have to be qualified to meet regulatory requirements.

Some typical guiding questions that can be employed to assess the system impact on product, process and cGMP are as follows:

- Does the system come into direct physical contact with the product or direct contact with a product contact surface?
- Does the system produce or distribute a material (e.g., an excipient) that has contact with the product?
- Does the system perform a critical process step or operation in manufacturing or processing or testing of the product?
- Is the system used in cleaning, sanitizing, or sterilizing?
- Does the system create or maintain a specified environmental condition required to preserve product quality (e.g., nitrogen purge for air-sensitive products)?
- Does the system produce, monitor, evaluate, store, or report data used to accept or reject product or CGMP materials, or data used to support compliance (e.g., electronic batch record system)?
- Does the system support a direct impact system without having a direct impact on product quality?

The answers to these questions will give an indication as to whether the system is critical to product quality or not and the scrutiny it should receive during qualification.

Table 11.2 illustrates the application of FMEA for a process impact assessment for the manufacture of a tablet produced by wet granulation. Only one attribute is shown in the example, whereas in the complete assessment there will be many more attributes that require evaluation for potential failure modes and should be documented. The next step in the process is to consider for each potential critical failure mode how the specific design and qualification testing will contribute to mitigating the potential failure.

Table 11.2 *Example of a process impact assessment for the manufacture of a tablet product.*
Attribute: Tablets will not be contaminated or compromised in any way by the manufacturing activity.

Potential product defect	Severity of product defect	Severity rationale	Potential failure	Mitigation
Contamination from contact with inappropriate materials used during construction of the process equipment.	10	Patient will be exposed to product with contamination that may cause harm.	Incompatible materials used during construction of processing equipment, e.g., non-TSE-free polishing compounds.	Verification of the materials used during construction of the processing equipment that are in contact with the product.
Contamination of the product by cross-contamination by other products.	10	Patient will be exposed to product with contamination that may cause harm.	Inadequate cleaning of equipment and cross-contamination of products.	1. Equipment is designed to be easily cleanable. 2. Cleaning procedure to be developed and validated.
			Airborne contamination.	Airflow directions in the processing areas should be appropriate.
			Foreign material in excipients or active pharmaceutical ingredient.	Quality control procedures, e.g., sampling and testing, letters of conformity.
			Inadvertent transfer of other products into the product process from operatives' garments.	Appropriate personnel change regimes and process room access discipline, procedures and operator training.
Contamination of the product from lubricants, etc., used in the process equipment.	6	Patient will be exposed to product with contamination that may cause harm.	Failure to isolate mechanical lubricants from the product.	Exclude lubricants by using dry bearings or using appropriate seal technology and where lubricants used are FDA approved.
Contamination of the product by use of wrong constituent materials.	10	Patient will be exposed to product with contamination that may cause harm, e.g., wrong API.	Incorrect material used in the formulation.	Verify identity in dispensing that the correct excipients and API are being used, and make a second check at charging(human or machine).

After establishing whether the system is direct or indirect, it is recommended that a component impact assessment also be performed to evaluate its criticality. This is best done after the URSs are in place and instrument drawings are available. The process is similar to that previously described for assessment of system impact, with similar typical guiding questions used to evaluate a particular component's impact on the operating, controlling, warning, and failure conditions of a system on the quality of a product. Any affirmative answer indicates a direct impact on product quality, thus qualifying the component as critical.

11.2.6 URSs for Systems

A URS is a detailed document used to specify the requirements of the user for individual aspects of the facility, equipment, utility, and system in terms of function, throughput, operation, documents, and qualifications required. The URS describes critical installation and operating parameters and performance standards that are required for the intended use of the system, and forms the basis for subsequent qualification and future maintenance. It can also be a useful document when purchasing a system, forming part of the purchasing specification.

11.2.7 Specification and Design

After the impact assessments have been performed and the URS developed, the qualification stages can be performed. First, the specification and design step is performed, where critical aspects should be captured, documented, and approved by the SMEs, using the risk and impact assessment processes and prior knowledge. *GEP*, defined as "those established engineering methods and standards that are applied throughout the lifecycle to deliver appropriate and effective solutions," should underpin and support this, and also leverage qualified *vendor expertise*. Vendor documentations may be used as part of verification documentation, but it is the responsibility of the pharmaceutical company to assess the vendor to assure quality and technical capability.

Functional specifications should be developed to provide acceptance criteria for functional tests specified in the verification plan. The approved critical aspects of the design, and associated acceptance criteria, should be linked to a structured verification plan through a traceability matrix, so that if a change control is required for a critical aspect of the manufacturing system, this traceability matrix provides an assessment of the impact by identifying all related specifications, design, and verification steps. The traceability matrix will summarize required testing and when it occurs.

11.2.8 Verification

The core concept of ASTM E2500 is described by the term "verification," which replaces the terms "commissioning,""qualification," and "validation" and is defined as

> a systematic approach to verify that manufacturing systems, acting singly or in combination, are fit for intended use, have been properly installed and are operating correctly.

SMEs should perform or oversee the activities and testing to execute the verification plan and to document the results. Testing should occur across all the traditional commissioning and validation stages (FAT, SAT, IQ, OQ, and PQ), with the more critical testing at the IQ/OQ stages to mitigate risk. Protocols may be combined into I/OQ or IQ/OQ/PQ for smaller, less complex system qualification, again based on risk assessment considerations.

ASTM E2500 requires a Summary Verification Report to provide an overview of the test results and highlight any non-conformances with the acceptance criteria or deviations or changes to the original Verification Protocol. This is achieved by a two-step process starting with a Verification Review performed by an independent SME to check the technical results and confirm the suitability for intended use. This is followed by an Acceptance and Release step, which is an approval by the SME and also by quality assurance (QA) for systems with critical aspects, to enable QA officially release the system for GMP operational use.

In summary, the key benefits of ASTM E2500 include reduced validation testing, building on installation testing, and prioritizing efforts on equipment and systems that directly impact product quality and patient safety. This latest approach reduces qualification costs by not repeating the same testing between commissioning and qualification, applying risk assessment involving SMEs to focus on what is important and reducing the qualification documentation, time, and resource. The main focus of the verification is on critical aspects of the manufacturing system that impact consistent product quality and patient safety. *Continuous improvement* can be incorporated into the process by seeking opportunities for improvements, based on review and evaluation, operational and performance data, and root-cause analysis of failures. By applying the eight concepts of ASTM E2500, organizations can benefit from a more flexible validation approach.

The essential differences between the ISPE Baseline® Guide Volume 5 and ASTM E2500 are summarized in Table 11.3.

Table 11.3 *Key differences between the ISPE Baseline® Guide and ASTM E2500.*

ISPE Baseline Guide Volume 5	ASTM E2500 Guide	Comments
Impact Assessment	Design inputs	Applying risk assessment by subject matter experts
Design Qualification	Design Review	Design review by subject matter experts
Commissioning	Risk Mitigation	Knowledge input from subject matter experts
IQ, OQ, PQ, and Acceptance Criteria	Verification Testing and Performance Testing	Commissioning and Qualification replaced with Verification
Good Engineering Practice (GEP) scope and QA scope overlapped	Clear boundaries between GEP and QA scope	Better focus and avoiding duplication of efforts
Focus on documentation deliverables	Focus on critical aspects that affect product quality and patient safety	Focus on what is important (critical). Not required to validate everything!
Rigid change control	Continuous process improvements	Lifecycle concept applied

The value of the ASTM E2500 standard has been acknowledged by the FDA, who reference it in a footnote in their latest guidance on PV because it can be used to support 2(1) of the PV guidance [3]. The FDA guide not only covers facilities and equipment qualification, but also PV, which is discussed in more detail later in this chapter.

In 2011, ISPE published another Baseline® Guide entitled "Science and Risk-Based Approach for the Delivery of Facilities, Systems, and Equipment" [4]. This guide reflects the latest industry and regulatory thinking and was introduced to provide direction to the pharmaceutical industry on how to implement a validation program based on ASTM E2500. The aim of the guide is to facilitate the translation of scientific knowledge about the product and process into documented specification, design, and verification of facilities, systems, and equipment. Specific implementation guidance is given on meeting the expectations of global regulators and is compatible with ICH documents (Q8 (R2) "Pharmaceutical Development" [5], Q9 "Quality Risk Management" [6], Q10 "Pharmaceutical Quality System" [7]), and ASTM E2500-07. The approach described in this guide focuses on QbD and establishing what is critical for the process, product, and patient, and recommends verification strategies for confirming these critical aspects. The activities described address the verification (or qualification) portion of the validation lifecycle upon which PV is built.

Finally, another ISPE guide titled "Applied Risk Management in Commissioning and Qualification [8]" is also well worth reading for organizations that are trying to change from the established traditional baseline approach to the more efficient science- and risk-based approach. It links the traditional terminology and approaches of commissioning and qualification as used in ISPE Baseline Guide Volume 5 to the newer approaches applied in the ISPE FSE Guide, ASTM E2500, and describes the application of QRM (ICH Q9). This document recognizes that it may take time and money for organizations to fully transition to the full science- and risk-based approach as described in the FSE Guide, but stresses the long-term benefit of doing so. Although it is not mandatory, it warns that there are regulatory expectations for QRM approaches to be adopted and the traditional approach is in danger of becoming non-compliant! ASTM E2500 was reviewed and revised in May 2013, but with no significant changes.

11.3 Drug Product Process Validation: A Lifecycle Approach

PV of the finished drug product is a strict GMP requirement, and manufacturers must comply with the current GMP regulations for finished pharmaceuticals (21 CFR 210 and 2011, FDA 2014 [9], EudraLex Volume 4, 2014 [10]). According to the cGMP guidelines, manufacturers must control the critical aspects of their particular operations through qualification and validation over the lifecycle of the product and process. Any planned changes to the facilities, equipment, utilities, and processes that may affect the quality of the product should be formally documented and the impact on the validated status or control strategy assessed.

The approach to PV is changing from the traditional approach, where the value of conducting three arbitrary PV batches as a "one-off" event to register a new commercial product has been challenged, to a new "lifecycle approach to process validation (PV)."

The traditional approach was compliance based and about following rules, whereas the new QbD is a science- and risk-based approach employed to understand product and process design, to identify, and focus on the critical aspects and to establish a control strategy for the manufacturing process. The application of the lifecycle approach to new and existing products is now an expectation from the major regulatory authorities, although it is not yet mandatory.

Annex 15 of the EU guideline on Qualification and Validation came into force from October 2015 [10]) to reflect the relevant concepts and guidance presented in ICH Q8, Q10, and Q11 and the most recent PV guidance updates discussed further below. The Annex 15 section on PV refers to CHMP's "Note for Guidance on Process Validation" and uses the same terminology to describe traditional, hybrid, and continuous verification approaches to product development. Products that have been developed by a QbD approach can follow a continuous process verification approach as an alternative to traditional PV. A hybrid approach using a mixture of the traditional approach and continuous process verification for different production steps can also be used. The Annex also includes a section on the qualification stages of equipment, facilities, and utilities including the URS as the starting point of reference, as discussed earlier in this chapter.

Both the European and the US regulatory authorities support the lifecycle approach and principles of QbD in their latest guidance documents on PV (EMA 2014 [11], FDA 2011 [3]). The FDA guidance covers the three different stages of the product lifecycle described further below in a single guidance document, where as the EU PV guidance covers "Stage 2 – Process Validation" only. The reason for this is that "Stage 1 – Product Development" is already covered in ICH Q8(R2) and "Stage 3 – Ongoing Process Verification" is covered in EU GMP Annex 15. A summary comparing the EU and US PV guidance and the key differences is given in Table 11.4.

The PV lifecycle approach links the process design/product development (Stage 1) closely with the qualification and validation of the commercial production processes (Stage 2), with Stage 3 being maintenance of the commercial production process in a state of control during routine production until the product is eventually discontinued:

- Stage 1 – Process design/product development
- Stage 2 – Process qualification
- Stage 3 – Continued process verification

The lifecycle approach to process validation recommended is consistent with a scientific, evidence- and risk-based approach led by the regulators through a series of QbD initiatives including ICH Q8 (R2), "Pharmaceutical Development"; ICH Q9, "Quality Risk Management"; ICH Q10, "Pharmaceutical Quality Systems," and the FDA guidance on process analytical technology (PAT) [12].

The US FDA definition of PV for pharmaceutical manufacturing tends to reflect the lifecycle approach and emphasizes that PV should no longer be considered to be a "one-off" event, but should involve "the collection and evaluation of data from the process design stage through commercial production that establishes scientific evidence that a process is capable of consistently delivering quality product" [3].

Comments made by senior FDA officials at public meetings indicate that the new concepts in their guidance are more important than the terminology used. Also, it is clear

Table 11.4 *Summary comparison of the EU and US process validation guidance.*

2011 US FDA Guidance	2014 EMA CHMP & Annex 15 Guidelines	Comments
Definition: "The collection and evaluation of data, from the process design stage through commercial production, which establishes scientific evidence that a process is capable of consistently delivering quality products"	Definition: "The documented evidence that the process, operated within established parameters, can perform effectively, and reproducibly to produce a medicinal product meeting its predetermined specifications and quality attributes"	Different definitions, but there are similarities in the intended meaning
All three stages of the lifecycle included in single document	Stage 1 not included but cross-refers to ICH Q8R2. Process Validation stage covered by new CHMP guidance and stage 3 covered by new Annex 15	The United States refers to "Process Design" for stage 1 whereas EU refers to "Product Development."The United States refer to Stage 3 as "Continued Process Verification," whereas the EU refers to "Ongoing Process Verification"
Intended to align with ICH Q8, Q9, Q10, and Q11	Intended to align with ICH Q8, Q9, Q10, and Q11	
Focus is on new lifecycle approach	Allows for traditional or a hybrid approach	
Emphasizes importance of statistics throughout lifecycle	Statistics only emphasized as an option	
Clearly endorses Continuous Process Verification	Ongoing Process Verification is referred to as an alternative approach	
No mention of number of validation batches required	Suggests "at least 3 validation batches" will be required	The United States goes further in emphasizing the use of all available data and statistics to support process validation, and this should dictate the number of batches

that the key focus of the guidance is on variation, understanding, detecting, responding, and controlling it from input to output. The FDA PV guiding principles clearly support a QbD approach:

- Quality, safety, and efficacy should be designed into the product.
- Quality cannot be adequately assured merely by in-process and finished-product testing or inspection.
- Each step of a manufacturing process is controlled to assure that the finished product meets all quality attributes including specifications.

The Parenteral Drug Association (PDA) has recently published Technical Report (TR) No. 60: "Process Validation: A Lifecycle Approach" [13], intended to provide practical guidance on the implementation of the three stages of the lifecycle approach to pharmaceutical PV. The TR follows the principles and general recommendations presented in current regulatory PV guidance documents from both the United States and Europe. A diverse group of SMEs from the pharmaceutical industry and from the regulatory bodies in the United States and Europe contributed to putting together this comprehensive view of validation theory and methodology. The TR also describes the use of enabling tools such as PAT, statistical analytical tools, QRM, technology transfer, and knowledge management as they relate to PV. Two examples are included in TR 60, one for a large molecule (biotech) and another for a small chemical (parenteral), respectively, illustrating the application of the three stages of PV to these product types.

One other validation guidance document to mention is a World Health Organization (WHO) publication still in draft, supplementary "Guidelines on Good Manufacturing Practices: Validation, Appendix 7: Non-Sterile Process Validation" [14]. This seems to be a fusion of the FDA's and EMA's PV guidance, but again supports the principles of QbD and the lifecycle approach.

The 2011 FDA PV guidance encourages QbD principles to be used when designing a manufacturing process (although it does not mention QbD by name). The three-stage lifecycle process illustrated in Figure 11.3 gives a recommended sequence of formulation and process design activities followed by PV activities to establish a robust product and manufacturing process with an associated design space and control strategy to assure product quality over the lifecycle of the product.

11.3.1 Stage 1: Process Design/Product Development

During Stage 1, the intended commercial manufacturing process is defined on the basis of knowledge gained through formulation and process development and scale-up activities as shown in Figure 11.3. The aim is to establish good product and process knowledge and understanding during the course of pharmaceutical development.

The principles of QbD recognize that quality cannot be tested into products, but should be built in by appropriate design and evaluated in a systematic way, for example, using design of experiments (DoE) as discussed in more detail in Chapter 7. Employing statistically designed multifactor (matrix) experiments (DoE), and multivariate analysis (see Chapter 8) where the relationship between several factors can be quantified simultaneously is usually the most time-efficient and resource-saving approach to product and process optimization. Applying QbD principles encourages pharmaceutical developers to apply modern statistical DoE and

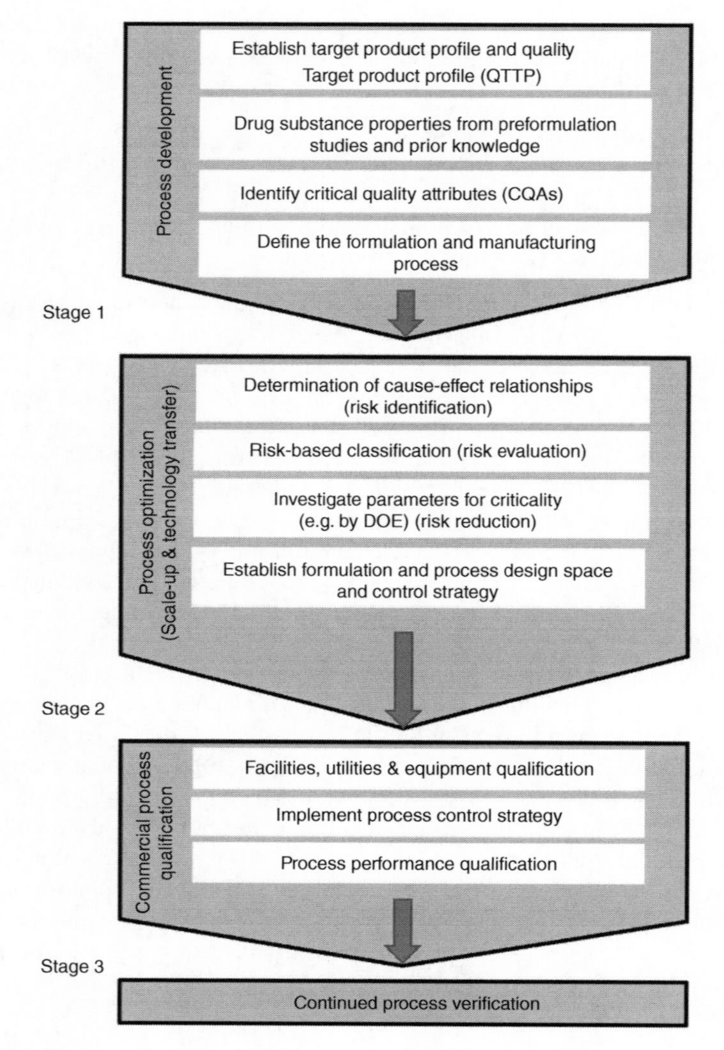

Figure 11.3 *Sequence of activities for formulation, process design, and optimization incorporating process validation activities according to the lifecycle approach.*

analytical procedures to define both the critical and noncritical sources of variability in the product and process, with the objective of demonstrating that the product is controlled within a broader "design space." Different types of DoE and recommended statistical methods, along with commercially available and user-friendly software tools, that can be applied throughout the QbD framework for the development of a pharmaceutical product to acquire formulation and process understanding can be found in the literature [15].

QRM and the application of risk management tools (discussed more thoroughly in Chapter 2) are also important and should be incorporated into product and process development in order to focus attention on the critical aspects as shown in the process flow given in Figure 11.3. According to ICH Q9: "Quality Risk Management is a systematic process for the assessment, control, communication and review of risks to the quality of the medicinal product across the product lifecycle." The initial risk assessment step is "Risk Identification," the use of information to identify hazards or potential risks considering any historical data available, prior knowledge, informed opinions, theoretical analysis, and the evaluation of cause-effect relationships from experimental design data. This is followed by a "Risk Evaluation" step to determine the most likely critical areas to monitor and control, but also to deduce where reduced scrutiny is required (Risk Reduction or Risk Acceptance). The risk evaluation process can be used to compare identified and analyzed risks against agreed criteria, considering probability, severity, and detectability. The output can be qualitative (high, medium, or low) or quantitative (probability × severity × detectability) to provide numerical values giving a relative ranking of the risks and enabling them to be prioritized. Parameters with the greatest likelihood of product impact should be evaluated first.

At the completion of process optimization, there should be a justification to support the drug product commercial manufacturing process steps and the CPPs, the limits of the process parameters, and confirmation that the product specifications can be met. Another key requirement for successful technology transfer from R&D to commercial production is that the product be "manufacturable" and that the manufacturing process be sufficiently "robust" during routine manufacture. Various metrics can be applied to give an indication of manufacturability such as the first time pass rate, the batch yield, the number of batch failures, batch cycle time, and resource and time costs, for example. A potential pitfall to avoid is to design a manufacturing process that can only be performed in the hands of "experts" or has to be nursed through because the operating limits for one or two CPPs are too narrow and cannot be consistently achieved. A measurement of process reproducibility and robustness as a function of the specification often used is the process capability index (C_{pk}). The C_{pk} is a simple number given by the lower of the two values calculated from the following equations:

$$C_{pk} = \frac{\text{upper limit of specification} - \text{mean}}{3 \times \text{standard deviation}} \quad (11.1)$$

$$C_{pk} = \frac{\text{mean} - \text{lower limit of specification}}{3 \times \text{standard deviation}} \quad (11.2)$$

There are some generally acceptable rules established on the basis of experimental data that relate the value of C_{pk} to process robustness. Values less than 1.0 indicate that the process is not capable, and further development work should be undertaken to improve it. C_{pk} values obtained between the range of 1.0 to 3.0 indicate a capable process meeting the specification target with acceptable variation, and a process with values greater than 3.0 is indicative of being very capable. However, aiming for C_{pk} values greater than 3.0 ("bomb-proof") is not recommended as it can lead to much wasted time and resource and is not necessary. Capability Indices data are useful for identifying the process variables that have the least or the most effect on product quality, to relax or tighten the control range, and to define any critical measures for in-process control and monitoring.

The final stage of product and process optimization is to generate sufficient stability data on one or more variants to select the best combination.

11.3.2 Stage 2: Process Qualification

During the process qualification (PQ) stage of PV, the process design output is evaluated to determine if it is capable of reproducible commercial manufacture. According to the current FDA PV guidance, this stage consists of two aspects, the qualification of the facility, equipment, and utilities already discussed in Section 10.2 of this chapter, and second, the PPQ. The major regulatory agencies such as the EMA and FDA stipulate that PPQ must be undertaken on the intended commercial product to fulfill regulatory and quality system requirements. Sufficient data must be available to demonstrate that the commercial-scale manufacturing process is reproducible, can be maintained within established parameters, and consistently produce product that meets product specifications. There has been a shift from the traditional approach to PPQ, where pharmaceutical companies were required to complete three successful validation runs to be able to claim that the process is validated. Although the EMA PV guidance does mention "at least 3 validation runs" for the regulatory submission, both the FDA guidance and PDA TR 60 do not stipulate the number of validation runs required. Instead, the onus is on a pharmaceutical company to scientifically determine the number of validation runs needed to acquire the necessary process understanding and control.

The expectation is that statistical considerations will be made to support the desired confidence level. TR 60 provides some useful guidance on statistical methods for determining the number of validation runs and appropriate sampling plans [13], and the EMA PV guidance discusses the extent of validation runs depending on the circumstances and whether the product type is of the standard or nonstandard type [11]. The PPQ will usually involve a higher level of sampling, testing, and greater scrutiny of process performance than typical routine commercial production. However, it is not typically necessary to explore the entire operating range at commercial scale if assurance can be provided by process design data. Previous credible experience with sufficiently similar products and processes can also be helpful. In any case, a justification for the PV studies will be required for the regulatory submission for the commercial product. This should explain how the data available from pilot-scale and commercial-scale studies support the design space for the proposed commercial manufacturing process.

The PPQ batches should be made to cGMP conditions employing trained personnel who are expected to manufacture routinely in the future, and the exercise should be appropriately documented. This means that there should be validation plans and protocols written prior to conducting the practical work, and validation reports written at the completion of the work. The PPQ Protocol should discuss:

- Manufacturing conditions such as operating parameters, processing limits, and raw material inputs.
- Data to be collected and when/how evaluated.
- Tests (in-process, release, characterization) and acceptance criteria for each critical process step.
- Sampling – plan, points, numbers, and frequency for each unit operation and quality attribute.

The PV Protocol and Report should discuss

- The process criteria and performance indicators to consistently produce a quality product, such as the statistical methods used to define batch variability (e.g., trending, evaluation of process stability and process capability, detection of process variability), and also to document how deviations and nonconforming data are addressed
- Equipment qualification.
- Personnel training.
- Analytical methods validation to support materials, in-process testing, and final product testing.
- Review and approval of the protocol.

Approval of the commercial process for routine use is conditionally based on a successful review of all the validation documentation outlined in a validation master plan, including cleaning validation, the results from all the sampling, and some stability testing of the manufactured product.

11.3.3 Stage 3: Continued Process Verification

The objective of the Continued Process Verification (CPV) stage is to provide a means of continually assuring that the manufacturing process remains in a state of control (the validated state) during routine commercial manufacture and meets GMP requirements. It involves an ongoing program of product and process data collection and evaluation, including the quality of the input materials and components used in the process, in-process testing, and finished-product testing. Therefore, a thorough understanding of functional relationships between process inputs and corresponding outputs gained during the previous stages is essential prior to starting the CPV program. It should be recognized that not all sources of potential variability may have been defined during Stages 1 and 2, and the CPV program is important for their evaluation. Typical examples of new potential sources of variability may include raw materials sourced from a different supplier or the use of different operatives who impact the manufacturing process. The root causes should be investigated and corrective actions implemented. In reality, there is likely to be unforeseen changes during the lifecycle that result in returning back to the process design stage and having to do more development work, as shown in Figure 11.4.

The FDA recommend that an equivalent level of monitoring and sampling of process parameters and product quality attributes to that used for the PPQ batches be continued until there are sufficient data to demonstrate that the process is being well controlled at the commercial scale. This will involve routine and non-routine sampling and testing employing off-line and on-line analysis (PAT) during manufacture to ensure that the necessary data set is collected. Acceptance sampling of the processed batch will also be considered to determine whether to accept or reject the batch for QC release. The limit of acceptability is referred to as the acceptance quality limit (AQL), defined in the ISO 2859-1 standard as the "quality level that is the worst tolerable" [16]. Suitable sampling plans and limits are also defined in a US standard: MIL- STD-105E. For pharmaceutical products, the limit will be 0% for critical defects (totally unacceptable as the patient might get harmed, or regulations are not accepted) and some relaxation of the limits for less serious defects. More recently, broader quality management systems have extended final inspection to

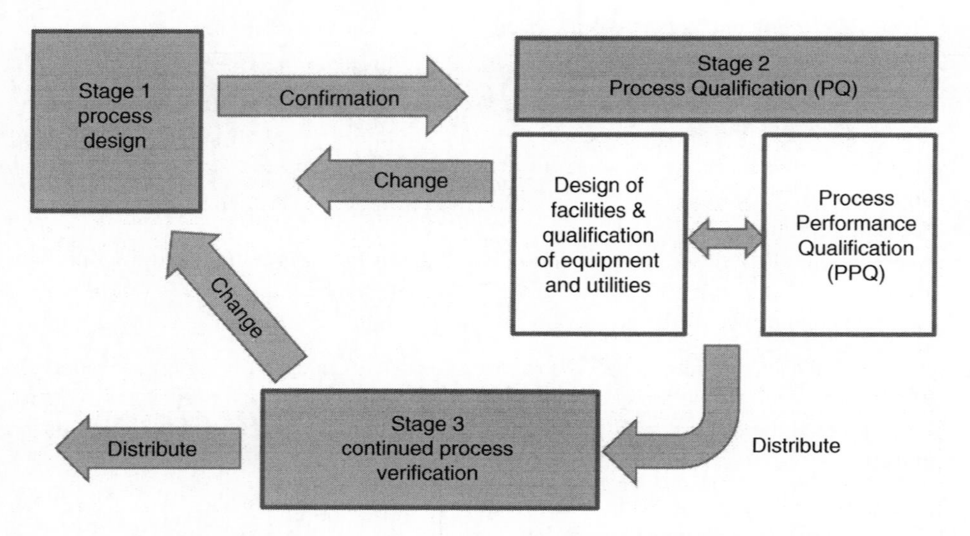

Figure 11.4 *Stages of process validation showing potential changes.*

include methodologies such as statistical process control, HACCP, 6 σ, and ISO 9000, and include all aspects of manufacture.

When sufficient confidence is obtained, the level of sampling and testing can be reduced. For certain parameters that appear to be already well controlled from the PPQ, enhanced sampling/testing may be omitted from the CPV program. Control charts are typically used to evaluate statistical process control over time and detect process trends. Where variables are found that are not being monitored adequately, changes to the monitoring methods may be required. The CPV should focus on the accuracy and reliability of the control methods used and if required consider potential improvements that can be made. However, any changes at this stage will need to be planned, justified, and approved by QA prior to implementation.

Finally, after knowledge and technology transfer from R&D to manufacturing operations has being undertaken successfully and routine manufacture is under way, pharmaceutical companies should be considering opportunities for continuous improvement. This might include making changes to improve the operational efficiency or moving from manual to automated controls, for example.

11.4 The Impact of QbD on Process Equipment Design and Pharmaceutical Manufacturing Processes

Modern pharmaceutical manufacturing on a large scale was first established during WWII (1939–1945), stimulated by the huge demand for penicillin and other discovered antibiotics. At this time, there were only a handful of large pharmaceutical companies with this capability. The number of companies proliferated during the postwar years as numerous new drugs were developed and mass-produced and marketed through the 1960s and 1970s.

There was a further proliferation of the pharmaceutical industry in the 1970s resulting from the incentive of stronger patent protection, to cover not only the drug substance, but also the drug product and the manufacturing process. As a result, Tagamet (cimetidine) and other "blockbuster" drugs were launched in the 1980s and 1990s, each blockbuster earning more than $1 billion a year in sales. During this golden period for the pharmaceutical industry with the associated huge profits, there was little incentive to change the way of working. So process equipment design and manufacturing technology remained fairly stagnant over this period compared to other industries.

Typically, for pharmaceutical processing in the manufacturing environment, the formulator or manufacturing personnel would manufacture one unique batch of drug product at a time, for example, according to the process flow shown in Figure 11.4 for a tablet formulation. In-process samples would be taken from the batch by the operatives during processing of a unit operation for onward testing to confirm that process was in control. A classical method of determining the wet granulation endpoint was for an experienced formulator or manufacturing operative to put their arm into the vessel and gently squeeze a handful of the wet granules between their fingers. Sometimes in-process samples could be tested locally in the manufacturing unit (e.g., tablet hardness and friability), but more often than not samples had to be sent away to the QC laboratory for testing located in another building. Manufacturing would often come to a standstill until the results were known and considered good enough to move to the next unit operation. Thus, manufacturing was stop-start due to the waiting times between unit operations. A significant proportion of the final product would also need to be tested in the QC lab at the end of manufacture to release the batch. So this traditional tried-and-tested approach was used for decades as it served well for both the industry and the regulatory bodies.

However, the climate changed in the 1990s as R&D drug pipelines started to dry up, and pharmaceutical companies struggled to replenish those blockbuster drugs on the market as the patent cliff took hold. In spite of increasing R&D investment, the number of new product launches decreased. This stimulated pharmaceutical companies to find a competitive edge to increase productivity, speed up R&D, and find drastic cost and efficiency savings during development and manufacture. Historical reports from PriceWaterhouseCoopers (PWC) [17] and IBM [18], respectively, compared some standard manufacturing benchmarks with other hi-tech industries and confirmed that there was much room for improvement for the pharmaceutical industry, with low process capability existing in pharmaceutical commercial manufacturing and associated high levels (5%–10%) of scrap and rework accepted, with the cost of quality typically exceeding 20%.

At the same time, some regulators were reporting dissatisfaction, and there was a paradigm regulatory shift, particularly led by the FDA, who wanted to accelerate the rate of change through the publication of extensive guidelines. Their initiative "Pharmaceutical Current Good Manufacturing Practices (cGMPs) for the 21st Century – A Risk-Based Approach" [19] was launched and its guidance on PAT that encouraged on-line, at-line, and in-line testing [12]. The FDA was openly critical of the current state of pharmaceutical manufacturing at the time, questioning not only the efficiency but also the outmoded production techniques and unacceptable levels of quality on the basis of the number of product recalls and drug shortages caused by manufacturing issues or production problems [20].

Another significant factor affecting process equipment design in recent years has been a host of new safety and environmental regulations introduced to address environmental and drug handling safety concerns. New drug molecules have generally become more potent and potentially hazardous. This has impacted the design of process equipment with an emphasis on increased containment and engineering controls built in, rather than relying on protecting the operatives with personal protective equipment (PPE). As a result of all these factors, for sampling and in-process testing, it was no longer acceptable for manufacturing operatives to access the process by opening the equipment and exposing themselves to potentially harmful materials. Alternative safer and timely means of sampling were required to be developed. Initially, pharma R&D process engineers attempted to retrofit the equipment by drilling holes and inserting probes, in-line sensors, and windows to enable collection of the in-process data. PAT devices were developed to measure changes and variability of the product, such as flow, pressure, and temperature. Later, various equipment suppliers came on board to design and automatically incorporate in-line sensors, windows, and analyzers into new process equipment, along with more sophisticated computer control and data capture capabilities to collect more complex data (e.g., using NIR spectroscopy). See Section 10.5 for more discussion about process control strategies and the use of sensors and analyzers used in pharmaceutical manufacturing.

All of these drivers for change have led to a concerted effort by equipment suppliers, academia, and the pharmaceutical industry to make great advancements in integrating QbD principles and PAT into pharmaceutical manufacturing processes and equipment design. The current pharmaceutical environment requires drug manufacturing processes that continuously improve product quality while reducing lead time in order to deliver more products on time every time. Therefore, many pharmaceutical companies are introducing continuous manufacturing to optimize their processes, increase their capacity, manage with variable demand, and reduce inventory costs. Applying continuous manufacturing and verification of product quality with advanced process control and real-time QA and release potentially result in faster times to market, as well as lowering operating costs and improving efficiency.

11.5 Introduction to Process Control in Pharmaceutical Manufacturing

A "process" is typically described as the act of taking something through an established set of procedures to convert it from one form to another. For pharmaceutical manufacturing processes, the intention is to convert a drug substance and formulation raw materials into a finished drug product. During the process design stage, the developer should apply QbD principles to establish a sound science-based understanding of the process in order to design the process correctly and identify what conditions drive the process in which direction; conditions that produce desired transformations are needed, and conditions that produce unwanted transformations (e.g., overprocessed material) are avoided. Also, the developer should aim to understand how to design the manufacturing process so that it either

- Naturally stops when the target is reached.
- Or it stops as a result of measurement of the target attributes and *control* of the process parameters.

The process equipment used will deliver the process conditions, and ideally the process parameters should directly express the process conditions without reference to any specific item of equipment. Equipment settings are applied to a specific piece of process equipment in order to achieve the desired process parameters (conditions) and may not give the same process parameters/conditions in any other piece of equipment. Different products will probably need different process parameters/conditions in order to achieve the target transformation.

Of course, you cannot control what you cannot measure, so there is a natural fit with PAT and QbD. Sensors and analyzers fitted to process equipment provide measurements, which can monitor equipment parameters, process parameters, or product attributes to determine whether the target conditions have been reached. Generally, a process sensor is a simple process analyzer that can only measure one variable at a time, for example, weight, pH, and pressure, whereas a process analyzer can measure many variables at the same time, directly or indirectly. There is now a regulatory expectation that process PAT tools will be used to generate measurements of critical quality and performance attributes during manufacture to control and gain a better understanding of the process. There are different possible measurement modes:

- *Off-line*: A sample is taken from the process and analyzed in a laboratory. It may take several days to get the results.
- *At-line*: The sample is taken from the process and analyzed in close proximity. It usually takes minutes to get the results.
- *On-line*: The sample is diverted from the manufacturing process and may be returned to the process stream. It usually takes seconds.
- *In-line*: The sample is not removed from the process stream. It can be invasive (probe in a process stream) or non-invasive (e.g., measurement through a window by a probe). This can take less than a second.

Typical measurements of common unit operations employed in drug product manufacture and the PAT that can be used are shown in Table 11.5.

Table 11.5 *Typical measurements of common unit operations.*

Unit operation	Measurements	Process analytics
Dispensing	Identity, particle size distribution	NIR (near infra red, at-line)
Milling	Particle size distribution	FBRM (focused beam reflectance measurements, on-line)
Granulation	Homogeneity, composition, endpoint	NIR, acoustics, on-line
Drying	Moisture content, endpoint	NIR, on-line
Blending/mixing	Homogeneity, endpoint	NIR, Raman, on-line
Compression	Identification, assay, content uniformity	NIR, on-line
Coating	Coating thickness, endpoint	NIR, on-line
Filling	Identification, head space	TDLS (tunable diode laser spectroscopy)
Lyophilization	Moisture content	NIR

NIR is probably becoming the most commonly used process analyzer because it can be widely used, both for gaining process understanding and for controlling a process, and it can be applied to gases, liquids, and solids such as powders, tablets, and packaging materials. It is a spectroscopic method that is fast and easy to use and requires no or little sampling. Quantitative analysis is possible through glass and plastic, and it is a nondestructive test. However, because it is not very sensitive and the very broad bands are difficult to assign, there is a need to use multivariate data analysis. Other techniques that can be used include Raman spectroscopy, x-ray fluorescence spectroscopy, x-ray diffraction, and chemical imaging. There are many factors to be considered for selection of the analyzer, but in summary it is important to understand the problem and what needs to be measured, and utilize the simplest technology that can do the job.

The benefit of employing PAT during process development and manufacture to achieve a *strategy for process control* is discussed later in this chapter, with the design of manufacturing equipment and PAT discussed in much more detail in Chapter 9 of this book.

Process control in QbD ensures that CPPs are kept within an operating space (or control space), and certainly within the design space, to ensure that the CQAs are ensured (see Chapter 6, Figure 6.8).

Process control actions can be human to bring a parameter as close as possible to a desired value, but ideally they should be automated actions. The latter can be through mechanical, hydraulic, pneumatic, electrical, or electronic actions to return a parameter as close as possible to a desired value. Closed-loop control has been used successfully for many years in other process industries, but the pharmaceutical industry is playing catch-up. Simple process control mechanisms are typically applied for controlling easy-to-measure CPPs such as temperature, flow rate, pressure, and pH. In advanced process control, discussed later in this chapter, process data and modeling software automatically control the process, which involves multivariable control problems. The elements of a simple, single-loop control system for a liquid process are shown in Figure 11.5.

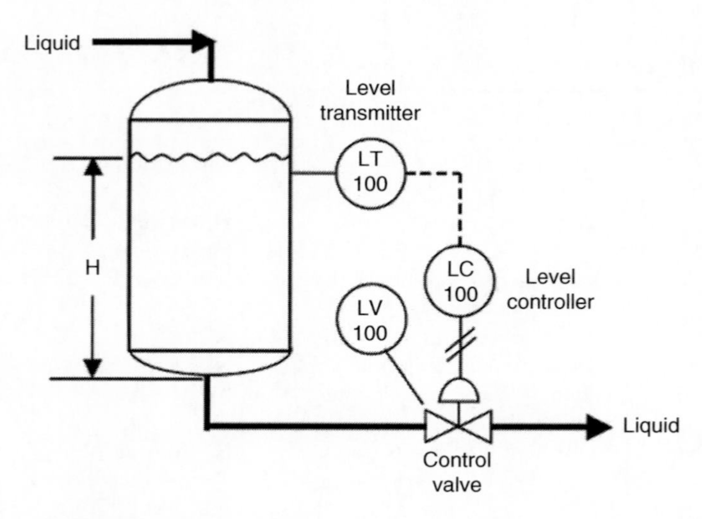

Figure 11.5 *The elements of a simple, single-loop control system for a liquid process. LT = sensor; LC = controller; LV = actuator.*

The diagram shows a sensor that measures the variable under control, that is, the liquid level in this example, and converts the physical property into a useful measurement signal. The controller determines the error between the actual and desired value of the variable and determines corrective actions. The actuator applies the corrective action by manipulating the process input, by opening or closing the control valve to maintain the correct liquid level.

The essential requirements of a process control system must be stable at all times, provide a fast speed of response with acceptable overshoot, reduce the set point error to zero, and reject any disturbances. The actuator elements in control systems are typically control valves. In the simple, single-loop liquid control system example, they modulate the flow of a liquid into or out of the process. The control valves have to be sized correctly. Ideally, the valve should be 40%–60% open under the normal steady state of the process, with a wide range of position to cater for the minimum to maximum set point range. The difference between the set point value and the measured value is referred to as the "set point error" and should be driven to zero by the control system. When the process is disturbed, the controlled variable will deviate from the set point. The control system should respond with corrective action to return the controlled variable back to the set point.

Controllers use a mathematical algorithm to determine the corrective action applied to the manipulated variable. One popular type of algorithm is "on/off," where the controller switches the manipulated variable either on or off in response to the set point error. Another popular type calculates the corrective action based on three types of control action: proportional, integral, and derivative (PID), and is hence referred to as *three term control*.

- P: Control effort is proportional to error.
- I: Control effort is proportional to the integral of error.
- D: Control effort is proportional to the rate of change of error.

There are many variations on the way the P, I, and D terms are calculated and implemented in process control systems. PID controllers used in pharmaceutical processes allow any combination of the three terms to be selected.

Process control is an integral part of the ICH control strategy and the PAT initiative. The overall control strategy will contain other controls such as SOPs, raw material controls, and so on, and is generally developed and initially implemented during clinical trial manufacture, but later refined for commercial manufacture as new knowledge is gained. In reality, the use of advanced processed control (APC) and the introduction of real-time release testing, discussed later in the next section, are more likely to be introduced after comprehensive product and process understanding have been obtained.

11.6 Advanced Process Controls (APC) and Control Strategy

A comprehensive pharmaceutical development approach will establish product and process understanding and identify sources of variability that need to be controlled. Understanding sources of variability and their impact on downstream processes or processing, in-process materials, and drug product quality can provide an opportunity to shift controls upstream and minimize the need for end-product testing (ICH Q8 (R2)). APC deals with multiple control objectives, interactions between process variables, process dynamics, and process disturbances. All of these characteristics allow APC to achieve tighter control of CPPs and CQAs. In addition, APC can ensure that design space limits are honored and move CPPs and CQAs within that space for economic advantage to the business.

There is no universally accepted definition of APC. The term is broadly used and applies to a variety of process control solutions for solving multivariable and/or complex control problems. An alternative broad definition is"a process control solution that is more complex than a collection of single-input, single output PID loops." With respect to QbD, process control ensures that CPPs are maintained within the operating space to ensure that theCQAs are ensured. Two key features of APC is that, in general, it deals with multivariable problems and utilize a model that describes the relationships between process inputs, CPPs, and CQAs. Process-modeling software compares real-time data to an ideal batch profile, identifies what parameters need to be changed to meet the ideal, and feeds this back into the control system.

According to ICH Q10 guidance, one of the main objectives is to establish and maintain a state of control of a pharmaceutical process by developing and using effective monitoring and control systems for process performance and product quality, thereby providing assurance of continued suitability and capability of processes. The ICH definition of a control strategy is "a planned set of controls, derived from current product and process understanding that ensures process performance and product quality. The controls can include parameters and attributes related to drug substance and drug product materials and components, facility and equipment operating conditions, in-process controls, finished product specifications, and the associated methods and frequency of monitoring and control."

ICH Q8 (R2) guidance explains that process controls can be in-process tests primarily for go/no-go decisions and off-line analysis, described as a minimal approach. Preferably, the use of PAT tools utilized with appropriate feedforward and feedback controls (APC) should be applied for an enhanced, QbD approach. Feedback is defined as "the modification or control of a process or system by its results or effects," and feedforward is defined as "the modification or control of a process using its anticipated results or effects."However, feedback/feedforward can be applied technically in process control strategies and conceptually in quality management (ICH Q10).

The control strategy options, at a minimum, should control the CPPs and material attributes, but a control strategy can include, but is not limited to, the following according to ICH Q8 (R2):

- Control of input material attributes (e.g., drug substance, excipients, and primary packaging materials) on the basis of an understanding of their impact on processability or product quality.
- Product specification(s).
- Controls for unit operations that have an impact on downstream processing or product quality (e.g., the impact of drying on degradation, particle size distribution of the granulate on dissolution).
- In-process or real-time release testing in lieu of end-product testing (e.g., measurement and control of CQAs during processing).
- A monitoring program (e.g., full product testing at regular intervals) for verifying multivariate prediction models.

A control strategy can include different elements, so, for example, one element could rely on end-product testing, whereas another could depend on real-time release testing. A well-developed control strategy will reduce risk but does not change the criticality of attributes. It plays a key role in ensuring that the CQAs are met, and hence that the QTTP is realized [21].

In practical terms, there are various potential control strategy options to ensure that the process remains within the design space:

1. Set each parameter at a fixed value.
2. Set one parameter at a fixed value, and allow a range for the second parameter.
3. Monitor parameter 1, and adjust parameter 2 to ensure both remain in the acceptable region.
4. Measure the CQAs, and adjust parameters 1 and 2 to ensure the CQA meets the acceptance criteria.

Consider the following practical example for a simple immediate release tablet to illustrate the potential options, where two CQAs and two CPPs with acceptable ranges have been established using risk assessment and DoE. The CQAs are tablet dissolution and tablet friability, respectively, and the derived design space for the two CPPs in the manufacturing process is illustrated in Figure 11.6. The control strategy potential options are described in Table 11.6 along with the pros and cons for each.

There are several challenges associated with establishing process control. For example, there can be complex interactions between process variables, CPPs, and CQAs, or there can be changing process dynamics throughout the batch. A key challenge can be the non-linearity of the process such that there are varying time constants and a varying magnitude in the relationships between variables. Reversals in relationships can be especially challenging for control systems. Another key challenge is the availability of CQA measurements. Ideally, the CQA parameters should be continuously measured during manufacture.

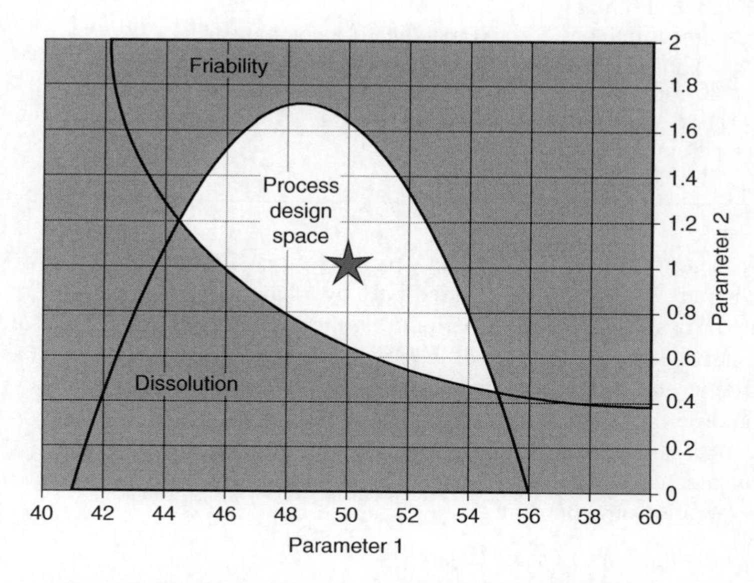

Figure 11.6 *Control strategy and design space for an IR tablet.*

Table 11.6 *Control strategy options for a simple IR tablet.*

	Potential options	Advantages	Disadvantages
1	Set each parameter at a target value: Parameter 1 = 50 Parameter 2 = 1 Verify dissolution and friability for each batch	Simple control strategy reliant on GMP – values specified in batch documentation	Variability System less responsive to variation, e.g., raw materials
2	Set each parameter at a target value: Parameter 1 = 50 Parameter 2 = 0.8–1.5 Verify dissolution and friability for each batch	Simple control strategy reliant on GMP – values specified in batch documentation Flexibility to accommodate some variation Can accommodate feedback control from batch to batch	Limited control of variability
3	Set each parameter at a fixed value: Constantly monitor parameter 1 in real time Adjust parameter 2 to remain within range, e.g., 0.8–1.5 Verify dissolution and friability for each batch	Process constantly monitored and kept under control Flexibility to accommodate variation in real time	More complex solution and costly
4	Measure CQA of dissolution using an on-line PAT: Probably using a surrogate or an indirect measurement Feedback; manually or using an automated feedback mechanism to adjust parameter 1 and 2 in real time	Process constantly monitored and kept under control Flexibility to accommodate variation in real time Can be registered as Real-Time-Release	More complex solution and costly

If no direct measure is possible, the CQA parameter may be inferred from a PAT sensor calibration model, or it may be measured only by a laboratory assay at the end of batch manufacture or from samples taken periodically during the batch. A further challenge can be the risk associated with scale-up of the process as the product is developed and maybe transferred between R&D and commercial sites. QRM tools can be used to assess the risks from using different processing equipment, facilities, personnel capability, and prior knowledge, for example [21].

Three fundamental characteristics of APC are enablers for delivering the benefit of QbD in the manufacturing environment:

- Reduced variability in CPPs and CQAs.
- A "leaner" manufacturing process.
- Economic optimization.

The key factors to consider when selecting a control strategy are to minimize risk to the patient, the effectiveness of the control system applied, and the cost of implementing a more advanced control system for the business.

Consideration should be given to continually improving the control strategy over the product lifecycle as more knowledge is gained and data trends have been assessed [21]. When multivariate prediction models are used, systems that maintain and update the models help to assure the continued suitability of the model within the control strategy. Continuous process verification is one approach that enables a company to monitor the process and make adjustments to the process and/or control strategy, as appropriate (see Chapter 10, Section 10.3)

11.7 The Establishment of Continuous Manufacture

Traditionally, a general pharmaceutical process involved in the manufacture of drug products consists of a series of unit operations. These separate process units when combined together are referred to as a "process train" that can be used to manufacture a unique batch of product. Batch scale-up is achieved by switching to a process train incorporating larger-scale equipment as the project progresses with the associated risk of introducing scale-related changes. For example, Figure 11.7 illustrates a typical batch process flow for a tablet formulation manufactured using a wet granulation process with the associated interruptions and waiting times. Preliminary development studies will typically start at

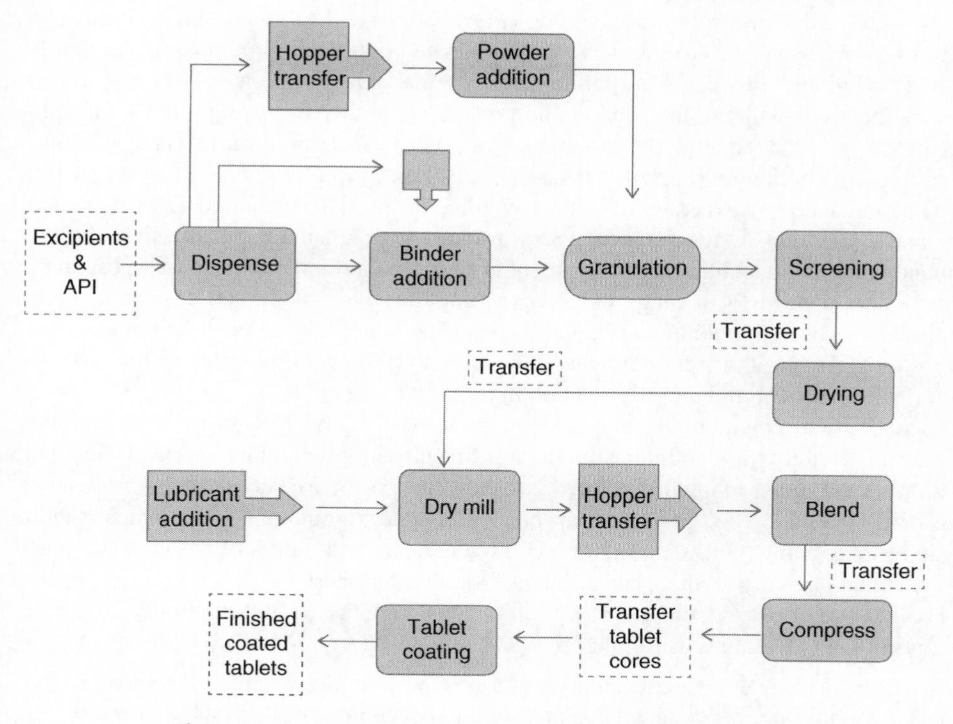

Figure 11.7 Typical batch process flow for a granulated tablet.

laboratory-scale such as 3–10 L, followed by pilot-scale studies at 65 to 150 L in R&D. If the project is successful, final scale-up batches to 500–1000 L or more will usually involve a technology transfer to the operations section.

Traditional batch processes do present some advantages such as the flexibility to re-arrange the unit operations and process equipment in whatever order necessary; also, individual pieces of equipment can be used for multiple drugs. However, there are significant disadvantages too, because the individual process steps are disconnected. There are typically many interruptions, with long waiting times and transport steps between the resulting in-process steps resulting in extended throughput times from start to finish of the manufacture. The batch size will be defined by the scale and output of the equipment used, and the quality of the manufacturing process is typically determined by in-process sampling and end-product testing.

An ASTM working group has suggested a definition for "continuous processing" as meeting one or more of the following criteria [22]:

- Materials are fed into a system at the same time as product is removed from the system.
- The process materials condition is a function of its position within the process as it flows from inlet to outlet.
- The quantity of product produced is a function of the duration for which the process is operated and the throughput rate of the process.

Thus, continuous processing for pharmaceuticals offers the possibility of an ultra-lean method of manufacturing. It involves process trains designed to run continuously with each unit operation linked through a computer-controlled system to enable timely transition from one processing step to the next. Well-designed continuous processes are generally more efficient and effective than traditional batch manufacture, because they are able to process materials without interruption and process variability is generally more immediately apparent, thus reducing the overall cost of quality. Scale-up of products should also be simpler for continuous processes, because the process is run for longer rather than when using larger-scale process equipment, thus avoiding the need to consider changes in geometry and equipment design. This provides a more flexible solution to meeting shifting demand patterns. In addition, there is potential for a smaller GMP footprint because there is no longer a need for a range of process trains to cover all the scales of operation. Continuous processing equipment is far smaller than conventional batch equipment, and estimates of footprint reduction in converting from batch to continuous vary from 20% to 35% (ISPE) to >50% and a 50% reduction in operating costs [23].

There are also potential advantages of improved and "real-time" process control and quality associated with continuous processes, linked with the application of on-line and in-line measurements to monitor the process and apply corrective action during processing (see Table 11.7). More experimental data can be generated, facilitating process understanding in a shorter time (QbD data), thus enabling the rapid exploration of design space with less effort and less drug, which can accelerate the R&D process.

The FDA regulatory definition of a "batch" which applies to both traditional batch or continuous manufacture is as follows:

> A specific quantity of drug or other material that is intended to have uniform character and quality, within specified limits, and is produced according to a single manufacturing order during the same cycle of manufacture.

Table 11.7 Comparing batch and continuous processes.

	Continuous process	Batch process
Process information	Simultaneous collection of information about all process steps is collected for small quantities of material (high number of data points per amount of product)	Sequential collection of information about each process step is collected for whole batch (fewer data points per amount of batch)
Detection and correction of deviations from desired set point	Potential to detect small deviations in small quantities of product and to apply corrective action to the specific transformation step before effect is significant to the quantity of the total production run	Only able to detect deviations that have already affected the whole batch. Any corrective must then be based on returning the whole batch to the desired condition
Elimination of nonconforming product	Potential to design systems which are able to reject small quantities of nonconforming products	Whole batch may be scrapped

So, for a continuous process, the batch size is more flexible because it is defined by time, although it has to be predetermined for a specific need. The QA release/approval of the product is made according to the defined batch.

Continuous processing has been employed extensively and for some time outside of the pharmaceutical industry, such as in the food, chemical, and petrochemical industries, driven by large-volume cost efficiencies and quality benefits. The pharmaceutical industry, however, has been relatively slow to adopt continuous processes due to several real and perceived barriers:

- Culture change – a reluctance to change from a tried-and-tested batch approach employed for many years.
- The time and cost of change – a reluctance to invest in new technologies, processes, and personnel training.
- A generally perceived concern about the amount of valuable product lost (waste) while reaching steady state associated with start-up and shut-down of continuous processes.
- A perception of limited use, for example, only for large quantities.
- A perception that there will be regulatory implications due to a lack of experience within the pharmaceutical industry and the different views from the various regulatory authorities.

However, the tide now seems to be turning as the pharmaceutical industry is finally realizing the overall benefits of continuous processing in providing significant quality, cost, time, and material benefits [24]. One of the key drivers for change has been a push from the US FDA, which recognized the need to eliminate the hesitancy for pharmaceutical companies to innovate and through their initiative entitled "Pharmaceutical cGMPs for the 21st Century: A Risk-Based Approach" to encourage the early adoption of new technological advances. The FDA's aim is to encourage the use of the latest scientific advances in manufacturing and technology, incorporating a series of QbD and PAT initiatives to reach the "desired state for pharmaceutical manufacturing" [19]. In recent years, the FDA has engaged with the pharmaceutical industry, expressing their full support to move in the direction of continuous processing and to encourage the inherent use of QbD and PAT in the development of continuous processes with the establishment of a control strategy. In fact, a senior FDA representative has made a bold statement that in the next 25 years, the pharmaceutical industry will shift from batch to continuous manufacturing and make current production methods "obsolete" [25]. Although there is some concern about the different attitudes from other regulatory authorities, there appear to be no current regulations or guidance anywhere that prohibits the use of continuous manufacturing and moving away from batch manufacture. A recently published article [26] discussing the status of continuous processing summarizes it quite nicely: "The benefits of continuous processing are without dispute. The path to introduction, however, is less stable and dependent on conviction. Continuous manufacture is a natural progression in the technology of production and, as techniques develop and systems improve, take up is inevitable."

11.8 The Tablet Press as Part of a Continuous Tableting Line

The urgent need to improve the efficiency and productivity within the pharmaceutical manufacturing sector to stay competitive and reduce costs has resulted in a range of initiatives involving pharmaceutical companies, academia, equipment suppliers, and government funding bodies to build considerable momentum in the field of continuous product manufacture for both drug substance and drug product manufacture [27]. There has been a significant focus on oral solid dose technologies, probably because they still account for the majority of dosage forms on the market today, and this is unlikely to change in the foreseeable future.

The manufacture of solid oral dosage form tablets is achieved be employing three main technologies: direct compression; dry granulation, for example, roller compaction; or wet granulation. Direct compression is the simplest and involves mixing/blending the drug with excipients followed by tableting and coating. Dry granulation is slightly more complex, where the drug substance and excipients are mixed/blended and processed through a roller compactor, for example. The output material is then milled and further mixed/blended with suitable excipients and tableted. Wet granulation is the most complex, with several unit operations required to manufacture a solid dosage form as shown in Figure 11.7.

Some established pharmaceutical unit operations are inherently continuous by design, but have tended to be utilized in batch mode. They can be operated continuously as long as they are fed with input material continuously. For example, milling or tablet compression on a rotary tablet press, and so on, are continuous, where after start-up the process equipment will operate in a continuous manner at steady state for the majority of time until it is finally closed down. Other unit operations are batch in nature, such as mixing or granulation. Table 11.8 gives a list of unit operations used in solid oral dose manufacture today that are either batch mode or inherently continuous. Another solid oral dosage form technology available is capsule filling (hard or soft-gel), where the unit operations involving blending/mixing the drug with excipients followed by filling, are also intrinsically continuous.

For wet granulation to be totally continuous, all the unit operations in the process would need to be conducted in a continuous manner: blending/mixing, wet granulation, drying, milling, blending/mixing, tableting, and coating. The design and commercial availability of

Table 11.8 *Unit operations currently used in oral solid dose manufacture.*

Batch mode	Inherently continuous
Mixing and blending	Milling
Wet granulation	Roller compaction
Drying	Hot melt extrusion
Coating	Tableting
	Spray drying
	Capsule filling

new continuous or semi-continuous processing solid dose equipment has enabled the pharmaceutical industry to finally exploit the potential advantages of continuous manufacture. Continuous equipment for the production of pharmaceutical granules has been available for some time, but these early models were fraught with shortfalls, such as high waste at start-up and shut-down, issues with plug flow, and were not easily cleanable. Only with plug flow can a product plug be identified and traced throughout the process line. More recently, GEA Pharma Systems has developed a new generation of continuous process equipment to overcome these problems. The ConsiGma™ continuous manufacturing solid dosage form platform is in compliance with the FDA's QbD initiative and is now being used by several pharmaceutical companies in both R&D and commercial production. The ConsiGma™ provides flexibility to produce products to meet demand with potentially a continuous output of 25 to 150kg/hr depending on the model used (there is a smaller model intended for R&D to minimize drug usage). The system can perform dosing and mixing of raw materials, wet or dry granulation, drying, and tableting along with integrated advanced process control and PAT tools to allow continuous monitoring during production.

For wet granulation, the full tablet production line illustrated in Figure 11.8 consists of a blender, twin-screw granulator, fluid-bed dryer, and granule conditioning unit, combined with a rotary tablet press and continuous coater. The in-line blenders mix in the external phase between the systems. The ConsiGma™ granulator and dryer section has three modules: a wet high-shear granulation module, a segmented dryer module, and an evaluation module. For the wet granulation process, the dry raw materials (drug substance and excipients) are initially premixed in the continuous granulator, and then granulation liquid is added. The resultant material is fed into a segmented fluid dryer in packages of 1.5kg, each being dried in a separate segment of the dryer to ensure plug flow. When each segment is dry, it is emptied and transferred to an evaluation module, where the dried granules can be measured for CQAs such as particle size distribution, content uniformity, and moisture. The dryer module is continually filled with another segment of wet granules. Dried granules are milled before being pneumatically transferred into an in-line blender and then introduced immediately into a tablet press, followed by the continuous coating system.

An interesting design feature of the GEA continuous rotary tablet press (GEA Courtoy MODUL™ P) is that it employs external rather than internal lubrication. Conventionally, a lubricant, for example, magnesium stearate, is blended into the formulation after the granulation and drying stage, but this can lead to overblending or overlubrication, resulting in a poorly compactable powder, insufficient tablet hardness, and friable tablets. As an alternative, GEA has designed a lubricant spray system to spray the press,

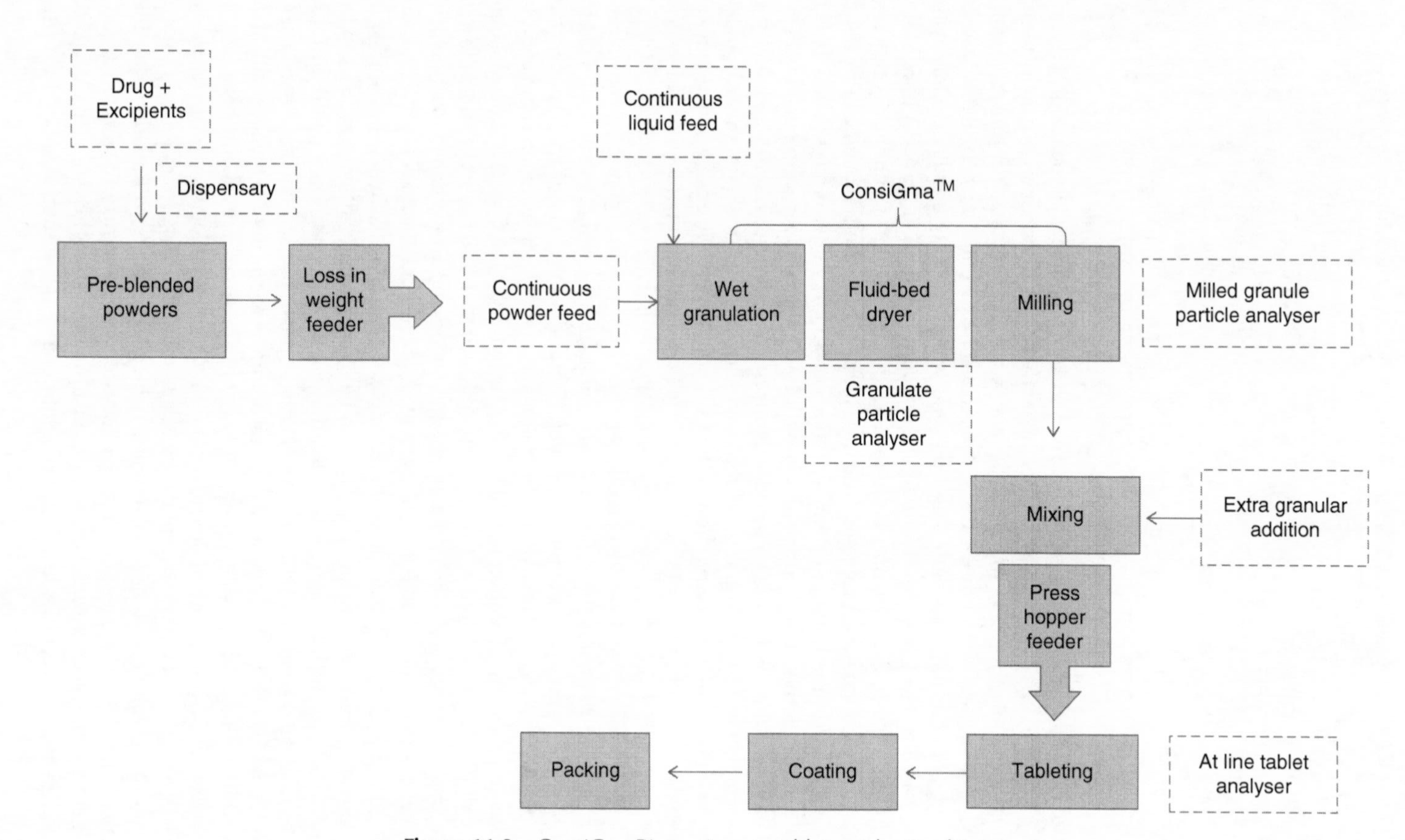

Figure 11.8 *ConsiGma™ continuous tablet production line.*

directly onto the components that require lubrication, the punch tips and die wall. As a result, the amount of lubricant used is significantly reduced. The complete GEA production line is very compact, being only one-third the size of a traditional granulation line, and GEA claims that it is possible to convert the raw materials into tablets in approximately 20–25 min.

PAT sensors have a significant role to play in the GEA system to allow monitoring of CQAs during operation. Moisture content and blend uniformity are measured using NIR, and particle size is measured using an on-line laser-diffraction system. Tablet content uniformity can be measured using Fourier transform (FT)-NIR transmission spectroscopy. There are PAT sensors at the inlet of the tablet press and also in the feeder of the tablet press directly above the dies. The latter effectively measures at the end of the powder line, just before the powder is compressed into tablets and can measure moisture, blend homogeneity, and content uniformity, for example. Process data can be generated for every individual tablet made on the press using special fast data acquisition hardware. Examples of CPPs like pre-compression force, pre-compression displacement, main compression force and displacement, ejection force, and the punch movements can be used as indirect measurements for CQAs, such as tablet weight and hardness. Consequently, they can be the control signals for the weight and hardness control system, so tablets with CQAs outside of the acceptance limits will be rejected by the press and into a reject bin.

If a displacement sensor is directly connected with a tablet punch, the punch movement while the tablet is made can be measured. If the punch movement is combined with the compression force, the compression energy can be calculated for one individual tablet, as well as the elastic recovery of the tablet. These are material-specific values that can be used to increase knowledge of the material and for material characterization. The higher the elastic energy, the higher the elastic expansion of the tablet after compression, and the more prone a tablet will be to defects such as capping or laminating. By correlating the compression energy with the CQAs, the compression energy can become a CPP and be used to optimize the process. Heckel plots can be produced from the punch movement and compression force data to give an indication of strength of the material, and which deformation mechanisms, for example, plastic deformation or fragmentation, are more pronounced during compression of the tablet.

All the data measurements from the entire production line are fed into a process control system that can apply calculations in real time (e.g., transformation of multivariate spectral data into specific CQA values) to provide closed-loop control. The software control system creates for every product plug a unique identifier, the product key (PK). Using a transfer function or residence time distribution function for every unit operation of the process line, the control system can trace every PK through the process line. For every unit operation, the control system time stamps the PK and adds the CPPs and CQAs to the PK. This allows the CPPs and CQAs generated for the different unit operations to be synchronized and correlated, enabling process understanding.

GEA Pharma Systems also produces a commercially available continuous dry blending process line linked to a rotary tablet press for direct compression tablet production. Other individual equipment suppliers can offer commercially available continuous processing equipment for blending, granulation, tableting, and coating, which if combined, could produce a continuous production line. So, the current status of continuous tablet manufacture is that most production lines today operate as a series of independent unit operations from a mixture of unit operations available, some of which are continuous as long as they are fed input material, and others which can be continuous in the context of an appropriately designed process and control strategy.

11.9 Real-Time Release Testing and Continuous Quality Verification

According to ICH Q8 (R2), real-time release testing (RTRt) is defined as "the ability to evaluate and ensure the quality of in-process and/or final product based on process data, which typically include a valid combination of measured material attributes and process controls." Although RTR is the ultimate goal for both the pharmaceutical industry and the regulators, not many companies have currently achieved it in practice.

In real-time release, product QA is based on on-line analysis. The product is released as it is produced rather than as a batch being held while waiting for QC testing. RTRt can be applied to any process: batch, continuous, existing, or new. For batch processing, the batch could be released if there were no deviations throughout the batch. In continuous processing, product could be released continually as long as there were no deviations. RTRt can be applied for some or all of the quality characteristics identified and can be applied to a unit operation or the entire manufacturing process. The main benefit of RTRt is a reduction in laboratory testing, and implementing closed-loop control of the manufacturing process will be the initial step toward achieving it. As mentioned earlier in this chapter (see Table 10.5), RTR data can be generated for most unit operations by means of direct measurements of the CQA during the manufacturing process, for example, on-line measurements. Alternatively, where there is a relationship between a CPP and a CQA and it is well understood, an appropriate model can be developed to monitor the CPP as a surrogate measurement of the CQA. For example, tablet disintegration may be used as a surrogate for tablet dissolution.

There is a strong link between RTR and CPV, discussed earlier in Section 10.3 as the best practice regarding business PV, and, Continuous Quality Verification (CQV), defined in ASTM E2537-08 as follows:

> An approach to process validation where manufacturing process (or supporting utility system) performance is continuously monitored, evaluated and adjusted as necessary.
>
> It is a science-based approach to verify that a process is capable and will consistently produce product meeting its predetermined critical quality attributes (CQAs) [28].

The key elements of CQV to ensure product quality are the following:

- Establishing thorough process understanding during process development.
- Implementing continuous quality monitoring and feedback during commercial manufacture on the basis of an established process control strategy.
- Establishing effective process analysis to confirm process performance and capability.
- Seeking opportunities for continuous process improvement.

The process control strategy is expected to include confirmation of product quality during manufacture such that each batch conforms to established CQAs, thereby enabling real-time release testing. Being able to take actions on a real-time basis is facilitated by systems that can capture process outputs and signal trends on a continuous basis during manufacture, and by use of control charting of CQA data.

The implementation of an appropriate data collection procedure is crucial for CQV. Multivariate tools can be used as part of a CQV approach for a new drug product relying on the information that summarizes the control strategy, as illustrated in a case study for a new tablet formulation [29,30].

Product quality reviews should be carried out on the basis of the statistical evaluation of the data collected to allow trend analysis of process consistency and capability, as well as intra-production quality analysis. Process information gained from manufacturing experience is used to make process improvements as applicable, with change management applied to implement the improvements.

Implementation of CQV represents a significant culture change for both the pharmaceutical industry and regulators. There can be a reluctance to change from traditional approaches, although many believe now that continuous manufacturing using on-line PAT with closed-loop process control and real-time release represents the future of pharmaceutical manufacturing.

11.10 Acknowledgments

I would like to thank the following for their contributions to this chapter:

Jurgen Boeck from GEA Pharma Systems for detailed information on the Consigma™ continuous production line and PAT measurements.

Trevor Page from GEA Pharma Systems for providing background materials on process engineering, process equipment, and continuous manufacturing.

Keith Smith from Perceptive Engineering for providing background materials on process controls and advanced process controls.

James Kraunsoe and David Holt from AstraZeneca for the background materials on Advanced Process Control.

Line Lundsberg from Lundsberg Consulting for the background materials on process analyzers and PAT tools.

11.11 References

[1] ISPE (2000) International Society for Pharmaceutical Engineering (ISPE) Baseline® Guide Volume 5: Commissioning and Qualification, *International Society for Pharmaceutical Engineering* (ISPE). www.ispe.org.

[2] ASTM E2500 (2007) Standard Guide for Specification, Design, and Verification of Pharmaceutical and Biopharmaceutical Manufacturing Systems and Equipment, July.

[3] FDA (2011) *Guidance for Industry, Process Validation: General Principles and Practices. US Food and Drug Administration*, http://www.fda.gov/downloads/Drugs/.../Guidances/UCM070336.pdf (accessed 30 August 2017).

[4] ISPE (2011) Baseline Guide Volume 12: Science and Risk-Based Approach for the Delivery of Facilities, Systems and Equipment. International Society for Pharmaceutical Engineering (ISPE).

[5] ICH Q8 (R2) (2009) *Pharmaceutical Development*, http://www.ich.org/fileadmin/Public_Web_Site/ICH_Products/Guidelines/Quality/Q8_R1/Step4/Q8_R2_Guideline.pdf (accessed 30 August 2017).

[6] ICH Q9 (2005) *Quality Risk Management*, http://www.ich.org/fileadmin/Public_Web_Site/ICH_Products/Guidelines/Quality/Q9/Step4/Q9_Guideline.pdf (accessed 30 August 2017).
[7] ICH Q10 (2008) *Pharmaceutical Quality System*, http://www.ich.org/fileadmin/Public_Web_Site/ICH_Products/Guidelines/Quality/Q10/Step4/Q10_Guideline.pdf (accessed 30 August 2017).
[8] ISPE, 2011. Good Practice Guide: Applied Risk Management in Commissioning and Qualification. International Society for Pharmaceutical Engineering (ISPE), October.
[9] 21CFR 210 and 211 (2014) Current Good Manufacturing Practice in Manufacturing Processing or Holding of Drugs: General.
[10] EudraLex (2015) EU Guidelines for Good Manufacturing Practice for Medicinal Products for Human and Veterinary use. Annexe 15, *Qualification and Validation by the European Commission*.
[11] EMA (2014) Guideline on Process Validation for Finished Products. Committee for Medicinal Products for Human Use (CHMP), *European Medicines Agency*, February.
[12] FDA PAT Guideline (2004) *Guidance for Industry, PAT – A Framework for Innovative Pharmaceutical Development, Manufacturing, and Quality Assurance*. (http://www.fda.gov/downloads/Drugs/GuidanceComplianceRegulatoryInformation/Guidances/ucm070305.pdf) (accessed 30 August 2017).
[13] PDA (2013) Technical Report No. 60. Process Validation: A Lifecycle Approach, Parenteral Drug Association, Bethesda, MD, USA.
[14] WHO (2014) Draft proposal for revision of the supplementary guidelines on Good Manufacturing Practices: Validation, Appendix 7: Non-Sterile Process Validation, *World Health Organization*.
[15] Gibson, M., Ashman, C.J. and Nelson, P. (2012) Optimization Methods in *Encyclopedia of Pharmaceutical Science and Technology*, Informa Healthcare, USA.
[16] ISO 2859-1 (1999) Sampling Procedures for Inspections by Attributes – Part 1: Sampling Schemes Indexed by Acceptance Quality limit (AQL) for Lot-by-Lot Inspection, *International Standard Organisation*.
[17] PWC (2001) Pharma Manufacturing – Unmet Performance Expectations, *Price Waterhouse Coopers*.
[18] IBM (2005) Report entitled "The Metamorphism of Manufacturing from Art to Science," http://www-935.ibm.com/services/us/imc/pdf/ge510-4034-metamorphosis-of-manufacturing.pdf (accessed 30 August 2017).
[19] FDA, (2004) *Pharmaceutical cGMPs for the 21st Century – A Risk-Based Approach*, http://www.fda.gov/downloads/drugs/developmentapprovalprocess/manufacturing/questionsandanswersoncurrentgoodmanufacturingpracticescgmpfordrugs/ucm176374.pdf (accessed 30 August 2017).
[20] Abboud, L. and Hensley, S. (2003) New Prescription for Drug Makers: Update the Plants, *Wall Street Journal*, 3rd September.
[21] ICH (2011) Quality Implementation Working Group Points to Consider (R2), *ICH-Endorsed Guide for ICH Q8/Q9/Q10 Implementation*.
[22] ASTM (2011) ASTM WK34349, New Guide for the Application of Continuous Processing in the Pharmaceutical Industry, Developed by Subcommittee E55.01, *ASTM International*.
[23] Ates, A. and Taliti, R. (2012) Poster entitled "Enablers and Barriers for Continuous Manufacturing in the Pharmaceutical Industry" presented at the *Annual EPSRC Centre Open Day, University of Strathclyde*.
[24] Plumb, K. (2005) Continuous Processing in the Pharmaceutical Industry – Changing the Mindset, *Chemical Engineering Research and Design*, **83**, 730–738.
[25] Woodcock, J. (2011) A statement by Janet Woodcock, Director, CDER, Food and Drug Administration, at the *AAPS Conference*, San Francisco, US.
[26] Bowen, R. (2015) Does Pharma Really Need Continuous Processing? *PDA Letter*, pp. 22–27, February.
[27] Gibson, M., de Matas, M. and Huang, Z. (2013) Making Continuous Processing a Reality in the Pharmaceutical Industry, *Bio Pharma Asia*, March/April.

[28] ASTM E2537-08 (2008) Standard Guide for Application of Continuous Quality Verification to Pharmaceutical and Biopharmaceutical Manufacturing, *ASTM International*, West Conshohocken, PA.

[29] Zomer, S., Gupta, M. and Scott, A. (2010) Application of Multivariate Tools in Pharmaceutical Product Development to Bridge Risk Assessment to Continuous Verification in a Quality by Design environment, *J. Pharm. Innov*, **5**,109–118.

[30] Conformia CMC-IM Working Group (2008) Pharmaceutical Development Case Study: "ACE Tablets," March.

12

Regulatory Guidance

Siegfried Schmitt and Mustafa A. Zaman

PAREXEL Consulting, PAREXEL International, United Kingdom

12.1 Introduction

From a regulatory perspective, submitting an application/dossier which has been developed from the outset using the principles of Quality by Design (QbD) to gain product and process knowledge is the ultimate goal. The condensed information from years of research and development can be submitted in one single document to the regulatory agencies in order to be granted a Marketing Authorisation (MA). As QbD aims to provide the most comprehensive insight into the relationship between quality attributes and process parameters, by definition, QbD submissions must be the most scientifically robust documents assembled. One might assume that industry would want to strive to deliver QbD submissions in all cases in order to stand the highest chance of approval. However, there are concerns in industry that are not easily overcome, both from a small molecule and large molecule perspective.

For example, [1] '*Pharmaceutical Technology*'s annual manufacturing and equipment survey found that while 28% of respondents stated they fully use QbD in process development/optimisation for solid-dosage manufacturing and 56% apply QbD to some extent, 16% said they do not use QbD. Respondents indicated that a lack of knowledge, training and clarification from regulators contributed to their reluctance to use QbD. Cost and time also factored into the decision to not incorporate QbD into their processes'. Despite the intrinsic advantages of QbD outlined in Figure 12.1, this does provide some pointers to the reasons why industry is not delivering as expected, but it does not give the full picture. One strong reason for not providing the full set of knowledge in a complete QbD submission is fear of loss of intellectual property (IP). As several agencies, such as the US FDA,

Pharmaceutical Quality by Design: A Practical Approach, First Edition.
Edited by Walkiria S. Schlindwein and Mark Gibson.

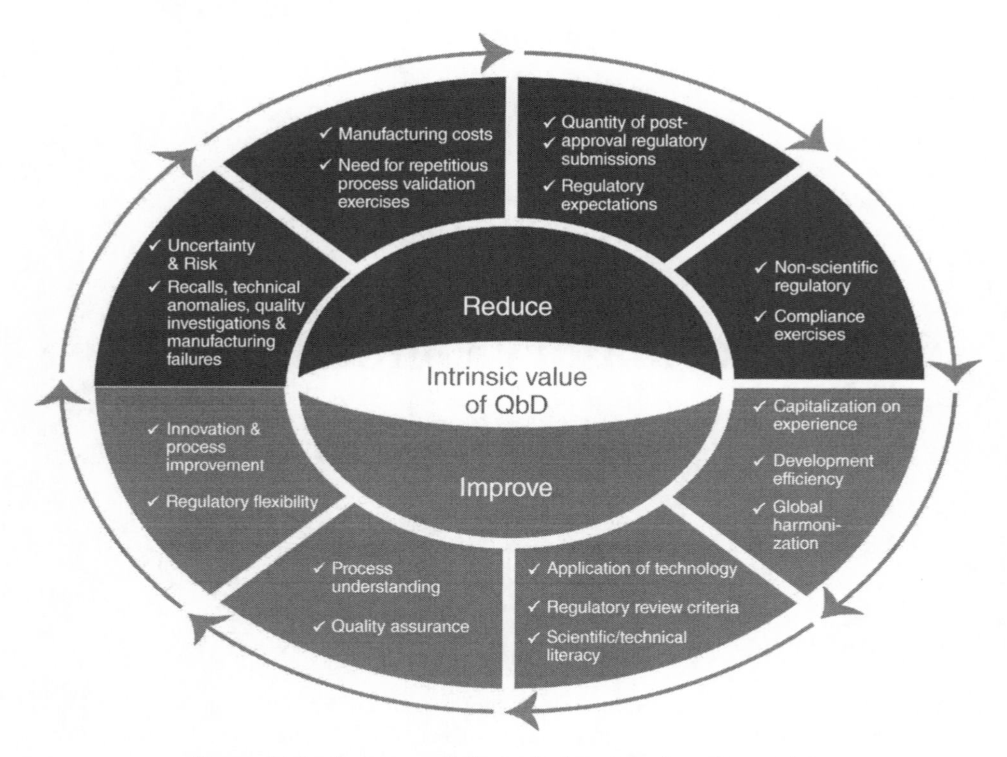

Figure 12.1 *Summary of the intrinsic value of applying QbD.*

actively encourages risk-based approaches, and require the incorporation of QbD elements in submissions/dossiers, industry does, of course, adhere to these requests. Nevertheless, there is a danger that applicants' not only incorporate QbD terminology without understanding the concepts, but they also include their own terminology, thus potentially confusing the agencies/reviewers further. It must be stated with clarity here that there is no agency in the world that mandates QbD.

This chapter will provide the reader with information on how to prepare submissions/dossiers in the Common Technical Document (CTD) format promulgated by the International Conference on Harmonisation (ICH) and mandated by a large number of regulatory agencies globally. Compared with traditional dossiers, QbD submissions contain much more detail on product development, risk management and knowledge management. When providing so much more information, it is essential to maintain a logical flow and provide coherent scientifically robust explanations for the selection of product attributes and process parameters.

12.2 The Common Technical Document (CTD) Format

This format is described in detail in ICH M4 [2]. It is assumed that readers are familiar with the format (Figure 12.2), so this chapter will only focus on QbD-specific issues relating to the CTD.

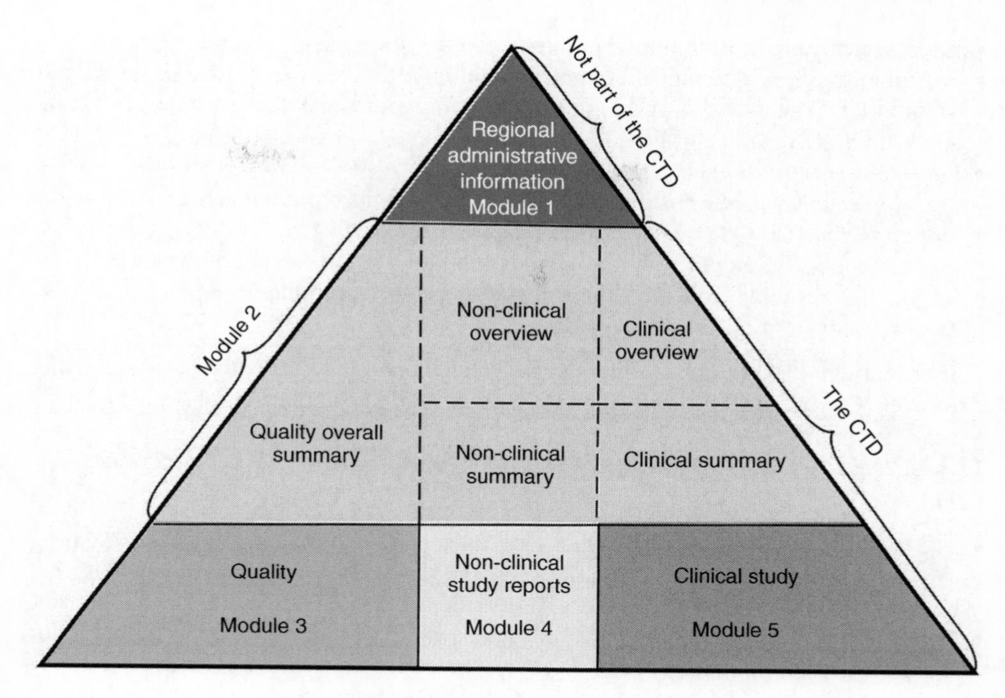

Figure 12.2 *The CTD triangle [4].*

Points to consider when submitting QbD-related information in CTD format:

- Submitted information should be organised in a clear manner and provide the regulators with sufficient understanding of the development approach; this information will be important for the evaluation of the proposed control strategy. It is advisable to highlight any pre-submission regulatory and/or scientific advice received from the agencies in the cover letter, as it is likely that the assessor reviewing the dossier is not the assessor who provided the advice at the respective meeting.
- It is highly advisable to use terminology that is stipulated in the relevant ICH Guidelines with regard to QbD [3].
- Companies generally perform, especially for QbD-containing submissions, an internal peer review process to assure quality, clarity and adequacy of the regulatory submission.
- It is important to realise that not all the studies performed and/or data generated during development need to be submitted.

However, sufficient supporting information and data should be submitted in the application to address the following:

- The scientific rationale for the studies conducted (in line with the Quality Target Product Profile (QTPP)).
- A concise description of methodologies used to conduct these studies and to analyse the generated data.
- The summary of results and conclusions drawn from these studies.

- The scientific justification for the proposed control strategy.
- For submissions containing additional/optional QbD elements (e.g. Real Time Release Testing (RTRT) and design space), it is helpful for regulators to have a statement by the applicant describing the proposed regulatory outcome and expectations.
- Drug substance (ICH Q11).
- Process development information should usually be submitted in Section S.2.6 of the CTD.
- Drug product (ICH Q8 (R2)).
- Pharmaceutical development information is submitted in Section P.2 of the CTD.
- Information resulting from development studies can be accommodated by the CTD format in a number of different ways.

Essential in any QbD submission is to identify/describe the QTPP, the Critical Quality Attributes (CQAs), the Critical Process Parameters (CPPs) and the control strategy.

12.2.1 Quality Target Product Profile (QTPP) and Critical Quality Attributes (CQAs)

- The QTPP forms the basis of design for the development of the product. Potential drug substance and drug product CQAs derived from the QTPP and/or prior knowledge are used to guide the product and process development.
- CQAs of the drug product and drug substance should be listed, and the rationale for designating these properties or characteristics as CQAs should be provided in the development sections of the application (S.2.6 and P.2). Essentially, these sections should provide a detailed rationale and explanation as to why the data that underpins the process is fit for purpose, and allows regulatory flexibility in the respective submission.
- Detailed information about structural characterisation studies that supports the designation of drug substance CQAs should be provided in, for example, S.3.1 'Elucidation of Structure and other Characteristics' and S.7 'Stability'.
- Discussion and evaluation of drug substance CQAs which may affect or are related to the attributes and/or process of the drug product (ultimately in patients). For example, factors such as drug substance impurities and physicochemical properties (e.g. BCS (Biopharmaceutics Classification System) class, solubility, permeability, partition coefficient, crystal/amorphous solid state, particle size distribution, and aerodynamic properties (inhalation products) should be provided in P.2.1 'Components of the Drug Product', and are best represented in a tabular summary, indicating if the drug substance attribute has a low or high impact on the drug product attribute or process.
- Functional characteristics of key excipients which may also affect the attributes and/or process of the drug product (ultimately in patients), and as they relate to drug product CQAs should be provided in P.2.1 'Components of the Drug Product'. For example, the intention behind the selection of the excipient, the role of the excipient and a discussion on drug substance and excipient compatibility should be provided.

12.2.2 Quality Risk Management (ICH Q9)

- Quality risk management tools can be used to rank and select quality attributes (including material attributes) and / or process parameters that should be further evaluated and/or controlled within appropriate ranges to ensure the desired product quality.

- The assessments used to guide and justify development decisions can be summarised in the development sections of the application (S.2.6 and P.2).
- Visual summaries of risk assessments at key stages of development can be useful in facilitating review.

12.2.3 Product and Process Development (S.2.6 and P.2)

- The applicant should clearly outline and summarise the intention behind the product design. For example, the following elements should be included:
  - Type of dosage form and posology.
  - QTPP.
  - Optimised formulation.
  - Optimised manufacturing method and process controls to ensure robust and reproducible product.
  - Optimised pharmacokinetic profile versus the reference product.
- The applicant should provide information of sufficient detail to demonstrate how the conclusions were reached:
  - The scientific rational behind the formulation selection should be fully discussed, and should include the formulations used in pivotal clinical trials, bioavailability or bioequivalence studies and primary stability studies. For example, why a particular dosage type was selected to produce a dosage form (e.g. direct compression used to manufacture an immediate-release tablet) in relation to dosage form uniformity, bioavailability, ease/difficulty of manufacture should be summarised. The latter principle could also be used to summarise why the final dosage form was coated or not.
  - The scientific rationale and basis for experiments that determined the final CQAs and process parameters. To aid this discussion, the CQAs (and other quality attributes) should be listed.
  - The linkage between CPPs, CQAs and the QTPP.
  - Identification of potential residual risk that might remain after the implementation of the proposed control strategy and discussion of approaches for managing the residual risk.
  - The applicant should comment on the impact of the following on risk assessment: (a) interaction of attributes and process parameters and (b) effect of equipment and scale. For both, the use of diagrams to identify risk are advisable (e.g. Ishikawa or fishbone diagrams; Figure 12.3).
- Inclusion of a full statistical evaluation of the experiments performed at early development stages are not expected; however, summary tables of the factors and ranges studied and the conclusions reached will be helpful.
- For design of experiments (DoEs) that are used to establish CPPs and/or to define a design space, the inclusion of the following information in the submission will greatly facilitate assessment by the regulators:
  - Rationale for selection of DoE variables (including ranges). Submitters should indicate if the factors are expected to be scale dependent
  - Any evidence of variability in raw materials (e.g. drug substance and/or excipients) that would have an impact on predictions made from DoE studies
  - Listing of the parameters/attribute that would be kept constant during the DoEs and their respective values, including comments on the impact of scale on these parameters

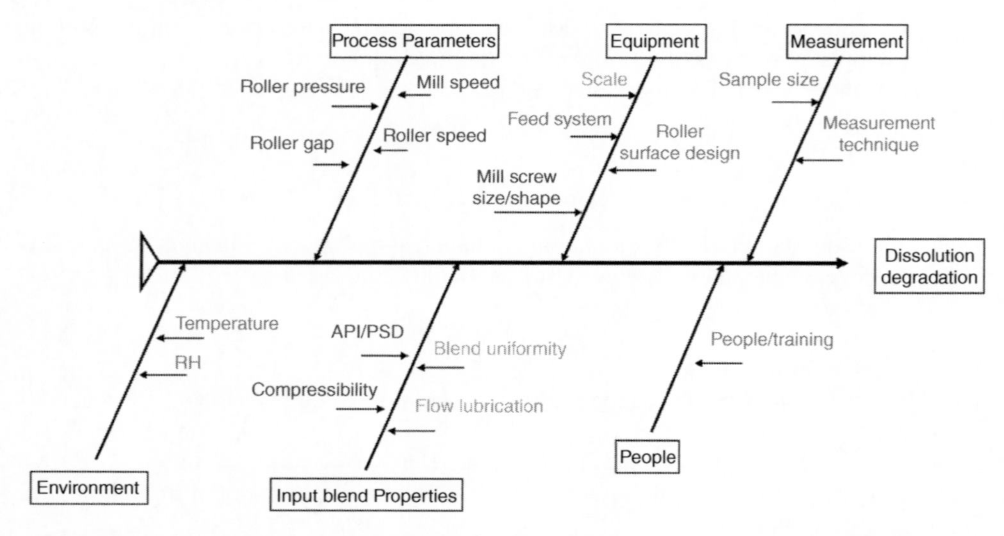

Figure 12.3 *A typical example of an Ishikawa or fishbone diagram. Key: red = potential high impact on CQA; Yellow = potential impact on CQA; green = unlikely to have significant impact on CQA. Example: roller compaction. (See insert for color representation of the figure.)*

(e.g. compression force, spray rate, temperature, milling speed, feed rate, mixer rate, water content and particle size of API)
- Type of experimental design used and a justification of its appropriateness, including the power of the design
- Reference to the type of analytical methods (e.g. HPLC and NIR) used for the evaluation of the data and their suitability for their intended use (e.g. specificity and detection limit)
- Results and statistical analysis of DoE data showing the statistical significance of the factors and their interactions, including predictions made from DoE studies relevant to scale and equipment differences

12.2.4 Control Strategy

- A planned set of controls, derived from current product and process understanding that assures process performance and product quality
- The controls can include parameters and attributes related to drug substance and drug product materials and components, facility and equipment operating conditions, in-process controls, finished product specifications and the associated methods and frequency of monitoring and control. Essentially, the control strategy could include three levels of control:
  - Level 1 Real-time automatic control and flexible process parameters to respond to variability in the input material attributes
  - Level 2 Reduced end-product testing, flexible material attributes and CPPs within the design space
  - Level 3 End-product testing and tightly constrained material attributes and process parameters

Table 12.1 Summary of where to present the control strategy in the CTD.

Control strategy element	Drug substance CTD module	Drug product CTD module
Manufacturing process development	—	P.2.4
Manufacturer(s)	S.2.1	P.3.1
Product composition/batch formula	—	P.1; P.3.2
Control of input materials	S.2.3	P.4 (excipients)
Manufacturing process description	S.2.2	P.3.3
Control of critical steps and intermediates	S.2.4	P.3.4
End-product specifications	S.4.1; S.4.5 (S.4.2; S.4.3; S.5)	P.5.1; P5.6 (P.5.2; P5.3; P.6)
Container closure system	S.6	P.7
Re-test period/shelf life	S.7.1	P.8.1

The control strategy is presented across the entire CTD format (see Table 12.1); however, justification of the drug substance and drug product specification and manufacturing process development (S.5.4, P.5.6 and P.2.4) is a good place to summarise the overall control strategy (Figure 12.2).

12.2.5 Design Space (Optional)

- As described in ICH Q8 (R2), when a design space is proposed by the applicant, it is also subject to regulatory assessment and approval. It is an optional element of the regulatory assessment process. To allow greater regulatory flexibility, the ideal design space should be independent of equipment used, scale employed and site utilised.
- Although drug product design space is discussed in more detail (with examples) in Chapter 6 of this book, different approaches can be considered when implementing a design space, for example, process ranges, mathematical expressions or feedback controls to adjust the parameters during processing.
- Regardless of how a design space is developed, it is expected that operation within the design space will result in a drug substance or product that meets the defined quality (within the parameters defined by the DSS (Drug Substance Specification) or FPS (Finished Product Specification).
- Justification that the control strategy ensures that the manufacturing process is maintained within the boundaries defined by the design space should be provided.
- The manufacturing process development (Sections S.2.6 and P.2 of the application) is the appropriate place to summarise and describe product and process development studies that provide the basis for the design space(s). It is also an ideal place to provide a description, together with reasons for using, novel technologies, such as Process Analytical Technology (PAT).
- As an element of the proposed manufacturing process, the design space(s) can be described in the section of the application that includes the description of the manufacturing process and process controls (S.2.2 and P.3.3).

- If appropriate, additional information can be provided in the section of the application that addresses the controls of critical steps and intermediates (S.2.4 and P.3.4).
- The relationship of the design space(s) to the overall control strategy can be discussed in the section of the application that includes the justification of the specification (S.4.5 and P.5.6).

Further information on what to consider is contained, for example, in the following documents:

- EMA ICH guideline Q8, Q9 and Q10 – questions and answers volume 4, EMA/CHMP/ICH/265145/2009, December 2010.
- US FDA Manual of Policies and Procedures MAPP 5016.1 Applying ICH Q8(R2), Q9 and Q10 Principles to CMC Review, Effective Date: 08 February 2011.
- US FDA Guidance for Industry Q8, Q9, and Q10 Questions and Answers (R4), November 2011.
- US FDA Guidance for Industry Q8, Q9, and Q10 Questions and Answers Appendix Q&As from Training Sessions, July 2012.

12.3 Essential Reading

As submissions/dossiers contain an organisation's IP (see earlier discussion), it is not easy to get hold of information on how companies assemble their QbD CTD documents. This has been recognised by the regulators and by industry. For that reason, several key documents have been published and are freely available for anyone's perusal. These are in no particular order:

- PhRMA Working Group Quality by Design for Biotech Products Parts 1 [5], 2 [6] and 3 [7], 2009.
- CMC Biotech Working Group, A-Mab: A Case Study in Bioprocess Development, 30 October 2009 [8].
- CMC-Vaccines Working Group, A-VAX: Applying Quality by Design to Vaccines, May 2012 [9].
- EMA Public Assessment Report (EPAR) Perjeta, EMA/17250/2013, 13 December 2012 [10].
- EMA-FDA Pilot Program for Parallel Assessment of Quality-by-Design Applications: Lessons Learnt and Q&A Resulting from the First Parallel Assessment, EMA/430501/2013, 20 August 2013 [11].
- PMDA Participation as an Observer in the EMA-FDA Pilot Program for Quality by Design [12].
- EMA-FDA Questions and Answers on Design Space Verification, 24 October 2013[13]
- Development and Verification of Design PDA: A Global Space Association, Tone Agasøster (Norwegian Medicines Agency) and Graham Cook (Pfizer), 28–29 January 2014 [14].
- Final Concept Paper ICH Q12: Technical and Regulatory Considerations for Pharmaceutical Product Lifecycle Management, 28 July 2014 [15].
- Quality by Design for ANDAs: An Example for Immediate-Release Dosage Forms, April 2012 [16].
- Quality by Design for ANDAs: An Example for Modified Release Dosage Forms, December 2011 [17].

12.4 What Is Not Written, or Hidden, in the Guidance Documents?

Though it is possible to create an n-dimensional design space, to date the few agencies that have reviewed full-fledged QbD submissions have not accepted more than three-dimensional design spaces, that is, limiting the observable interactions to two process parameters. It is simply a fact, and it is left to anyone's guess why this is so.

Risk assessments can be of high complexity and be extremely granular. Experience shows that very complex risk assessments are not acceptable to the reviewers. Furthermore, it should be noted that there is a combined EMA/FDA statement [11] that limits criticality levels to critical/non-critical, effectively dismissing the term 'key critical parameter'.

What happens when historically, the CQA has never been out of specification and the CPPs have also never been out of specification? If a company never had a problem with a CQA, do they still need to do further small-scale experiments to find out where the boundaries lie? Can they use the existing data to demonstrate that extensive QbD experimentation is not required for this CQA? Here it should be noted that a process can well be within specifications, without ever being within statistical control [18]. As the design space within the QbD realm is a statistical concept, there is nothing that can be done from a statistical perspective –that is, more experimentation may not reap any benefits.

Regulators want to have rationales from the submitting company as to why certain parameters, attributes or boundaries are sufficiently characterised by small-scale experiments and do not need large-scale ones. For example, most companies still perform three validation batches at one specific setpoint only; that is, they do not vary manufacture within the design space at all, and thus cannot provide as much scientific evidence as possible. Indeed, regulators have argued that if a narrow range of process parameters has been employed to manufacture large-scale batches, then wider-range process parameters cannot be used to manufacture and/or justify large-scale batches.

During the product lifecycle, moving from one area to another within the design space (i.e. change in the normal operating ranges (NORs)) may represent higher or unknown risks not previously identified during initial establishment of the design space. For this reason, regulators may expect 'design space verification' of the suitability of the design space, to thus confirm that all product quality attributes are still being met in the new area of operation within the design space. It is equally important to rationalise why a parameter or attribute is not critical, that is, why it does not influence quality attributes. QbD without real-time monitoring of processes is not an option.

All full-fledged QbD submissions have been submitted via the centralised procedure in the EU.

Acceptance of specific QbD elements in a submission by one agency (e.g. EMA) does not automatically lead to acceptance of the same by other agencies (e.g. US FDA).

Regulators often reiterate and clarify that the design space should not be confused with proven acceptable range (PARs). In line with ICH Q8(R2), PARs are defined as'a characterised range of a process parameter for which operation within this range, while keeping other parameters constant, will result in producing a material meeting relevant quality criteria'. The guideline clearly states that a combination of PARs *does not* constitute a design space. However, PARs based on univariate experimentation can provide useful knowledge about the process.

Although it is recognised that design space corroboration is expected for large/commercial-scale batches (particularly if the number of batches is limited), if a design space has not been verified using large/commercial-scale batches, then regulators would expect to see a comprehensive discussion on the ability of the manufacturing process to cope with scale-dependent CQAs – in such cases, regulators would also expect to refer to a protocol to verify the latter.

Analytical design spaces are currently unacceptable in submissions, as these are not defined in any of the applicable regulations. This has been reiterated time and again by regulatory agencies as part of published reviews of QbD submissions and at various conferences.

12.5 Post-Approval Change

Figure 12.4 summarises the types of post-approval changes available to applicants. Although the changes are applicable from a EU perspective, the schematic can also be applied to other regions, where the level of regulatory risk (e.g. in terms of data required or assessment review time and approval) increases as the procedure/changes become more complex. Importantly, it also illustrates that if a submission (initial or subsequent post-approval) includes an (approved) design space, then no further respective variations are required.

Conversely, when a dossier does not include verification of a design space with an initial application, a Post-Approval Change Application can be submitted. A key document of this

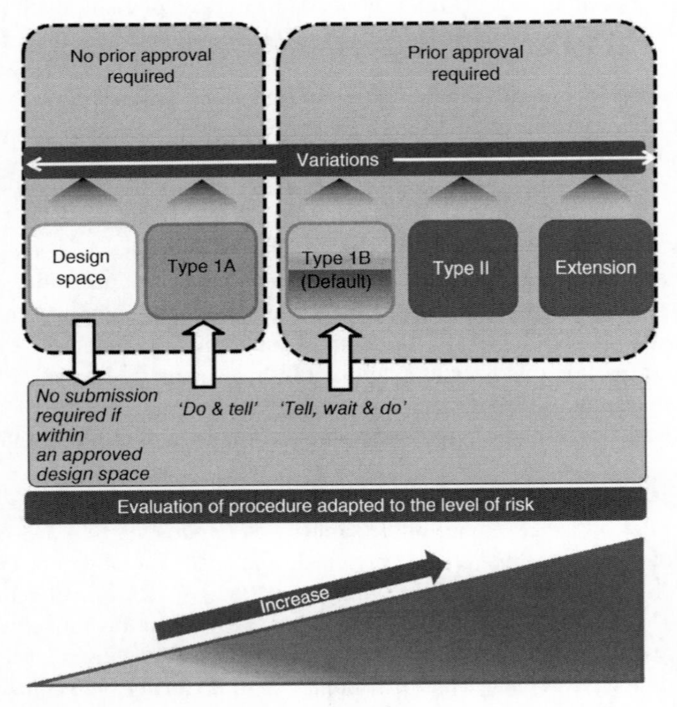

Figure 12.4 *Types of post-approval changes available in the EU relative to the adopted level of risk during the evaluation of the procedure.*

application is the Design Space Verification Protocol; the protocol is not expected to be exhaustive and should comprise the following basic elements [14]:

- The protocol should embody the spirit of process validation guideline.
- All the respective operations within the design space that are in scope.
- Verification of the initial NORs with regard to clinical, stability and pharmacovigilance batches).
- How the processes that are outside the NORs are assessed and monitored.
- Change management approach to evaluate the impact on control strategy and the API/ product quality.
- For scale-dependant processes, verification of the design space should be conducted at the intended production scale. The protocol should also list the scale-dependant parameters and discuss whether the control strategy can address the respective risk.

Although Section 12.2 of this chapter provides a summary of what should be presented in the dossier (this is equally applicable to Post-Approval Applications), how the Design Space Verification Protocol is submitted to the authorities can vary from country to country. For instance, for the USFDA, the protocol will be reviewed by the Office of Pharmaceutical Quality (OPQ) using an integrated review–inspection process; from a EU perspective, the protocol is submitted within the post-approval submission. Intriguingly, the EU variations regulations do not require filing of changes to non-critical process parameters. If these parameters are registered as part of a design space, however, changes have to be filed, although 'Qualified Person (QP) regulatory discretion' can be employed when isolated minor deviations occur for parameters in the design space [19].

One of the tools that is increasingly being employed to ease the regulatory burden is Post Approval Change Management Protocols (PACMP) (from an EU perspective) [20] or Change Protocols (CP) (from a US perspective) [21], and essentially describes specific step-wise changes that a company would like to implement during the lifecycle of the product and how these would be strategised, prepared and (later) verified. From a Japanese perspective, although post-approval change management is possible and the fundamental understanding of QbD is not different, there is currently no system to accept PACMP, nevertheless, there is a real drive to introduce this philosophy and harmonise the concept [22].

Managing post-approval changes from a global perspective is costly, time consuming and often requires good knowledge of the local country requirements. With this in mind, it makes sense that companies working on both established and new products transition to a QbD-based quality risk management process, which allows for more regulatory flexibility and is thus a possible solution to better supply-chain management [23].

12.6 Summary

Understanding what QbD elements to select and where to place them in a CTD (pre- and post- approval) is essential for successful submission and acceptance for assessment by the regulatory agencies. In addition, with concepts such as continuous manufacturing now becoming a reality, it is of paramount importance to provide a logical, science-based and comprehensive summary of the development process that ultimately leads to the present validated process, providing a product of the desired quality – always.

12.7 References

[1] Haigney, S. (2015) Quality by Design in Solid-Dosage Manufacturing, *Pharmaceutical Technology*, **39**(9), 26–32.

[2] ICH M4. *The Common Technical Document*, http://www.ich.org/products/ctd.html (accessed 5 March 2016).

[3] ICH Q8 (R2) (2009) *Pharmaceutical Development*, http://www.ich.org/fileadmin/Public_Web_Site/ICH_Products/Guidelines/Quality/Q8_R1/Step4/Q8_R2_Guideline.pdf (accessed 30 August 2017).

[4] ICH M4, *The Common Technical Document – The CTD Triangle*, http://www.ich.org/fileadmin/Public_Web_Site/ICH_Products/CTD/CTD_triangle.pdf (accessed 30 August 2017).

[5] Rathore, A.S., Look, J., Atkins, L.et al. (2009) Quality by Design for Biotechnology Products – Part 1: A PhRMA Working Group's Advice on Applying QbD to Biotech, *BioPharm International*, **22**(11).

[6] Rathore, A.S., Look, J., Smock, P.et al. (2009) Quality by Design for Biotechnology Products – Part 2: Second in a three-part series that discusses the complexities of QbD implementation in biotech development, *BioPharm International*, **22**(12).

[7] Rathore, A.S., Look, J., Atkins, L.et al. (2010) Quality by Design for Biotechnology Products – Part 3: Guidance from the Quality by Design Working Group of the PhRMA Biologics and Biotechnology Leadership Committee on how to apply ICH Q8, Q8R1, Q9, and Q10 to biopharmaceuticals, *BioPharm International*, **23**(1).

[8] CMC Biotech Working Group (2009) A-Mab: A Case Study in Bioprocess Development, http://c.ymcdn.com/sites/www.casss.org/resource/resmgr/imported/A-Mab_Case_Study_Version_2-1.pdf (accessed 30 August 2017).

[9] CMC Vaccines Working Group (2012) A-VAX: Applying Quality by Design to Vaccines, http://www.google.co.uk/url?sa=t&rct=j&q=&esrc=s&source=web&cd=1&cad=rja&uact=8&ved=0ahUKEwjDlcz1_v7JAhWCXhoKHQYBCDMQFggjMAA&url=http%3A%2F%2Fwww.ispe.org%2F2013-biotech-conference%2Fa-vax-applying-qbd-to-vaccines.pdf&usg=AFQjCNGg78eSyT59zzBL0WlBOz5sXsS1Bw&bvm=bv.110151844,d.ZWU (accessed 30 August 2017).

[10] European medicines Agency (EMA) Public Assessment Report (EPAR) (2012) Perjeta (International Non-Proprietary Name: Pertuzumab), Procedure No. EMEA/H/C/002547/0000 (EMA/17250/2013), http://www.ema.europa.eu/docs/en_GB/document_library/EPAR_-_Public_assessment_report/human/002547/WC500141004.pdf (accessed 30 August 2017).

[11] European Medicines Agency (2013) EMA-FDA Pilot Program for Parallel Assessment of Quality-by-Design Applications: Lessons Learnt and Q&A Resulting from the First Parallel Assessment (EMA/430501/2013).

[12] Pharmaceuticals and Medical Devices Agency (PMDA) (2011) PMDA Participation as an Observer in the EMA-FDA Pilot Program for Quality by Design, http://www.pmda.go.jp/english/rs-sb-std/standards-development/cross-sectional-project/0002.html (accessed 30 August 2017).

[13] European Medicines Agency (2013) Questions and Answers on Design Space Verification (EMA/603905/2013).

[14] PDA Joint Regulators/Industry QbD Workshop (2014) Case Study 2: Development and Verification of Design PDA: A Global Space Association, Agasøster, T. (Norwegian Medicines Agency) & Cook, G. (Pfizer), http://www.ema.europa.eu/docs/en_GB/document_library/Presentation/2014/02/WC500162148.pdf (accessed 30 August 2017).

[15] International Council on Harmonisation of Technical Requirements For Registration Of Pharmaceuticals For Human Use (2014) Final Concept Paper ICH Q12: Technical and Regulatory Considerations for Pharmaceutical Product Lifecycle Management.

[16] US Food and Drug Administration (FDA) (2012) Quality by Design for ANDAs: An Example for Immediate-Release Dosage Forms, http://www.fda.gov/downloads/Drugs/DevelopmentApprovalProcess/HowDrugsareDevelopedandApproved/ApprovalApplications/AbbreviatedNewDrugApplicationANDAGenerics/UCM304305.pdf (accessed 30 August 2017).

[17] US Food and Drug Administration (FDA) (2011) Quality by Design for ANDAs: An Example for Modified Release Dosage Forms, http://www.fda.gov/downloads/Drugs/DevelopmentApprovalProcess/HowDrugsareDevelopedandApproved/ApprovalApplications/AbbreviatedNewDrugApplicationANDAGenerics/UCM304305.pdf (accessed 30 August 2017).
[18] Torbeck, L. (2015) Why Life Science Manufacturers Do What They Do in Development, Formulation, Production and Quality: A History, PDA/DHI, Bethesda, USA.
[19] Busse, B., Rönninger, S., and Rössling, G. (2014) Industry, Regulators Meet to Drive QbD Implementation. *PDA Letter*, 30–31.
[20] European Medicines Agency (2012) Questions and Answers on Post Approval Change Management Protocols (EMA/CHMP/CVMP/QWP/586330/2010).
[21] Moore, C.M.V. (2013) -Approval Change Management: Challenges and Opportunities – An FDA Perspective. Speech at the DIA 49th Annual Meeting in Boston, 23–27th June 2013, http://www.fda.gov/downloads/AboutFDA/CentersOffices/OfficeofMedicalProductsandTobacco/CDER/UCM406881.pdf (accessed 5 March 2016).
[22] Matsuda, Y. (2015) Current Status and Perspectives on Pharmaceutical Products in Japan. Speech at the KFDC Annual Meeting, 13th November 2015, https://www.pmda.go.jp/files/000208435.pdf (accessed 30 August 2017).
[23] Avenatti, J. (2016) Managing Post-Approval Changes in a Global Environment. *PDA Letter, January* **2016**, 20–23.

Index

Pharmaceutical Quality by Design: A Practical Approach, First Edition.
Edited by Walkiria S. Schlindwein and Mark Gibson.